# Progress in Nonhistone Protein Research

## Volume II

Editor

**Isaac Bekhor, Ph.D.**
Professor
Laboratory for Molecular Genetics
University of Southern California
Los Angeles, California

Associate Editors

**Carol J. Mirell, Ph.D.**
Laboratory for Molecular Genetics
University of Southern California
Los Angeles, California

**C. C. Liew, Ph.D.**
Professor
Department of Clinical Biochemistry
The Banting Institute
University of Toronto
Toronto, Ontario, Canada

CRC Press
Taylor & Francis Group
Boca Raton London New York

CRC Press is an imprint of the
Taylor & Francis Group, an informa business

First published 1985 by CRC Press
Taylor & Francis Group
6000 Broken Sound Parkway NW, Suite 300
Boca Raton, FL 33487-2742

Reissued 2018 by CRC Press



**Library of Congress Cataloging in Publication Data**
Main entry under title:

Progress in nonhistone protein research.

Bibliography.
Includes indexes.
1. Nonhistone chromosomal proteins. I. Bekhor, Isaac. II. Mirell, Carol J. III. Liew. C. C.
[DNLM: 1. Chromosomal Proteins, Non-Histone.
QU 58 P964]
QP552.N62P76 1985 574.19'245 84-9408
ISBN 0-8493-5528-1 (v. 1)
ISBN 0-8493-5529-X (v. 2)

A Library of Congress record exists under LC control number: 84009408



ISBN 13: 978-1-315-89695-3 (hbk)
ISBN 13: 978-1-351-07605-0 (ebk)

Visit the Taylor & Francis Web site at http://www.taylorandfrancis.com and the
CRC Press Web site at http://www.crcpress.com

# FOREWORD

Nonhistone chromosomal (NHC) proteins are assumed to include regulatory proteins which modulate genetic expression by activating or repressing specific genes. This assumption has lured many investigators to search for sequence-specific DNA-binding proteins.

At the present time, however, the field of nonhistone protein research has been sharply curtailed. Many investigators have shifted their efforts to DNA cloning and sequencing. Molecular biologists have cloned hundreds of genes in the past several years, and the data for nucleotide sequences of genes and gene fragments have grown at a dashing pace. Sequencing data continue to provide valuable insight as to how genes are structured, but have given little information as to how gene expression is manipulated. In particular, the nucleotide base sequences of gene control regions, from higher organisms, have provided little insight as to how gene control is accomplished. It is possible that DNA sequencing has been overemphasized, and that the secondary structure of DNA, which may be due to the DNA sequence itself or possibly due to the intimate association of DNA with special proteins, is responsible for the ultimate mechanism of control.

Why have so many well qualified and highly creative investigators abandoned their search for the function of nonhistones? These proteins constitute one of the least understood groups of proteins whose functions are extremely difficult to define. At one time the genome was not understood. Now many investigators may claim that by close examination of the DNA we will learn how to control the gene. How realistic is this belief? From studies on prokaryotes, it is clear that DNA-binding proteins play a vital role in determining the various functions necessary for genetic survival of that organism. Bacteria provide an excellent system for the determination of a specific genetic function of a defined protein; the genome can be modified by the production of mutants, and the functional changes can be monitored. Chromosomal proteins of higher organisms are not amenable to such a study. Clearly the struggle to find defined functions of nonhistones requires the skills of the creative investigator, who is willing to expend his efforts even though the field does not give way to easy answers and breakthroughs. Many investigators have thus been discouraged. We believe that more encouragement should be given to those who are willing to continue in this difficult area of research, and that more rapid progress would be made as a reflection of this positive attitude.

In this spirit we have asked various investigators to contribute information recently obtained in their laboratories; we have also encouraged them to express their sentiments as to how these proteins functionally interact with DNA. This undertaking has not been easy since many investigators, discouraged by the inherent difficulties in studying these proteins, have left the field. These highly heterogeneous proteins are not readily extractable with physiological buffers, nor are they easy to isolate or purify. Then, even if our efforts at purification are successful, how are we to ascertain their biological activity? Clearly, these proteins must do more than merely fill in a nuclear space. It is scientifically correct to assume that each protein must possess a specific function. What is that function? These functions must be investigated and tabulated. The task is obviously frustrating and formidable, yet exciting and promising. Resolving the questions of gene control is the reward offered to those scientists whose persistence and pioneering attitudes allow them to continue in this complex study; this is the reward offered to the scientists contributing to this volume.

These volumes are rich with information on HMG proteins (Bhorjee, Bustin, Reeck, Teller, Isackson, Cox, Manning, Teng, Teng), on specific nuclear antigens (Chiu, Epstein, Clevenger, Pettijohn, Sculley, Spelsberg), on a specific phosphoprotein (Halikowski, Zhao), on salt-soluble nonhistones (Horst, Rackowicz-Szulczyńska, Wang, Chamberlin), on salt-insoluble nonhistones (Duerre, Traynor, Leatherbury, Horst, Rackowicz-Szulczyńska, Halikowski, Zhao, Pettijohn, Wang, Chamberlin, Hentzen, Sculley, Spelsberg), on hormone interactions with DNA (Wang, Chamberlin), on nonhistones with enzyme functions (Miyazaki, Yamashita, Kihira, Hagiwara, Nagao, Yamashita, Horio, Ruchirawat, Becker, Andoh, Ishii, Nagasawa, Seno), on nonhistones binding carcinogens (Trafzer, Rho), on Epstein-Barr nuclear antigen (Sculley, Spelsberg), on nuclear matrix salt-insoluble DNA (Vogelstein, Small, Robinson, Nelkin, Pogo, Procyk, Razin, Chernokhvostov, Yarovaya, Georgiev), on nuclear matrix salt-insoluble nonhistones (Pogo, Procyk, Razin, Chernokhvostov, Yarovaya, Georgiev), on methods for analysis of nonhistones in both normal and malignant cells (Rodriguez, Ramagli, Johnston), and on the use of monoclonal antibodies to study functions of nonhistones (Epstein, Clevenger, Pettijohn). Clearly, these two volumes encompass many of the difficult areas encountered in research on nonhistones. The scope is rather broad, yet it is focused on one theme: How do these nonhistones participate in the control of gene function?

Our purpose is to encourage research on nonhistones and to stimulate the imagination of other investigators whose future efforts might result in new discoveries as to the significance of these proteins. We hope that future volumes in this series will contain data from many more laboratories, which have been "turned-on" to NHC protein research.

**I. Bekhor**
**C. J. Mirell**
**C. C. Liew**

## THE EDITOR

**Isaac Bekhor, Ph.D.,** is Chief, Laboratory for Molecular Genetics, University of Southern California, and Professor of Biochemistry and Molecular Biology, School of Dentistry, University of Southern California, Los Angeles, California.

Dr. Bekhor graduated from the University of California, Los Angeles, with a B.S. in Chemistry, and obtained his Ph.D. in 1966 in Biochemistry, University of Southern California, Los Angeles, where the Graduate School presented him with the Harry J. Deuel, Jr. Award in Biochemistry in recognition of outstanding research. Between 1966 and 1969 Dr. Bekhor was a postdoctoral Fellow at the California Institute of Technology, Pasadena. In 1970 he joined the University of Southern California School of Dentistry Faculty. In 1971 he won a 5-year Career Development Award from the National Institutes of Health.

Dr. Bekhor is a member of the American Association for The Advancement of Science, American Chemical Society, American Society for Cell Biology, International Biochemical Society, International Society for Developmental Biologists, Southern California Craniofacial Group and West Coast Chromatin Group. He has been a consultant to *Industry on Enzyme Technology* for over 15 years.

His major research interest is the participation of nonhistone chromosomal proteins in gene regulation in eukaryotes. He has published more than 50 papers in *Molecular Biology,* and is currently cloning specific genes to facilitate his studies on the possible regulatory roles of specific nonhistones.

CONTRIBUTORS

**Jaswant S. Bhorjee, Ph.D.**
Assistant Professor
Division of Biomedical Sciences
Meharry Medical College
Nashville, Tennessee

**Michael Bustin, Ph.D.**
Laboratory of Molecular Carcinogenesis
National Cancer Institute
National Institutes of Health
Bethesda, Maryland

**Linda L. Chamberlin**
Department of Biological Sciences
State University of New York
Buffalo, New York

**Victor V. Chernokhvostov**
Postgraduate Student
Department of Nucleic Acids Biosynthesis
Institute of Molecular Biology
U.S.S.R. Academy of Sciences
Moscow, U.S.S.R.

**David J. Cox, Ph.D.**
Head
Department of Biochemistry
Kansas State University
Manhattan, Kansas

**Georgii P. Georgiev**
Head
Department of Nucleic Acids Biosynthesis
Institute of Molecular Biology
U.S.S.R. Academy of Sciences
Moscow, U.S.S.R.

**Paul J. Isackson, Ph.D.**
Postdoctoral Fellow
Department of Microbiology and Molecular Genetics
California College of Medicine
University of California
Irvine, California

**Dennis A. Johnston, Ph.D.**
Associate Professor
Associate Biomathematician
Department of Biomathematics
The University of Texas System Cancer Center
M.D. Anderson Hospital and Tumor Institute
Houston, Texas

**David Manning**
Research Assistant
Department of Biochemistry
Kansas State University
Manhattan, Kansas

**Carol J. Mirell, Ph.D.**
Department of Biochemistry
School of Medicine
University of California
Los Angeles, California

**Barry D. Nelkin**
Instructor
The Oncology Center
Johns Hopkins University School of Medicine
Baltimore, Maryland

**A. Oscar Pogo, M.D., D.M.Sci.**
Senior Investigator and Head
Laboratory of Cell Biology
The Lindsley F. Kimball Research Institute of the New York Blood Center
New York, New York

**Roman Procyk, Ph.D.**
Research Associate
Laboratory of Cell Biology
The Lindsley F. Kimball Research Institute of the New York Blood Center
New York, New York

**Louis S. Ramagli**
Research Associate
Section of Experimental Pathology
The University of Texas System Cancer Center
M.D. Anderson Hospital and Tumor Institute
Houston, Texas

**Sergei V. Razin**
Junior Research Worker
Department of Nucleic Acids Biosynthesis
Institute of Molecular Biology
U.S.S.R. Academy of Sciences
Moscow, U.S.S.R.

**Gerald R. Reeck, Ph.D.**
Professor
Department of Biochemistry
Kansas State University
Manhattan, Kansas

**Joon Rho**
Professor
School of Pharmacy
University of Southern California
Los Angeles, California

**Sabina I. Robinson**
Graduate Student
Department of Pharmacology and Oncology
Johns Hopkins University School of Medicine
Baltimore, Maryland

**Lewis V. Rodriguez, Ph.D.**
Associate Professor
Associate Pathologist
Section of Experimental Pathology
The University of Texas System Cancer Center
M. D. Anderson Hospital and Tumor Institute
Houston, Texas

**Thomas B. Sculley, Ph.D.**
Research Associate
Queensland Institute of Medical Research
Brisbane, Australia

**Donald Small**
Graduate Student
The Oncology Center
Johns Hopkins University School of Medicine
Baltimore, Maryland

**Thomas C. Spelsberg, Ph.D.**
Professor
Section of Biochemistry
Mayo Clinic and Graduate School of Medicine
Rochester, Minnesota

**David C. Teller, Ph.D.**
Professor
Department of Biochemistry
University of Washington
Seattle, Washington

**Ching Sung Teng, Ph.D.**
Professor
Department of Anatomy, Physiology, and Radiology
North Carolina State University
Raleigh, North Carolina

**Christina T. Teng, Ph.D.**
Associate Professor
Department of Anatomy, Physiology, and Radiology
North Carolina State University
Raleigh, North Carolina

**Rebecca Trafzer**
School of Pharmacy
University of Southern California
Los Angeles, California

**Bert Vogelstein**
Associate Professor
Department of Oncology
Johns Hopkins University School of Medicine
Baltimore, Maryland

**Tung Y. Wang, Ph.D.**
Professor
Department of Biological Sciences
State University of New York
Buffalo, New York

**Olga V. Yarovaya**
Institute of Molecular Biology
U.S.S.R. Academy of Sciences
Moscow, U.S.S.R.

TABLE OF CONTENTS

Volume II

Chapter 1

# HIGH MOBILITY GROUP PROTEINS: PURIFICATION, PROPERTIES, AND AMINO ACID SEQUENCE COMPARISONS

**Gerald R. Reeck and David C. Teller**

## TABLE OF CONTENTS

# I. BACKGROUND INFORMATION

The high-mobility group (HMG) proteins are a small set of distinctive nonhistone chromatin (NHC) proteins which, without doubt, have been studied more intensively than any other nonhistones. That is reflected by the inclusion in this volume of several chapters that deal with the HMG proteins. The attention that has been focused on these proteins is in large part due to practical considerations. After the histones, the HMG proteins are the chromatin proteins most amenable to investigation with currently available techniques. They are among the most abundant nonhistone proteins of chromatin, they are readily extracted from chromatin, and are soluble under nondenaturing conditions. Hence, the HMG proteins are relatively easy to purify, in comparison to other NHC proteins. Further, the acid solubility of the HMG proteins allows them to be easily identified in complex mixtures of proteins with at least some degree of confidence. As a result of the comparative ease with which they can be studied, a great deal of structural information has been accumulated on HMG proteins. In addition, studies aimed more directly at elucidating their functions have raised intriguing possibilities for the biological roles of the HMG proteins, with the effect of greatly heightening interest in them. That interest is fully justified. At the same time, there is no reason to believe that many other NHC proteins will be any less interesting when they have been investigated as thoroughly as the HMG proteins.

The first portion of this chapter is intended to provide a modicum of background information on the HMG proteins. This brief overview is by no means a comprehensive review, and we have cited only a few of the many articles that deal with HMG proteins. Much more extensive information on these proteins can be found in a recent monograph by E. W. Johns.[1]

We define the HMG proteins to be the calf thymus prototypes (HMG-1, -2, -14, and -17) and proteins closely related in amino acid sequence to one of those four proteins. By "closely related in sequence" we mean possessing extensive regions of statistically significant similarity. In the last section of this chapter we will justify our use of this restricted definition of HMG proteins.

## A. Characteristic Properties of the HMG Proteins

In 1973, Goodwin et al.[2] reported a simple procedure for obtaining a relatively small set of distinctive, NaCl-dissociated calf thymus NHC proteins. They called the set of proteins the high mobility group, and showed that the calf thymus HMG proteins had the following properties:

1. They could be extracted from chromatin with 0.35 *M* NaCl.
2. They were soluble in 2% trichloroacetic acid (TCA).
3. They had relatively high mobilities in an acid/urea polyacrylamide gel electrophoresis system (hence the name "high mobility group").
4. They had strikingly high contents of both acidic and basic amino acid residues.

A corollary to the second property listed that emerged in later work from the same laboratory[3] is that the HMG proteins can be extracted from chromatin with acids, such as 5% perchloric acid.

These properties can serve as a useful guide in attempting to identify HMG proteins from other sources, but particularly when applied to proteins from invertebrates, they fall short of allowing definitive identification. (Refer to Section III for nomenclature.)

The 2% TCA-soluble portion of the 0.35 *M* NaCl extract of calf thymus chromatin was resolved into 17 bands by acid/urea gel electrophoresis, and the bands were assigned the names HMG-1 through -17.[4] It became apparent, however, that many of the bands seen in the early preparations of calf thymus HMG proteins were degradation products of histones and of intact HMG proteins. With careful control of proteolysis, Goodwin et al.[3] identified four bands in calf thymus HMG protein preparations (HMG-1, -2, -14, and -17) that represented intact proteins and not the results of proteolytic degradation of larger proteins.

Sequence homologs of those four proteins (and, therefore, bona fide members of the set of HMG proteins) have since been identified in other species. HMG-E was discovered in avian erythrocytes[5] and was originally thought to occur only in that source. It is now recognized as a minor component in other tissues and other species.[6] HMG-T from trout testis[7] is, like HMG-E, a homolog of HMG-1 and -2. Another HMG protein from trout testis, H6, is a homolog of HMG-14 and -17.[8] Results have also been obtained in many laboratories (for the most part not including amino acid sequence data) that indicate that true HMG proteins quite probably occur in chromatins from all nucleated vertebrate cells as relatively abundant NHC proteins.

### B. Isolation and Purification

There are two approaches that are commonly used at an analytical level (i.e., in preparation for gel electrophoresis). In Method 1, chromatin or nuclei are extracted with 0.35 *M* NaCl. The extract is adjusted to 2% TCA and the soluble proteins are recovered by centrifugation. In Method 2, chromatin, nuclei, cells, or organs are extracted with acid (5% perchloric acid, 5% TCA, and 0.4 *M* sulfuric acid have been used) and the extracted proteins are separated from insoluble material by centrifugation.

The first method, particularly when applied to chromatin, minimizes the risk of contamination since 0.35 *M* NaCl extracts only a portion of the nonhistone proteins of chromatin[9] and, of the histones, only a small portion of H1. On the other hand, this extraction, which is typically carried out near neutrality, exposes the protease-sensitive HMG proteins to conditions ideal for their degradation (refer to Section II.C). Neither method can be assumed to provide a preparation that has bona fide HMG proteins as the only nonhistones. This is true as a first approximation in calf thymus (if proteolysis is controlled) and in chicken erythrocytes, the two most common sources of HMG proteins, but thymocytes and avian erythrocytes are rather peculiar in the simplicity of their NHC protein complements and in the nature of their cytoplasms, as well.

An identification is most convincing when a protein is compared by two electrophoretic systems with calf thymus or chicken erythrocyte standards. Even then, our experience is that identification, particularly of HMG-14 and -17, can be problematical. A peculiarity of those proteins that can be a useful supplementary criterion for identification is the red tinge they develop upon prolonged destaining after exposure to Coomassie blue. In at least one type of electrophoresis (nonequilibrium pH gradient electrophoresis) the presence of histones precludes the analysis of HMG-1, -2, and -E, presumably as a result of complex formation between the HMG proteins and histones (L. Wen and G. R. Reeck, unpublished observations).

Goodwin and his colleagues[10] worked out the first methods to purify individual HMG proteins at a preparative level for structural studies, including amino acid sequence determinations. NaCl or perchloric acid extracts were treated with TCA and/or acetone to yield preparations of HMG proteins at various levels of purifications. Those preparations were dissolved in aqueous solvents and chromatographed on CM-Sephadex® or CM-cellulose to yield pure proteins. Similar methods have been used in other laboratories to purify HMG-E, -T, and H6. Essentially pure preparations of HMG-14 and -17 (as a mixture) can be obtained without column chromatography by exploiting the extraordinary solubility of those proteins in concentrated TCA.[11]

The preparative procedures just mentioned are entirely suitable for studies in which the physical state of the purified proteins is irrelevant (e.g., amino acid sequence determination). Those purification methods are, however, less than ideal for studies in which it is desirable or essential that the proteins be in their native states. An assumption, usually implicit, in studies carried out with HMG proteins after their exposure to TCA and/or acetone is that the proteins regain their native states upon dissolution under nondenaturing solvent conditions. In the case of HMG-14 and -17, exposure to denaturing conditions may well be entirely without effect, since the proteins appear to be devoid of secondary or tertiary structure in their native states (refer to Section I.F). HMG-1 and -2, however, have globular domains and are clearly susceptible to denaturation. In the absence of a functional assay for HMG-1 or -2, it has been necessary to rely on indirect evidence for renaturation. This evidence includes the solubility of the proteins in nondenaturing, aqueous solvents, and the substantial content of helix in HMG-1 and -2 preparations.[12,13] Clearly, neither of these criteria provides an incontrovertible demonstration that a protein is in its native state. That exposure of basic proteins to acid may have subtle but potentially important effects that persist after return to nondenaturing conditions has been demonstrated by Brand et al.[14] They have shown that acid-extracted histone H1 differs in conformation from H1 that had not been exposed to acid.

In the absence of a functional assay for a protein and therefore of a fully reliable means of determining if the protein is in its native state, the safest course is to not expose the protein to overt denaturing conditions during its purification. Reeck's laboratory has developed methods for purification of HMG proteins in which denaturing conditions are avoided. This has been done simply as a precaution since there is as yet no proof that HMG proteins that have been exposed to TCA or acetone fail to fully renature upon dissolution in a nondenaturing solvent. (We would note, however, that in our hands, preparations of HMG-1, -2, and -E do not fully dissolve under nondenaturing conditions after having been exposed to TCA. Preparations of the same proteins that have never been exposed to denaturing conditions are fully soluble, however. It therefore seems likely that the HMG-1, -2, and -E which remain insoluble after exposure to TCA are in a denatured state. Whether all the molecules that do dissolve in the nondenaturing solvent are fully renatured is an open question.)

In our purification schemes, chromatin is treated with NaCl, usually 0.35 *M*. The extracted proteins are obtained as a supernatant after centrifugation and applied to columns, either in 0.35 *M* NaCl or after dilution or dialysis to a lower ionic strength. Phosphocellulose has proven to be a particularly useful matrix for column chromatography of the HMG proteins.[15] All of the proteins bind sufficiently tightly to it that they are retained in 0.35 *M* NaCl. By judicious choice of elution conditions (increased NaCl concentrations), a good separation of HMG proteins can be achieved. We have also found chromatography on DNA-cellulose columns to be useful,[16,17] but this is not as readily scaled up as is phosphocellulose chromatography. Particularly for purification of HMG-1, chromatography on DEAE-cellulose columns or columns containing immobilized histone H5 promise to be valuable additions to the arsenal of purification methods (L. Wen and G. R. Reeck, unpublished observations).

### C. Degradation of the HMG Proteins

The HMG proteins are all quite sensitive to degradation by proteases. In the case of the calf thymus proteins, this presents a truly vexing problem because chromatin from that tissue has avid protease(s) which, to our knowledge, no one has been able to adequately control without resorting to the use of denaturing conditions. Degradation of calf thymus HMG proteins can be minimized by extracting fresh or carefully stored tissue directly with acids.[3] Extracting calf thymus nuclei or chromatin with NaCl, however, inevitably allows substantial proteolysis of the HMG proteins. The degradation products uniformly bind more tightly to double-stranded DNA than do the parent proteins, and chromatography of NaCl extracts on immobilized DNA affords a simple way of separating the intact proteins from degradation products.[18] The intact proteins nonetheless continue to be hydrolyzed after separation from the degradation products.

To our knowledge, no other source approaches calf thymus in the seriousness of the problem of chromatin protein degradation. The HMG proteins from any source, however, are sensitive to attack by proteases, and care must be taken to control endogenous activities (even if low) and to not adventitiously introduce exogenous protease activity. HMG-1, -2, and -E are all degraded by trypsin to a relatively stable intermediate that resembles calf thymus HMG-3, a degradation product shown to be the N terminal two thirds of HMG-1.[3] The presence of HMG-3, readily ascertained by sodium dodecyl sulfate (SDS) or acid/urea gel electrophoresis, is a useful diagnostic indicator of proteolysis.

### D. Occurrence of HMG Proteins in Phylogenetically Diverse Species

Results obtained in many laboratories taken together suggest that chromatin from any nucleated vertebrate cell population will contain bona fide HMG proteins as some of the most abundant NHC proteins. Only limited information is available on the possible occurrence of HMG proteins in other eukaryotes. Spiker et al.[19] obtained proteins from wheat and yeast chromatins that had several of the features characteristic of calf thymus HMG proteins, including high contents of both acidic and basic amino acid residues. Whether those wheat and yeast proteins are HMG proteins in the restricted sense in which we use the term cannot be determined until sequence information becomes available.

Several laboratories have worked with insect proteins that display some of the features of HMG proteins.[20-22] The situation is a bit confused at this moment because of apparent discrepancies between laboratories, but it seems safe to conclude that TCA fractionation methods, when applied to insect nuclei or chromatin, do not yield proteins similar in all their properties to the calf thymus HMG proteins. In the absence of sequence information, it is not clear whether the insect proteins studied to this point are HMG proteins. We have screened proteins from an insect cell line for nonhistone that display DNA binding properties similar to those of vertebrate HMG proteins.[16,23] Sequential chromatography on columns containing immobilized double- and single-stranded DNAs revealed no proteins that displayed preferential affinity for single-stranded DNA and that had molecular weights near those of HMG-1 and -2. Instead, a protein of substantially higher molecular weight (65,000 to 70,000) was found as the major protein with those DNA binding properties.[24] Whether that protein corresponds to the *Drosophila* A63 protein of Bassuk and Mayfield[22] has not been determined.

### E. Multiplicity of the HMG Proteins

Covalent, posttranslational modifications of several sorts have been described for HMG proteins. HMG-14 and -17, but apparently not HMG-1 or -2, are phosphorylated.[25] Acetylation of all four HMG proteins[26,27] and methylation of HMG-1 and -2[28] have been reported. H6 from trout testis has been shown to be ADP-ribosylated,[29] and it has been reported that HMG proteins from Friend erythroleukemic cells are glycosylated.[30]

The products of in vitro translation of HMG-1 and -2 mRNAs appear to have the same molecular weights as purified HMG-1 and -2.[31] Hence, there appears to be no proteolytic processing in the synthesis of those proteins.

Generally, heterogeneity in any of the HMG proteins (HMG-1, for instance) is not apparent by SDS or acid/urea gel electrophoresis. Multiple forms of HMG-1, -2, and -E have, however, been separated by isoelectric focusing[32] or a nonequilibrium variant of it.[33] Recently, improved resolution in two-dimensional gel electrophresis has revealed that chicken erythrocyte HMG-14 occurs as several species (L. Wen and G. R. Reeck, unpublished observations), whereas only two such species had previously been supposed to exist.[15] The relationship, if any, between the multiple forms of HMG proteins observed by isoelectric focusing and heterogeneity produced by posttranslational modifications is not yet clear.

### F. Physical Properties

Molecular weights of HMG-1 and -2 determined by sedimentation equilibrium under

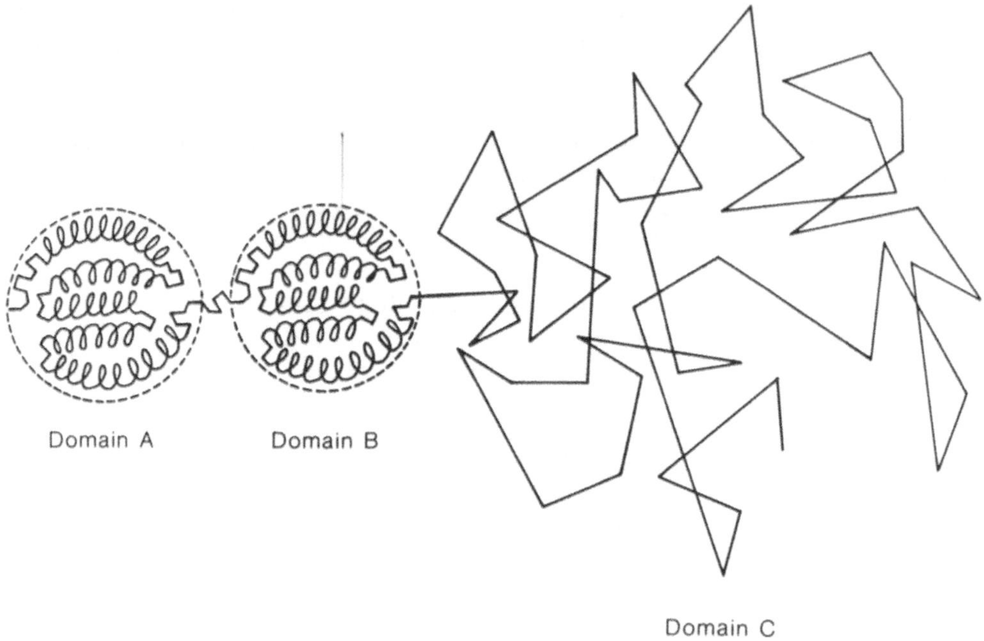


FIGURE 1. Model of three-dimensional organization of HMG-1 and its homologs. Domains A and B were assumed to fold as compactly as possible into spheres, for which we calculated radii of 14.5 and 13.7 Å, respectively. Domain C was assumed to occur in a randomly coiled state comparable to that assumed by polypeptides in 6 *M* guanidine hydrochloride. Under that assumption, using Equation 5 of Tanford et al.[37] we calculated a sedimentation coefficient from domain C as if it were a separate polypeptide. From the sedimentation coefficient we calculated a frictional coefficient, which in turn allowed us to calculate a Stokes radius (41.9 Å). In drawing the diagram, the randomly coiled polypeptide of domain C was confined to a circle of radius 41.9 Å, although in actuality in a random coil with that Stokes radius, the chain would frequently extend outside a sphere of radius 41.9 Å. Since domain C has an extremely high negative charge density, charge repulsion would likely cause it to be more extended than is a random coil.

nondenaturing conditions (26,500 and 26,000, respectively)[10] approximate the polypeptide molecular weights calculated from amino acid sequence data.[34] Thus, the proteins occur predominantly as monomers. Calf thymus HMG-1 and -2 have pIs of about 6.5 and 8.5, respectively.[32] Each has been estimated by circular dichroism measurements to have about 40 to 50% of its amino acid residues in helical conformations.[12,13] HMG-T, on the other hand, has been reported to contain little if any helix.[35] Since HMG-T is very similar in amino acid sequence to HMG-1 and -2,[7] its low helix content suggests that it refolds incompletely after exposure to the acidic conditions used in its isolation.

Analysis of the amino acid sequences of HMG-1 and -2 (presented in Section II.C below) indicates that each of these proteins is organized into three domains, each of approximately 80 to 90 residues. The N terminal and central domains, which we call domains A and B, occur as independent globular units connected by a several-residue portion of the polypeptide chain. Each of these domains is expected on the basis of the Chou and Fasman rules[36] to have a high content of helix. Domain C is the highly acidic C terminal third of the molecules. Because it is nearly devoid of hydrophobic amino acid residues and because it possesses a very large net negative charge, it can be presumed to occur essentially as a random coil, without organized folding. Our current view of the three-dimensional organization of HMG-1 and -2 is depicted schematically in Figure 1.

A comparison of ultracentrifuge data[10] and amino acid sequence[38] indicates that HMG-17, like HMG-1 and -2, does not self-associate to a significant extent. Circular dichroism and NMR studies have shown that HMG-14 and -17 are devoid of organized folding.[39,40]

This is not unexpected in light of the essential lack of hydrophobic amino acid residues in these proteins. Amino acid sequences[38,41] and nonequilibrium pH gradient electrophoresis[33] show HMG-14 and -17 to be highly basic proteins.

### G. Postulated Functions

The functions of the HMG proteins are unknown. That is indeed true for all NHC proteins other than those with recognized enzymatic activities. Several proposals have been raised on the functions of the HMG proteins. We will do little more than list some of them. For a far more detailed presentation of relevant data and a thorough assessment of the significance of those data, see the recent chapter by Goodwin and Mathew.[42]

One apparent certainty amidst the current abundance of uncertainties is that the HMG proteins are incapable of recognizing DNA sequences on purified DNA. Thus, they do not appear to be among those NHC proteins that bind to DNA with sequence-specificity.

HMG-1, -2, and -T have been proposed to be involved in either DNA replication or in transcription. Since neither of these is proven, the possibility that the proteins are involved in neither process is certainly open. Our suggestion that HMG-1 and -2 might participate in DNA replication stemmed from our finding that they possess preferential affinity for single-stranded DNA and that they are enriched in rapidly proliferating cell populations.[16,23,43] There is, however, no direct evidence at this time for participation of HMG-1 or -2 in DNA replication. A case for localization of this class of proteins in regions of chromatin containing transcribed DNA sequences has been made most persuasively by Levy-Wilson et al.[44] (in the case of HMG-T). That sort of localization, which has been argued largely on indirect evidence, would presumably imply a generalized participation of the protein in transcription. Some direct evidence for such an involvement has recently been provided by Stoute and Marzluff.[45] They showed that RNA synthesis by endogenous polymerase in NaCl-extracted nuclei could be stimulated by addition to the nuclei of preparations that contained HMG-1 and -2.

Considerable interest has been focused on the possible participation of HMG-14 and -17 in transcription. Weisbrod and Weintraub[46] have presented evidence that HMG-14 and -17 are responsible for the greater sensitivity to DNase I of transcribed DNA sequences (in chromatin) than of bulk DNA (in chromatin). The proteins appear to bind preferentially to nucleosomes that contain transcribed DNA sequences.[11,47] There are two binding sites for HMG-14 or -17 per nucleosomal core particle.[11,48] Levy-Wilson et al.[44] have shown that the trout homolog H6 is enriched in nucleosome preparations that are enriched in transcribed DNA sequences. The preferential association of HMG-14 and -17 (or H6) with nucleosomes that contain transcribed DNA sequences suggests that these proteins play a generalized role in transcription. Although there has been a tendency to consider the association of HMG-14 and -17 with transcribed chromatin as specific, such is apparently not the case. Chromatin greatly enriched in presumably nontranscribed satellite DNA sequences has been found to contain HMG-14 and -17 at only modestly lower levels than does bulk chromatin.[49] The available data appear to be consistent with preferential binding of HMG-14 and -17 to nucleosomes that contain transcribed DNA (at a level of preference that is not yet determined) and qualitatively different effects of their binding on the susceptibility to DNase I of that DNA, on the one hand, and of bulk DNA sequences, on the other hand.

The studies reported to this point on HMG-14 and -17 have been very stimulating, but the story is far from complete. Among the important questions that remain to be answered are by what mechanism HMG-14 and -17 bind more tightly to transcriptionally active nucleosomes than to bulk nucleosomes and even more fundamentally, what role, if any, HMG-14 and -17 have in transcription. Our recent finding that HMG-14 and -17 exhibit preferential affinity for single-stranded DNA[50] may have some bearing on one or both of those issues.

## II. AMINO ACID SEQUENCE COMPARISONS

### A. Methods

We have used computerized procedures to analyze the sequences of HMG proteins for similarities with each other and with other proteins. Computations were performed on a PDP-12 or a VAX 11/780 computer using programs developed in Teller's laboratory. The methods have for the most part been previously described rather fully,[51,52] so we will present them here in summary form only.

There are three principal elements to the procedures used in the current work: a diagonal search procedure, the Sankoff algorithm for optimizing alignments, and Monte Carlo approaches to evaluate the statistical significance of sequence similarities.

The *diagonal search* is used as a preliminary way of finding extensive regions of sequence similarity or as a means of finding all short regions of similarity between two sequences. The approach was introduced by Fitch[53] and has since been used widely. All stretches of a prescribed length (typically up to 25 residues) from one protein are compared to all stretches of the same length in a second sequence. For two 100-residue sequences, there are 6561 comparisons of stretches of length 20 in an exhaustive diagonal search. The score for the comparison of any two stretches is the summation of the scores at each position for the amino acid residues paired at that position. We most frequently use the scoring scheme of McLachlan[54] and supplement it most commonly with the scores of Dayhoff.[55] Both of these sets of scores are based on the frequencies with which pairs of amino acids have been interchanged in the course of evolution of homologous proteins. Occasionally we use minimum base differences[56] as a scoring scheme. In our experience and others' as well,[55] minimum base differences are considerably less sensitive in detecting sequence similarities.

The mean and standard deviation for comparisons of all stretches of the prescribed length are provided as output, as are identifications of the comparisons with the highest scores. The mean and standard deviation can give some feel for the statistical significance of the score of a comparison of potential interest, but a rigorous estimate of statistical significance is obtained by a Monte Carlo method described below. With slight modification, the diagonal search procedure is used to examine a single sequence for internal similarities.

The *Sankoff algorithm* was devised to find the best alignment between two nucleic acid sequences under a constraint imposed by the investigator on the number of gaps used in making the alignment.[57] We have adapted and used the procedure for comparison of protein sequences. Details are provided elsewhere.[51] The power of this approach lies in its objectively producing optimized alignments and in doing so under a specified restriction on the use of gaps. Our programs for both the diagonal search and the Sankoff algorithm can be used to compare families of proteins, with many sequences in each family.

We use *Monte Carlo methods* to evaluate the statistical significance of similarities revealed by the diagonal search or by the Sankoff algorithm. A Monte Carlo method compares the result of an analysis of real data with the results from analogous analyses of randomized data produced by a digital computer. This allows one to estimate the probability of obtaining a score by chance that is as high or higher than the score obtained from the real data (amino acid sequences, in our case). Such a randomization method was suggested by Sankoff and Cedergren[58] for evaluating the statistical significance of gaps in alignments of nucleic acid sequences. The power of the method is that it makes no assumptions whatsoever about the distribution of scores from comparisons of random sequences. Instead, the distribution appropriate for each problem is produced by the computer, and the score for an alignment between the actual sequences is compared directly to scores for alignments between many pairs of randomized sequences. Depending upon the degree of similarity between the actual sequences, the use of the Monte Carlo method (as opposed to a calculation based on the assumption of a Gaussian distribution of scores) can be quite important in ascertaining the statistical significance of an alignment, since in general the scores from comparison of randomized sequences do not have a Gaussian distribution.[51]

To test the significance of a short stretch of similarity revealed by a diagonal search, the computer generates 100 pairs of randomized sequences of the same lengths and amino acid compositions as the actual sequences between which the similarity was found. A diagonal search is then performed on each of the 100 pairs of randomized sequences. The computer counts the number of pairs of randomized sequences in which, by the diagonal search, a similarity is found to equal or exceed the similarity found between the real sequences. From this we calculate an estimated level of significance as described below.

The randomization approach is used in an analogous way to evaluate a similarity revealed by the Sankoff algorithm. In so doing, we are also able to determine the number of gaps that are justifiably introduced into an alignment. This is the only nonarbitrary method that we are aware of to deal with the longstanding problem of how many gaps to use in aligning protein or nucleic acid sequences. The score for comparison of any two sequences can essentially always be improved by introducing more gaps into the alignment. In proposing any given alignment, one needs to know how many gaps are justified. We determine this by using the Sankoff algorithm in combination with Monte Carlo randomizations. Many pairs of randomized sequences (typically 100) are generated by the computer. The randomized sequences have the same lengths and amino acid compositions as the particular pair of actual sequences between which an alignment was found by the Sankoff algorithm. The computer performs the Sankoff algorithm on each pair of randomized sequences at numbers of gaps ranging up to and beyond the number of gaps being considered for aligning the actual sequences. The scores for the best alignment obtained at each number of gaps for every pair of randomized sequences are provided as output, as are the increments in scores for alignments obtained in allowing each larger number of gaps.

We use a formula of Birnbaum[59] to calculate an estimated level of significance, $\alpha$, for a score obtained by comparison of actual sequences:

$$\alpha = \frac{1 + N\gamma}{1 + N}$$

In this formula, N is the number of randomizations performed and $\gamma$ is the fraction of randomizations in which a score was observed that is equal to or greater than the score obtained from comparison of the actual sequences; $\alpha$, the level of significance of the real score, is the probability that the degree of similarity of the real sequences (as indicated by their score) could be due simply to chance. Very similar sequences will have low $\alpha$. Even if a score from comparison of the actual sequences exceeds all of the scores from comparisons of the randomized sequences (i.e., $\gamma = 0$), we are limited by the Birnbaum formula to concluding that $\alpha = 0.01$ when $N = 100$. This limitation holds regardless of how much the score for the actual sequences exceeds the highest score obtained in the 100 randomizations. More precise estimates of $\alpha$ can in principle be obtained by performing more randomizations, but in many cases the number of randomizations required to produce an alignment as good or better than that between two actual sequences of substantial similarity exceeds all reasonable bounds on the use of computer time. Under such circumstances we fall back on rough estimates of probabilities calculated under the assumption of Gaussian distributions.

The number of gaps justified on statistical grounds is determined using the Birnbaum formula on incremental scores. For example, the increment in score achieved in going from two to three gaps in aligning the actual sequences is compared to the increments obtained in going from two to three gaps in aligning each of the 100 pairs of randomized sequences. Typically, the probability of achieving the incremental score by chance ($\alpha_{inc}$) decreases up

```
                                   10                  20                  30                  40
BOVINE  HMG-14:  P K R K V S S A E G A A - - - - - - - - K E E P K R R S A R L S A K P A P A K V E
CHICKEN HMG-14:  P K R K - A P A E G E A - - - - - - - - K E E P K R R S A R L S A K P A P P K P E
BOVINE  HMG-17:  P K R K - - - A E G D A K G D K A K V K D E P Q R R S A R L S A K P A P P K P D
CHICKEN HMG-17:  P K R K - - - A E G D T K G D K A K V K D E P Q R R S A R L S A K P A P P K P D
TROUT H6:        P K R K - - - S A - - T K G - - - - - - D E P A R R S A R L S A R P V P - K P A

                                   50                  60                  70                  80
BOVINE  HMG-14:  T K P K K A A - G K D K S S D K K V Q T K G K R G A K - - G K Q A E V A N Q E T
CHICKEN HMG-14:  P K P K K A A P P K E K A A D D K
BOVINE  HMG-17:  P K P K K A P A K K - - - - - G E K V P K G K K G K A D A G K E - - - - - - - -
CHICKEN HMG-17:  P K P K K A A P K K - - - - - S E K V P K G K K G K A D A G K D - - - - - - - -
TROUT H6:        A K P K K A A A P K - - - - - - - K A V K G K K - - - - - - - - - - - - - - - -

                                   90                  100                 110
BOVINE  HMG-14:  K E D L P A E N G E T K N E E S P A S D E A E E K E A K S D
BOVINE  HMG-17:  - G N N P A E N G D A K T N Q A E K A E - - G A G D A K
CHICKEN HMG-17:  - G N N P A E N G D A K T N Q A E K A E - - G A G D A K
TROUT H6:        - - - - A A E N G D A K A E A K V Q A A G D G A G N A K
```

FIGURE 2. Alignment of the amino acid sequences of the HMG-14/HMG-17/H6 family. A-Ala; C-Cys; D-Asp; E-Glu; F-Phe; G-Gly; H-His; I-Ile; K-Lys; L-Leu; M-Met; N-Asn; P-Pro; Q-Gln; R-Arg; S-Ser; T-Thr; V-Val; W-Trp; Y-Tyr.

to a certain number of gaps and then precipitously increases as more gaps are allowed. The number of gaps that we deem justified is that number which has associated with it the lowest $\alpha_{inc}$. The behavior of $\alpha_{inc}$ with increasing numbers of gaps and the relationship between that behavior and the level of significance for the alignment can be seen in graphical form elsewhere.[51]

## B. Alignment of HMG-14, -17, and H6

The sequences of HMG-14,[41] -17,[38] and trout H6[8] have been determined in the laboratories of Johns and Dixon. Those investigators recognized that these proteins for a family of homologs. We have reinvestigated the alignment of the set of sequences to optimize it with our computerized procedures.

The alignment shown in Figure 2 was constructed as follows. We first aligned the most closely related sequences and then sequentially added more distantly related sequences into the family. Bovine and chicken HMG-17s, which differ at only five residues,[60] were aligned visually without gaps. Bovine HMG-14 was then aligned with the 49-residue N terminal sequence of chicken HMG-14.[61] In doing so, we used the Sankoff algorithm to optimize the alignment, and randomizations to determine how many gaps were justified. In visually aligning these sequences, Walker et al.[61] included only one gap (that at position 5 in our Figure 2). We find that two gaps are actually clearly justified (at positions 5 and 48). Using McLachlan and Dayhoff scores, we found $\alpha_{inc}$ for including two gaps to be 0.01 and 0.02, respectively. In examining 100 pairs of randomized sequences with the two scoring schemes, only one case was found in which the increment in going from one to two gaps was equal to or greater than the increment in going from one to two gaps in the actual sequences. Thus, two gaps are justified. A third is clearly not, however, since $\alpha_{inc}$ for including three gaps were found to be 0.91 and 0.99 with McLachlan and Dayhoff scores, respectively. In allowing three gaps in aligning the randomized sequences, nearly all such alignments improved more over the two-gap alignments than did the alignment between the real sequences in going from two to three gaps.

In this family of HMG proteins, there is a highly conserved region (positions 20 to 50 in Figure 2) that has been previously identified by other investigators by visual inspection of the sequences. On either side of that region, however, the similarities among HMG-14, -17, and H6 are substantially less. It is in those portions of the sequences where the Sankoff algorithm for objectively finding optimized alignments and the Monte Carlo method for determining how many gaps are justified are particularly useful. By pairwise comparisons among the three families formed thus far (HMG-14s, -17s, and the one-member H6 family), we found that H6 and the HMG-17 family form the most closely related pair of families in the region C terminal to position 50. We therefore proceeded in our constuction of the overall alignment by optimizing the alignment between H6 and the two HMG-17s. Finally, that three-member family was aligned with the family of two HMG-14s, with the result shown in Figure 2. That alignment contains all the gaps that are justified on statistical grounds. The number of gaps is substantially lower than the number included by other workers in aligning the same set of sequences. This is consistent with our rather general experience that in forming alignments visually many investigators use more gaps than are actually justified.

At each step in constructing the alignment in Figure 2, use of the McLachlan and Dayhoff scores gave results that were in good agreement. By examining in isolation the portions of the molecules N terminal to position 20 and C terminal to position 50, we found that statistically significant sequence similarity exists over the entire sequences and is not limited to the highly conserved stretch at positions 20 to 50.

The HMG-14/HMG-17/H6 family is notable for an extraordinarily low content of hydrophobic amino acid residues. The proteins, as a result, are presumably unable to form an interior and assume globular conformations. Physical measurements (Section I.F) are consistent with the three proteins' existing as random coils. In globular proteins, a major constraint that operates to conserve sequence similarity is the requirement that an amino acid substitution not disrupt the folding of a protein. That constraint would appear to be absent in the evolution of HMG-14/HMG-17/H6. The striking conservation of sequence in positions 20 to 50 therefore presumably results solely from constraints on this region that are needed to preserve the functions of the proteins. The very high content of basic amino acid residues in positions 20 to 50 suggests that this is a DNA-binding region of the molecule. NMR studies have in fact indicated that portions of the proteins corresponding rather closely to positions 20 to 50 are involved in DNA binding.[39,40] There would appear, therefore, to be very strong constraints on amino acid substitutions in the DNA-binding portions of HMG-14/HMG-17/H6.

The formation of a complex between one of these HMG proteins and DNA would result in the system losing a great deal of configurational entropy since the portion of the protein that binds to the DNA would be restricted in its conformation in the complex and largely unrestricted when the protein is not bound. Configurational entropy would therefore strongly favor dissociation. The loss in configurational entropy upon binding is probably minimized in these HMG proteins by the high proline content of their DNA-binding regions. The covalent bond between the $C_\alpha$ and the N atoms of proline would allow far less conformational freedom in the DNA-binding regions than the same regions would have if alanine, for example, were substituted for every proline. Since the unstructured DNA-binding regions of some of the histones, particularly H1 and H5, are similarly rich in proline, this may be a frequently used strategy in DNA-binding polypeptides that lack defined three-dimensional structure. It is also worth noting that a protein such as these HMG proteins, which lack organized folding and for which configurational entropy strongly favors dissociation, must have more lysine and arginine residues in their DNA-binding regions than would be required to achieve equally tight binding in other DNA-binding proteins that have similar three-

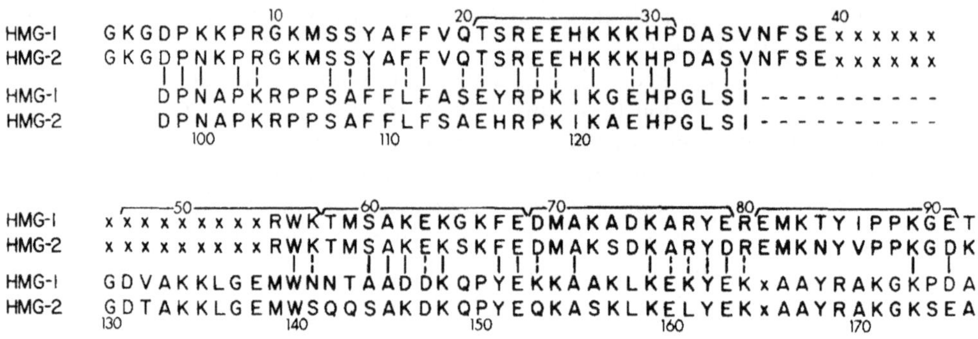


FIGURE 3. Alignment showing the internal sequence similarity in HMG-1 and -2. X indicates positions at which the identity of the amino acid residue is still unknown.[34] Brackets indicate five short (11 residue) stretches that exhibit similarity to each other (see text). For the one-letter abbreviations for amino acids see the legend to Figure 2.

dimensional structures when bound to DNA and when not bound to DNA. This would ensure that that conformation which binds to DNA binds sufficiently tightly to more than offset the strong effect of conformational entropy to promote dissociation.

### C. Internal Sequence Similarities in HMG-1 and -2

We have recently reported the existence of internal similarity in the sequences of HMG-1 and -2.[62] Here we further document that finding and take note of smaller repeating units in the sequences of the proteins.

In our work with HMG-1 and -2, we combined the sequences reported by Walker et al.[34] into a two-member family that we call HMG-1,2. In Figure 3 we show an alignment that depicts the internal sequence similarity in HMG-1,2. We have used X to indicate positions at which residues are believed to occur but whose identities are not yet established. Based on the sequence information currently available, the first portion of the sequence of HMG-1,2 that is involved in the sequence similarity (residues 1 to 92) is somewhat longer than the second portion (residues 98 to 176). Thus, when the sequences are completed it appears that one gap will be needed to optimize the alignment. In assessing the statistical significance of the alignment shown in Figure 3, we have therefore compared the score of that alignment with scores of alignments of randomized sequences in which one gap was allowed. Regardless of the scoring scheme used, the alignment in Figure 3 was produced by the Sankoff algorithm as the optimal alignment.

The internal similarity in HMG-1,2 is sufficiently strong that it is not feasible to estimate $\alpha$ precisely by Monte Carlo methods. In either the McLachlan or Dayhoff scheme, the score of the alignment is much higher than the highest score obtained in the randomizations (Table 1). Hence it would require an unreasonable amount of computer time to perform enough randomizations to produce an alignment by chance that had a score as high or higher than the score of the alignment of the actual sequences. Apparent probabilities calculated under the assumption of Gaussian distributions serve to emphasize the statistical significance of the internal similarity in HG-1,2 (Table 1). That significance results from long stretches of only moderate levels of similarity rather than from short stretches of high similarity.

The internal similarity in the sequences of HMG-1 and -2 divides the amino acid sequences of the proteins into three sections, each of approximately the same length: residues 1 to 92, 98 to 176, and 177 to 259. We call those sections domains A, B, and C, respectively. A region that corresponds approximately to what we call domain C has been previously recognized to be distinct from the rest of the molecule on the basis of its peculiar amino acid composition:[34] in HMG-1, over 50% of the residues in domain C are Glu or Asp, and very few hydrophobic residues occur in that portion of the molecule.

**Table 1**
**STATISTICAL ANALYSIS OF INTERNAL SIMILARITY IN HMG-1,2**

| Scoring scheme | Data from 100 randomizations: Mean | S.D. | Top score | Score of alignment of actual sequences | Distance from mean (in σ) | $P_{app}$ |
|---|---|---|---|---|---|---|
| McLachlan scores | 625 | 66.3 | 862 | 1072 | 6.74 | $1 \times 10^{-11}$ |
| Dayhoff scores | 820 | 85.6 | 1028 | 1440 | 7.24 | $3 \times 10^{-13}$ |

*Note*: $P_{app}$ was calculated under the assumption of Gaussian distributions of scores of alignments of randomized sequences. $P_{app}$ is the apparent probability of obtaining by chance an alignment with a score equal to or greater than the score of the alignment between the actual sequences.

We believe that the sequence domains also occur as separate domains in a three-dimensional sense. Digestion of HMG-E with trypsin releases fragments that correspond to domains A and B.[62,63] Each is a water-soluble, DNA-binding protein with no apparent tendency to interact with the other. We therefore believe that in the intact protein, domains A and B are independent globular units joined by a short connecting peptide (postulated to be Lys-Lys-Lys-Phe-Lys in HMG-1 and Lys-Gly-Lys-Lys-Lys in HMG-2). In general, polypeptides with similar amino acid sequences are believed to fold similarly. That is known from X-ray crystallographic structure determinations to be true for proteins with substantially less sequence similarity than that exhibited by domains A and B of HMG-1,2 (for instance, distantly related globin chains[64]). On the basis of similarity in predicted secondary structures,[62] domains A and B seem likely to obey the generalization and possess similar three-dimensional structures. Domain C presumably exists as a random coil. It is therefore expected to occupy a much greater volume than does either domain A or domain B (see Figure 1).

HMG-1 and -2 contain several shorter regions of internal sequence similarity that we have not previously reported. We discovered these by diagonal searches at short stretch lengths on the family of 4 sequences aligned as in Figure 3. We found several stretches of length 11 that exhibited similarity in pairwise comparisons. Those stretches are identified in Figure 3 by brackets above the appropriate portions of the sequence of domain A of HMG-1. The similarities among these bracketed regions are not overwhelmingly strong. Designating the most N terminal region as segment 1, the two most similar pairs of segments are numbers 2 and 4 and numbers 3 and 4. Each of these pairs has a probability of about 0.1 associated with it, as assessed by our Monte Carlo method. That is, one would find two stretches of length 11 with similarity as great or greater in about one of every 10 pairs of randomized sequences with length and composition of the family of four sequences shown in Figure 3. The probability of finding five similar stretches of length 11 in a single sequence (or a single family of four sequences) would be quite low, however. We have not yet attempted to evaluate that probability. Study of these small repeats is hampered by the fact that the sequences of HMG-1 and -2 are incompletely known. We will reexamine the occurrence and significance of the small repeats when more complete sequence information becomes available.

The sequence similarity between domains A and B of HMG-1,2 is most readily explained in the usual manner — an internal gene duplication followed by divergent evolution. It is not possible with currently available information to speculate meaningfully on whether the duplication occurred before or after the DNA coding for domain C had become part of the gene. The existence of short repeats within domains A and B suggests that the gene that coded for the single domain ancestral to domains A and B arose in part by internal duplication.

### D. Comparisons of HMG Proteins with Other Proteins

**HMG-14 and HMG-17 with Lysine-Rich Histones.** Several short stretches of similarity

### Table 2
### SHORT REGIONS OF SEQUENCE SIMILARITY BETWEEN LYSINE-RICH HISTONES AND HMG-14 OR -17

| Proteins | Sequences (starting positions) | α |
|---|---|---|
| Bovine HMG-14 | (35) KPKKAAGKDK | |
| Rabbit H1.3 | (139) KPKKAAGAKK | 0.02 |
| Bovine HMG-14 | (35) KPKKAA | |
| Trout H1 | (175) KPKKAA | 0.07 |
| Bovine HMG-14 | (34) TKPKKAA | |
| Chicken H5 | (116) TSPKKAA | 0.14 |
| Bovine HMG-17 | (38) PKPKKAPAKK | |
| Rabbit H1.3 | (119) PKPKKAGAAK | 0.03 |
| Bovine HMG-17 | (38) PKPKKAPAKK | |
| Rabbit H1.3 | (197) PKPKAAKAKK | 0.03 |
| Bovine HMG-17 | (38) PKPKKAPAKK | |
| Trout H1 | (115) PKAKKVPAKK | 0.20 |
| Bovine HMG-17 | (9) AKGDKAK | |
| Chicken H5 | (96) AKSDKAK | 0.13 |

*Note*: This table lists all similarities found to have a $\alpha \leqq 0.2$. α is the probability (estimated by Monte Carlo methods described in the text) of finding a similarity as great or greater between random sequences of the same lengths and amino acid compositions as the two proteins between which a similarity was found.

between the sequences of HMG-14 or -17 and lysine-rich histones (H1 or H5) have been pointed out by other investigators.[38,40-42,61] We have asked whether there are statistically significant similarities between those HMG proteins and H1 or H5. With our Monte Carlo method, we have found that a stretch of similarity must be at least 6 residues long to begin to approach statistical significance. For pairs of randomized sequences of the lengths and compositions of chicken H5[65] and bovine HMG-17, α for a 6-residue *identity* is 0.05. That is, approximately 1 of every 20 pairs of randomized sequences will contain 1 or more 6-residue identities. A 6-residue alignment with 1 nonidentity or stretches shorter than 6 residues falls well short of statistical significance. For example, α for a 4-residue identity is 0.8.

We have used the diagonal search procedure to see whether regions of statistically significant similarity exist between bovine HMG-14 or -17 and three lysine-rich histones: rabbit thymus H1.3,[66] trout H1,[67] or chicken H5.[65] All stretches that have $\alpha \leqq 0.2$ are shown in Table 2. The strongest alignment, which occurs between HMG-14 and rabbit H1.3, is ten residues long and contains nine identities. That alignment is of modest statistical significance ($\alpha = 0.02$). All but one of the similarities in Table 2 contain the tetrapeptide sequence KPKK from HMG-14 or -17 (positions 42 to 46 of the alignment in Figure 2). In every case, the similarities break down on both sides of the alignments shown in the table.

The portions of H1 or H5 involved in the alignments in Table 2 are the C terminal, lysine-rich tails of those molecules. We used the Sankoff algorithm to find optimal alignments between HMG-14 or the family of sequences aligned in Figure 2 and the C terminal tail of trout H1 (Residues 103 to 194). With the family of HMG proteins, we found no alignment that was close to statistical significance. With HMG-14 alone, a more promising alignment was found (not shown), but it nonetheless fell well short of significance ($\alpha$ = 0.17).

There clearly are regions of similarity between HMG-14 or -17 and the lysine-rich histones. All are quite short, however, and even the longest does not provide an adequate basis for suggesting that it arose as a result of divergence from a common ancestral sequence. The term "homology" is therefore not appropriate to describe these regions of similarity, since "homology" is taken by many investigators, ourselves included, to imply a divergence from a common ancestor.

The similarities shown in Table 2 (some of which are among those pointed out by other investigators) occur in portions of the HMG proteins and the lysine-rich histones that appear to have similar functions, in a broad sense. The sequence similarities are between small portions of larger DNA-binding regions that are devoid of organized folding. These short stretches of sequence similarity could well reflect similarities in the details of DNA binding in those limited stretches of the polypetides. These DNA-binding regions in the HMG proteins and the lysine-rich histones are peculiar in amino acid composition (being rich in lysine, proline, and alanine). It is possible that this makeup is required for a randomly coiled DNA-binding polypeptide. Thus, the similarities between HMG-14 or -17 and the lysine-rich histones may have resulted from convergent evolution to compositional similarity. The similarity in composition would be expected to produce short stretches of sequence similarity. We have previously discussed this possibility in considering the N terminal regions of the core histones.[52] Even if extensive stretches of sequence similarity were found in regions of distinctive and similar compositions (and this is not the case with the HMG proteins and lysine-rich histones), it would be difficult to argue convincingly that the sequence similarities resulted from divergent evolution.

**HMG-14 and HMG-17 with the Core Histones.** Diagonal searches between the sequences of bovine HMG-14 or -17 and the sequences of the four core histones (H2A, H2B, H3, and H4) revealed few similarities that approached statistical significance. The following alignment between HMG-17 and histone H3[68] had $\alpha$ = 0.12:

```
HMG-17    (41) KKAPAKKGEK
H3        (27) KSAPATGGVK
```

A more interesting similarity (because it extends over a longer region) is the following one between HMG-17 and H2B:[69]

```
HMG-17    (31) PAPPKPDPKPKKAPAKKGEKVPKGKKGK
H2B        (1) PEPAKSAPAPKKGSKKAVTKAQKKDGKK
```

This alignment of length 28 had $\alpha$ = 0.01. (In 100 pairs of randomized sequences, no stretches of length 28 were found with similarity as great as that of the alignment shown above.) The similarity does not extend further, however, and it is between lysine- and proline-rich regions of two DNA-binding polypeptides. We feel it is best interpreted as a short stretch of sequence similarity that results from compositional similarity.

**HMG-1 and HMG-2 with Histones.** We searched for similarities between domain A or domain B of HMG-1,2 and the histones. A stretch of strong similarity was found between HMG-1,2 and trout H2B:[70]

```
HMG-1      (1) GKGDPKKPRGKMSSYAFFV
H2B       (21) GKGGKKRKRSRKESYAIYV
```

Monte Carlo estimation of the significance of this similarity gave $\alpha = 0.01$. The strength of the alignment is dependent, however, upon peculiarities of the trout H2B sequence. Alignment of the corresponding portion of bovine H2B with the first 19 residues of HMG-1,2 is not statistically significant. The Sankoff algorithm, applied to domain A of HMG-1,2 and the trout H2B sequence, revealed an alignment that contained the above stretch and extended over the entire sequence of domain A (not shown). While not statistically significant ($\alpha = 0.10$), this alignment merits further attention when the HMG-1 and HMG-2 sequences are completed or when the sequences of homologous proteins become available.

**Other Comparisons.** Based on comparisons of amino acid compositions, Spiker et al.[19] suggested that certain bacterial DNA-binding proteins (HTa from *Thermoplasma acidophilum*[71] and HU from *Escherichia coli*[72]) might be more appropriately considered HMG proteins than histone-like proteins. We used the Sankoff algorithm to search for sequence similarities between HTa or HU and domain A or domain B of HMG-1,2. No statistically significant similarity was found. We therefore believe it is inappropriate to consider HTa and HU to be HMG proteins.

We examined all entries in the Protein Sequence Database[73] for possible similarity to HMG-14, -17, or domain A or B of HMG-1,2. No similarities of apparent statistical significance were found other than those already discussed above. An interesting pattern emerged, however, in searching for sequences with similarities to the C terminal portions of HMG-14 or -17. Those regions of these HMG proteins (particularly HMG-14) have many Asp and Glu residues and few basic residues[38,41] (see Figure 2). Diagonal searches of those acidic regions of HMG-14 and -17 against all sequences in the Database produced a set of alignments that were based primarily on matches between acidic residues. Although none of the alignments was statistically significant, the group of alignments with the greatest similarities was interesting because of the preponderance of nucleic acid-binding proteins among the sequences selected from the Database. Excluding comparisons to other HMG proteins, of the 26 stretches most similar to residues 76 to 100 of bovine HMG-14, 15 were from DNA- or RNA-binding proteins. Thus, it appears that stretches rich in Asp and Glu are rather common features of proteins that interact with nucleic acids. These negatively charged stretches occur most commonly near the C terminus of the proteins. HMG-1 and -2 probably represent the extreme form of this phenomenon with their rather long domain C of extraordinarily high negative charge density. HMG-14 and -17, with their shorter stretches of lower charge density, are more representative of the negatively charged regions in nucleic acid-binding proteins as revealed by our search of the Database.

## III. NOMENCLATURE CONSIDERATIONS

The difficulties of assigning names to previously unstudied proteins have been emphasized recently by Johns.[1] The term "HMG protein" was devised for a heterogeneous protein preparation about which only rudimentary physical and chemical information was available and for which nothing about function was known.[2] Given the paucity of information available when the name originated, it is not surprising that the term presents some difficulties today. This is true even when it is taken to include only calf thymus HMG-1, -2, -14, and -17 and proteins related to them in amino acid sequence. We will briefly consider some of those difficulties here to support our belief that the term "HMG protein" should be used only in that restricted sense.

As has been pointed out by several authors including Johns,[1] the canonical HMG proteins fall into two distinct families: HMG-1 and -2 form one family, and HMG-14 and -17 form another. The two families are quite different in amino acid compositions and are unrelated in sequences or three-dimensional organizations. They bear no apparent similarities in their interactions with nucleosomes. Thus, even in its restricted usage, "HMG protein" refers to two unrelated families of proteins. The effects of using a single name for unrelated proteins are far from trivial. In order to identify one of the families, one must use a cumbersome qualifier. For instance, we have used "high molecular weight HMG proteins" to limit the reference to HMG-1, -2, and -E. In addition to its awkwardness, this phrase suffers from the fact that HMG-1, -2, and -E are not high molecular weight at all — they are simply larger than the other HMG proteins. Nonetheless, a more suitable qualifier has not been suggested. Perhaps because it is easier to write and to read sentences about "the HMG proteins" than about "the high molecular weight (or low molecular weight) HMG proteins," many authors simply omit such qualifiers.* One therefore becomes accustomed to reading properties attributed to the entire set of HMG proteins even when authors provide references to support the attribution to only one of the two families. This expediency is no doubt used with the intention of writing directly and simply, but when it is repeated frequently the simplification can subtly and intrusively distort one's logic to the point where one thinks of the HMG proteins as all being interrelated. There is no doubt that people outside the field and people who work on its periphery usually fall into that trap. So powerful are the effects of a name, however, that thoroughly informed people working directly with the HMG proteins must actively resist the tendency to think of them as a completely interrelated group.

In retrospect, it would have been preferable to give 17 unrelated names to the 17 electrophoretic bands in the original HMG protein preparations. As information accumulated about the proteins, groupings based on biologically meaningful information such as sequence similarities could have given rise to a nomenclature that reflected those groupings. The problems that have resulted to this point in using the term "HMG protein" have been relatively minor and are probably not worth trying to remedy by a change in nomenclature. We should be careful, however, to not compound those problems by including still other families of unrelated proteins under the rubric of "HMG protein". That will surely happen if we do not restrict the use of the term to mean proteins related in amino acid sequence to the calf thymus HMG proteins.

The pitfalls in using less stringent criteria can be seen if we examine a recently revised set of properties of calf thymus HMG proteins that Johns[1] has suggested can be used to tentatively evaluate whether a candidate protein is an HMG protein. We emphasize that Johns states that a final decision must be reserved until an amino acid sequence is known.[1] Given the substantial obstacles to obtaining an amino acid sequence, it would be quite easy to overlook Johns' conclusion that sequence is the definitive criterion for membership in the high mobility group and to use less restrictive criteria based on more readily acquired information.

In two of the five criteria suggested for tentative evaluation of a candidate protein, properties that are continuous are treated as if they were all or none. Johns states that calf thymus HMG proteins are extractable from chromatin with 0.35 *M* NaCl.[1] To test the extractability of a protein, the most common approach is to examine only the extract, since it can be far more difficult to analyze the proteins that remain bound to DNA. At any NaCl concentration, there will be an equilibrium between the free and bound forms of any chromatin protein. Thus, all chromatin proteins would be extracted to some extent at 0.35 *M* NaCl (or any other NaCl concentration), and how many of them one visualizes in the extract depends upon the sensitivity of the detection method. To what extent should a protein be extracted at 0.35 *M* NaCl? At what chromatin concentration should its extraction be characterized, since the extent of dissociation will depend upon that concentration? (By the law of mass action, high concentrations promote complex formation.)

* A related tendency is to refer to the proteins as the "HMGs". Since only one high mobility group has been defined, the term "high mobility groups" (HMGs) is meaningless. In almost all contexts, "HMG proteins" should be used instead of "HMGs".

The TCA-solubility criterion is similarly oversimplified. Johns states that HMG proteins are not precipitated from a 0.35 *M* NaCl extract in 2% TCA.[1] With such an extract of chicken erythrocyte chromatin, readily detectable portions of each of HMG-1, -2, and -E are in fact precipitated at 2% TCA. Our tentative conclusion to that effect[33] has now been confirmed by radioimmunochemical detection after two-dimensional gel electrophoresis of the material precipitated by 2% TCA (L. Wen and G. R. Reeck, unpublished observations). Thus, chicken erythrocyte HMG-1, -2, and -E have been found in the so-called low mobility group (LMG) fraction. With a naive interpretation of a TCA-solubility criterion, they could therefore be called low mobility group proteins. Conversely, proteins that are only slightly soluble in 2% TCA may be detectable in the 2% TCA-soluble fraction, depending upon the sensitivity of one's detection method. The HMG proteins are indeed more soluble in 2% TCA than are most other proteins. Differences in solubility are, however, differences in degree. No protein is infinitely soluble or totally insoluble in any given solvent. What must the solubility of a candidate protein be in 2% TCA to meet the solubility criterion?

Johns[1] has proposed the following as distinctive features of the amino acid compositions of the calf thymus HMG proteins: Asx + Glx $\geqq$ 20%; Lys + Arg $\geqq$ 25%; Pro $\geqq$ 7%. In earlier work,[2,10] a value of 30% for Ask + Glx was used, but all of the individual calf thymus HMG proteins actually fall short of that. Moving the lower limit for acidic amino acids to 20% was necessary to include HMG-17, for which Asx + Glx = 22.5%. A lower limit of 20% is sufficiently low, however, to make it rather nondiscriminating. In a sample of 207 proteins chosen without regard to function or source, we found average Asx and Glx contents of 10.7 and 10.6 mol%, respectively.[74] Since Asx and Glx are not statistically correlated,[74] half of all proteins will have Asx + Glx $\geqq$ 21.3%. Many proteins that meet Johns' criteria for the contents of Pro and of Lys + Arg will therefore exceed the requirement of 20% set on Asx + Glx. This requirement is therefore quite weak. Nevertheless, it is still too strong to include all known HMG proteins. As Johns points out,[1] H6 is quite decisively excluded since it has Asx + Glx = 13%.[8] Thus, the amino acid composition criteria as currently formulated are seriously flawed. We suggest that a criterion be used that compares the entire compositions of a candidate protein and a calf thymus HMG protein. We have found such comparisons to be good indicators of sequence similarity.[75]

The five criteria listed by Johns[1] are clearly less than rigorous in their current form. Nevertheless, they have legitimate uses. HMG proteins have been thoroughly characterized from a wide range of vertebrates. With that backlog of information, the five criteria of Johns can be used to rather confidently identify HMG proteins from new vertebrate sources. We would recommend, however, that the criteria be supplemented with electrophoretic studies. In studying proteins from invertebrate sources, the criteria are useful as only rough guidelines and are not an adequate basis for naming a protein an HMG protein. Our position seems to be in substantial agreement with that of Johns, who has also indicated that a final determination must be based on amino acid sequence data.[1]

Johns has made it clear that "HMG protein" was simply a piece of ad hoc laboratory jargon for proteins that had been placed under investigation in his laboratory and that therefore needed a name.[1] This name has seemed to assume greater significance to other investigators that its originators intended. Some workers have been treating the name as if it were part of a classification scheme that divided chromatin proteins into groups based on fundamentally important shared properties. As we have emphasized, however, even in its restricted usage "HMG protein" refers to two unrelated families whose greatest similarity is relatively high solubility in 2% TCA. This is presumably not of fundamental biological significance. We believe it is important to continue using "HMG protein" only with its limited interpretation. To fail to do so would result in calling other unrelated families of proteins by the same name and generating a good deal of confusion.

## ACKNOWLEDGMENTS

Work in our laboratories on HMG proteins and comparisons of their sequences has been supported by NIH grants CA-17782 and GM-29203 to Gerald R. Reeck and GM-13401 to David C. Teller. Gerald R. Reeck is the recipient of NIH Research Career Development Award CA-00425. This is publication number 82-681-B, Kansas Agricultural Experiment Station.

## REFERENCES

1. **Johns, E. W., Ed.,** *The HMG Chromosomal Proteins*, Academic Press, New York, 1982.
2. **Goodwin, G. H., Sanders, C., and Johns, E. W.,** A new group of chromatin associated proteins with a high content of acidic and basic amino acids, *Eur. J. Biochem.*, 38, 14, 1973.
3. **Goodwin, G. H., Walker, J. M., and Johns, E. W.,** Studies on the degradation of High Mobility Group nonhistone chromosomal proteins, *Biochim. Biophys. Acta*, 519, 233, 1978.
4. **Goodwin, G. H., Nicolas, R. H., and Johns, E. W.,** An improved large scale fractionation of High Mobility Group nonhistone chromatin proteins, *Biochim. Biophys. Acta*, 405, 280, 1975.
5. **Sterner, R., Boffa, L. C., and Vidali, G.,** Comparative structural analysis of High Mobility Group nonhistone chromosomal proteins from a variety of sources. Evidence for a High Mobility Group protein unique to avian erythrocyte nuclei, *J. Biol. Chem.*, 253, 3830, 1978.
6. **Mathew, C. G. P., Goodwin, G. H., Gooderham, K., Walker, J. M., and Johns, E. W.,** A comparison of the high mobility group nonhistone chromatin protein HMG-2 in chicken thymus and erythrocytes, *Biochem. Biophys. Res. Commun.*, 87, 1243, 1979.
7. **Watson, D. C., Peters, E. H., and Dixon, G. H.,** The purification, characterization and partial sequence determination of a trout testis non-histone protein, HMG-T, *Eur. J. Biochem.*, 74, 53, 1977.
8. **Watson, D. C., Wong, N. C. W., and Dixon, G. H.,** The complete amino acid sequence of a trout testis nonhistone protein, H6, localized in a subset of nucleosomes and its similarity to calf thymus nonhistones proteins HMG-14 and HMG-17, *Eur. J. Biochem.*, 95, 193, 1979.
9. **Bidney, D. L. and Reeck, G. R.,** Analysis of the effectiveness of sodium chloride in dissociating non-histone chromatin proteins of cultured hepatoma cells, *Biochim. Biophys. Acta*, 521, 753, 1978.
10. **Goodwin, G. H., Walker, J. M., and Johns, E. W.,** The High Mobility Group (HMG) nonhistone chromosomal proteins, in *The Cell Nucleus*, Vol. 6, Busch, H., Ed., Academic Press, New York, 1978, 181.
11. **Sandeen, G., Wood, W. I., and Felsenfeld, G.,** The interaction of High Mobility proteins HMG-14 and HMG-17 with nucleosomes, *Nucleic Acids Res.*, 8, 3757, 1980.
12. **Baker, C., Isenberg, I., and Johns, E. W.,** Physical studies of the nonhistone chromosomal proteins HMG-1 and HMG-2, *Biochemistry*, 15, 1645, 1976.
13. **Cary, P. D., Crane-Robinson, C., Bradbury, E. M., Javaherian, K., Goodwin, G. H., and Johns, E. W.,** Conformational studies of two non-histone chromosomal proteins and their interactions with DNA, *Eur. J. Biochem.*, 62, 583, 1976.
14. **Brand, S. H., Kumar, N. M., Walker, I. O.,** The properties of salt-extracted histone H1, *FEBS Lett.*, 133, 63, 1981.
15. **Isackson, P. J., Debold, W. A., and Reeck, G. R.,** Isolation and separation of chicken erythrocyte High Mobility Group non-histone chromatin proteins by chromatography on phosphocellulose, *FEBS Lett.*, 119, 337, 1980.
16. **Isackson, P. J., Fishback, J. L., Bidney, D. L., and Reeck, G. R.,** Preferential affinity of high molecular weight High Mobility Group non-histone chromatin proteins for single-stranded DNA, *J. Biol. Chem.*, 254, 5569, 1979.
17. **Isackson, P. J., Clow, L. G., and Reeck, G. R.,** Comparison of the salt dissociations of high molecular weight HMG non-histone chromatin proteins from double-stranded DNA and from chromatin, *FEBS Lett.*, 125, 30, 1981.
18. **Isackson, P. J. and Reeck, G. R.,** Removal of degradation products from calf thymus High Mobility Group non-histone chromatin proteins by chromatography on double-stranded DNA, *Biochim. Biophys. Acta*, 697, 378, 1982.
19. **Spiker, S., Mardian, J. K. W., and Isenberg, I.,** Chromosomal HMG proteins occur in three eukaryotic kingdoms, *Biochem. Biophys. Res. Commun.*, 82, 129, 1978.

20. **Alfageme, C. P., Rudkin, G. T., and Cohen, L. H.,** Isolation, properties and cellular distribution of D1, a chromosomal protein of *Drosophila*, *Chromosoma*, 78, 1, 1980.
21. **Marquez, G., Moran, F., Franco, L., and Montero, F.,** Cl proteins: a class of High-Mobility-Group non-histone chromosomal proteins from the fruit fly *Ceratitis capitata*, *Eur. J. Bochem.*, 123, 165, 1982.
22. **Bassuk, J. A. and Mayfield, J. E.,** Major High Mobility Group like proteins of *Drosophila melanogaster* embryonic nuclei, *Biochemistry*, 21, 1024, 1982.
23. **Bidney, D. L. and Reeck, G. R.,** Purification from cultured hepatoma cells of two nonhistone chromatin proteins with preferential affinity for single-stranded DNA: apparent analogy with calf thymus HMG proteins, *Biochem. Biophys. Res. Commun.*, 85, 1211, 1978.
24. **Fishback, J. L., Tallman, G., Reeck, G. R., and Kramer, K. J.,** Non-histone chromatin proteins from an insect cell line: preferential affinity for single-strand DNA of a high molecular weight protein, *Insect Biochem.*, 12, 251, 1982.
25. **Saffer, J. D. and Glazer, R. I.,** The phosphorylation of High Mobility Group proteins 14 and 17 and their distributon in chromatin, *J. Biol. Chem.*, 257, 4655, 1982.
26. **Sterner, R., Vidali, G., and Allfrey, V. G.,** Studies of acetylation and deacetylation in High Mobility Group proteins. Identification of the sites of acetylation of HMG-1, *J. Biol. Chem.*, 254, 11577, 1979.
27. **Sterner, R., Vidali, G., and Allfrey, V. G.,** Studies of acetylation and deacetylation in High Mobility Group proteins. Identification of the sites of acetylation in High Mobility Group proteins 14 and 17, *J. Biol. Chem.*, 256, 8892, 1981.
28. **Boffa, L. C., Sterner, R., Vidali, G., and Allfrey, V. G.,** Post-synthetic modifications of nuclear proteins. High Mobility Group proteins are methylated, *Biochem. Biophys. Res. Commun.*, 89, 1322, 1979.
29. **Wong, N. C. W., Poirier, G. G., and Dixon, G. H.,** Adenosine diphosphoribosylation of certain basic chromosomal proteins in isolated trout testis nuclei, *Eur. J. Biochem.*, 77, 11, 1977.
30. **Reeves, R., Chang, D., and Chung, S.-C.,** Carbohydrate modifications of the High Mobility Group proteins, *Proc. Natl. Acad. Sci. U.S.A.*, 78, 6704, 1981.
31. **Bustin, M., Neihart, N. K., and Fagan, J. B.,** mRNA of chromosomal proteins HMG-1 and HMG-2 are polyadenylated, *Biochem. Biophys. Res. Commun.*, 101, 893, 1981.
32. **Walker, J. M., Goodwin, G. H., and Johns, E. W.,** The similarity between the primary structures of two nonhistone chromosomal proteins, *Eur. J. Biochem.*, 62, 461, 1976.
33. **Tyrell, D., Isackson, P. J., and Reeck, G. R.,** Two-dimensional gel electrophoresis of nonhistone chromatin proteins with nonequilibrium pH gradient electrophoresis as the first dimension, *Anal. Biochem.*, 119, 433, 1982.
34. **Walker, J. M., Gooderham, K., Hastings, J. R. B., Mayes, E., and Johns, E. W.,** The primary structures of non-histone chromosomal proteins HMG 1 and 2, *FEBS Lett.*, 122, 264, 1980.
35. **Cary, P. D., Crane-Robinson, C., Bradbury, E. M., and Dixon, G. H.,** Structural studies of the non-histone chromosomal proteins HMG-T and H6 from trout testis, *Eur. J. Biochem.*, 119, 545, 1981.
36. **Chou, P. Y. and Fasman, G. D.,** Prediction of the secondary structure of proteins from their amino acid sequences, *Adv. Enzymol.*, 47, 45, 1978.
37. **Tanford, C., Kawahara, K., and Lapanje, S.,** Proteins as random coils. I. Intrinsic viscosities and sedimentation coefficients in concentrated guanidine hydrochloride, *J. Am. Chem. Soc.*, 89, 729, 1967.
38. **Walker, J. M., Hastings, J. R. B., and Johns, E. W.,** The primary structure of a non-histone chromosomal protein, *Eur. J. Biochem.*, 76, 461, 1977.
39. **Abercrombie, B. D., Kneale, G. G., Crane-Robinson, C., Bradbury, E. M., Goodwin, G. H., Walker, J. M., and Johns, E. W.,** Studies on the conformational properties of the High-Mobility-Group chromosomal protein HMG 17 and its interaction with DNA, *Eur. J. Biochem.*, 84, 173, 1978.
40. **Cary, P. D., King, D. S., Crane-Robinson, C., Bradbury, E. M., Rabbani, A., Goodwin, G. H., and Johns, E. W.,** Structural studies on two High-Mobility-Group proteins from calf thymus, HMG-14 and HMG-20 (ubiquitin), and their interaction with DNA, *Eur. J. Biochem.*, 112, 577, 1980.
41. **Walker, J. M., Goodwin, G. H., and Johns, E. W.,** The primary structure of the nucleosome-associated chromosomal protein HMG 14, *FEBS Lett.*, 100, 394, 1979.
42. **Goodwin, G. H. and Mathew, C. G. P.,** Role in gene structure and function, in *The HMG Chromosomal Proteins*, Johns, E. W., Ed., Academic Press, New York, 1982, 193.
43. **Shastri, K., Isackson, P. J., Fishback, J. L., Land, M. D., and Reeck, G. R.,** The influence of nonhistone chromatin protein HMG-1 on the enzymatic digestion of purified DNA, *Nucleic Acids Res.*, 10, 5059, 1982.
44. **Levy-Wilson, B., Connor, W., and Dixon, G. H.,** A subset of trout testis nucleosomes enriched in transcribed DNA sequences contains High Mobility Group proteins as major structural components, *J. Biol. Chem.*, 254, 609, 1979.
45. **Stoute, J. A. and Marzluff, W. F.,** HMG-proteins 1 and 2 are required for transcription of chromatin by endogenous RNA polymerase, *Biochem. Biophys. Res. Commun.*, 107, 1279, 1982.
46. **Weisbrod, S. and Weintraub, H.,** Isolation of a subclass of nuclear proteins responsible for conferring a DNase I sensitive structure on globin chromatin, *Proc. Natl. Acad. Sci. U.S.A.*, 76, 630, 1979.

47. **Weisbrod, S. T.,** Properties of active nucleosomes as revealed by HMG 14 and 17 chromatography, *Nucleic Acids Res.*, 10, 2017, 1982.
48. **Mardian, J. K. W., Paton, A. E., Bunick, G. J., and Olins, D. E.,** Nucleosome cores have two specific binding sites for nonhistone chromosomal proteins HMG 14 and HMG 17, *Science*, 209, 1534, 1980.
49. **Mathew, C. G. P., Goodwin, G. H., Igo-Kemenes, T., and Johns, E. W.,** The protein composition of rat satellite chromatin, *FEBS Lett.*, 125, 25, 1981.
50. **Isackson, P. J. and Reeck, G. R.,** Nonhistone chromatin proteins HMG-14 and HMG-17 bind preferentially to single-stranded DNA, *Nucleic Acids Res.*, 9, 3779, 1981.
51. **de Haën, C., Swanson, E., and Teller, D. C.,** The evolutionary origin of proinsulin. Amino acid sequence homology with the trypsin-related serine proteases detected and evaluated by new statistical methods, *J. Mol. Biol.*, 106, 639, 1976.
52. **Reeck, G. R., Swanson, E., and Teller, D. C.,** The evolution of histones, *J. Mol. Evol.*, 10, 309, 1978.
53. **Fitch, W. M.,** An improved method of testing for evolutionary homology, *J. Mol. Biol.*, 16, 9, 1966.
54. **McLachlan, A. D.,** Tests for comparing related amino acid sequences. Cytochrome *c* and cytochrome $c_{551}$, *J. Mol. Biol.*, 61, 409, 1971.
55. **Dayhoff, M. O.,** *Atlas of Protein Sequence and Structure*, Vol. 5, National Biomedical Research Foundation, Silver Springs, Md., 1972.
56. **Cantor, C. R. and Jukes, T. H.,** The repetition of homologous sequences in the polypeptide chains of certain cytochromes and globins, *Proc. Natl. Acad. Sci. U.S.A.*, 56, 177, 1966.
57. **Sankoff, D.,** Matching sequences under deletion/insertion constraints, *Proc. Natl. Acad. Sci. U.S.A.*, 69, 4, 1972.
58. **Sankoff, D. and Cedergren, R. J.,** A test for nucleotide sequence homology, *J. Mol. Biol.*, 77, 159, 1973.
59. **Birnbaum, Z. W.,** Computers and Unconventional Test Statistics, Contract N-onr-477(38), Tech. Rep. No. 62, Office of Naval Research, Department of the Navy, Washington, D.C., 1973.
60. **Walker, J. M., Stearn, C., and Johns, E. W.,** The primary structure of non-histone chromosomal protein HMG-17 from chicken erythrocyte nuclei, *FEBS Lett.*, 112, 207, 1980.
61. **Walker, J. M., Brown, E., Goodwin, G. H., Stearn, C., and Johns, E. W.,** Studies on the structures of some HMG-like non-histone chromosomal proteins from trout and chicken tissues. Comparison with calf thymus proteins HMG 14 and 17, *FEBS Lett.*, 113, 253, 1980.
62. **Reeck, G. R., Isackson, P. J., and Teller, D. C.,** Domain structure in high molecular weight High Mobility Group nonhistone chromatin proteins, *Nature (London)*, 300, 76, 1982.
63. **Isackson, P. J.,** Isolation of High Mobility Group Nonhistone Chromatin Proteins and Studies of Their DNA Binding Properties, Ph.D. thesis, Kansas State University, Manhattan, 1980.
64. **Lesk, A. M. and Chothia, C.,** How different amino acid sequences determine similar protein structures: the structure and evolutionary dynamics of the globins, *J. Mol. Biol.*, 136, 225, 1980.
65. **Briand, G., Kmiecik, D., Sautiere, P., Wouters, D., Borie-Loy, O., Biserte, G., Mazen, A., and Champagne, M.,** Chicken erythrocyte histone H5. IV. Sequence of the carboxy-terminal half of the molecule (96 residues) and complete sequence, *FEBS Lett.*, 112, 147, 1980.
66. **Cole, R. D.,** Special features of the structures of H1 histones, in *The Molecular Biology of the Mammalian Genetic Apparatus*, Vol. 1, Ts'o, P.O.P., Ed., North-Holland, Amsterdam, 1977, 93.
67. **MacLeod, A. R., Wong, N. C. W., and Dixon, G. H.,** The amino acid sequence of trout-testis histone H1, *Eur. J. Biochem.*, 78, 281, 1977.
68. **DeLange, R. J., Hooper, J. A., and Smith, E. L.,** Histone III. III. Sequence studies on the cyanogen bromide peptides; complete amino acid sequence of calf thymus histone III, *J. Biol. Chem.*, 248, 3261, 1973.
69. **Iwai, K., Hayashi, H., and Ishikawa, K.,** Calf thymus lysine- and serine-rich histone. III. Complete amino acid sequence and its implication for interactions of histones with DNA, *J. Biochem.*, 72, 357, 1972.
70. **Koostra, A. and Bailey, G. S.,** Primary structure of histone H2B from trout (*Salmo trutta*) testes, *Biochemistry*, 17, 2504, 1978.
71. **DeLange, R. J., Williams, L. C., and Searcy, D. G.,** A histone-like protein (HTa) from *Thermoplasma acidophilum*. II. Complete amino acid sequence, *J. Biol. Chem.*, 256, 905, 1981.
72. **Mende, L., Timm, B., and Subramanian, A. R.,** Primary structures of two homologous ribosome-associated DNA-binding proteins of *Escherichia coli*, *FEBS Lett.*, 96, 395, 1978.
73. **Orcutt, B. C. and Dayhoff, M. O.,** *Protein Sequence Database*, NBR Report 820535-08710, National Biomedical Research Foundation, Washington, D.C., 1982.
74. **Reeck, G. R. and Fisher, L.,** A statistical analysis of the amino acid compositions of proteins, *Int. J. Peptide Protein Res.*, 5, 109, 1973.
75. **Harris, C. E. and Teller, D. C.,** Estimation of primary sequence homology from amino acid compositions of evolutionarily related proteins, *J. Theor. Biol.*, 38, 347, 1973.

Chapter 2

# STUDIES ON THE INTERACTIONS OF HMG-1 AND ITS HOMOLOGS WITH DNA

**Paul J. Isackson, David J. Cox, David Manning, and Gerald R. Reeck**

## TABLE OF CONTENTS

# I. INTRODUCTION

HMG-1, -2, and -E are a set of sequence homologs from the family of high molecular weight high mobility group nonhistone chromatin (NHC) proteins.[1] A number of interesting observations have been made on the DNA-binding properties of these proteins. We reported that the proteins from several sources possess a preferential affinity for single-stranded DNA.[2,3] That conclusion was drawn from chromatography of the proteins on double- and single-stranded DNAs immobilized on cellulose. Javaherian et al.[4] reported that calf thymus HMG-1 and -2 are capable of changing the helical structure of DNA — specifically, of reducing the linking number of nicked circular DNA that is covalently closed in the presence of one of the proteins. That effect could be produced by any of several mechanisms, including unwinding due to preferential binding to single-stranded DNA.[4] Javaherian et al.[5] later reported that calf thymus HMG-1 and -2 were capable of lowering the $T_m$ of poly(dA-dT). Duguet and de Recondo and their colleagues have been studying a single-stranded DNA-binding protein (called S25) from rat liver and a protein (called HD25, indistinguishable in most respects from S25) from regenerating rat liver. Recently that group has identified S25 and HD25 as cytosolic forms of rat HMG-1.[6] HD25 (hence, cytosolic HMG-1 from regenerating rat liver) has been reported to be capable of lowering the $T_m$ of poly(dA-dT).[7,8]

Several prokaryotic proteins that bind preferentially to single-stranded DNA are known to be required for DNA replication.[9] The gene 32 protein of bacteriophage T4, the most thoroughly studied of those proteins, is capable of lowering the $T_m$ of poly(dA-dT)[10] but does not lower the $T_m$s of naturally occurring DNAs.[9] We have suggested[2,3,11,12] that HMG-1 and -2 might play a role in DNA replication analogous to that of the gene 32 protein, which is believed to promote the formation of single-stranded DNA regions at replication forks.[9] Our suggestion was based on the preferential affinity of HMG-1 and its homologs for single-stranded DNA[2,3] and on the increased amounts of HMG-1 and -2 in rapidly proliferating hepatoma cells compared to the amounts in adult rat liver.[2,12] In light of that proposal, we are interested in the details of the interactions of HMG-1 and its homologs with DNA and with chromatin, and we are particularly interested in whether the proteins can lower the melting temperature of DNA or of synthetic polynucleotides.

Chicken erythrocyte chromatin is an attractive source of HMG proteins. HMG-1, -2, and -E are the most abundant nonhistone proteins in that chromatin[3] and the total complement of nonhistones in chicken erythrocyte chromatin seems simpler than that in other sources.[13] Erythrocytes can be obtained in large quantities, and HMG proteins can be isolated from erythrocyte chromatin without much proteolytic degradation. We have therefore chosen to purify HMG proteins on a large scale from chicken erythrocytes and to study the interactions of those purified proteins with DNA. The usefulness of such studies depends in large part on obtaining the purified proteins in their native states. We have developed a large-scale purification procedure for chicken erythrocyte HMG proteins that purposefully avoids overt denaturing conditions.[14] In that regard, it differs from more commonly used methods for purifying HMG proteins. (See Reference 1 for a discussion of that point.)

In this chapter, we report studies designed to determine whether chicken erythrocyte HMG proteins are capable of lowering the $T_m$ of poly(dA-dT). We also report binding studies from which we obtain estimates of the affinities of HMG-E for single- and double-stranded DNAs and of the binding site sizes for the two types of DNA.

# II. RESULTS

## A. Thermodynamic Parameters for the Interaction of HMG-E with DNA

Upon binding to DNA, the intrinsic fluorescence of a protein commonly decreases, either as a result of a conformational change in the protein, the positioning of fluorescent side chains next to DNA, or both. The decrease in fluorescence can be used to generate a binding curve from which estimates of binding parameters can be extracted using a suitable model.[15] In the case of the HMG proteins binding to DNA, the suitable model is the overlapping sites formulation of McGhee and von Hippel.[16] Their approach takes into account the fact that potential binding sites overlap for proteins that bind without sequence specificity to sites larger than one base in single-stranded DNA or one base pair in double-stranded DNA. This overlap results in a form of negative cooperativity that makes it difficult to fully saturate a DNA lattice with the protein ligand. (The system maintains a state of high entropy when not all bases, or base pairs, are covered, because of the many possible arrangements of the bound protein molecules on the less-than-saturated lattice.) The McGhee and von Hippel model can also accommodate more typical forms of cooperativity in the interaction of a protein with DNA. With positive cooperativity, a protein molecule is more likely to bind immediately adjacent to a previously bound molecule than to leave bases unoccupied between protein molecules. Strongly positive cooperative interactions (such as the binding of the gene 32 protein to single-stranded DNA)[10,17] can overcome the entropic negative cooperativity inherent in the nonspecific binding of a protein to DNA and readily produce saturation of the DNA. In the work that follows, however, there is no evidence for positive cooperativity, and the interaction of HMG-E with DNA is fully described by two parameters:[15] K, the affinity of the protein for a site on the DNA, and n, the binding site size; that is, the number of bases (or base pairs) occluded by the binding of one molecule of protein to the DNA lattice.

Using fluorescence titration and the overlapping binding sites model, we have studied the binding of HMG-E to DNA. HMG-E was chosen because it binds more tightly to DNA than does HMG-1[3,18] and because it is more abundant (and therefore more readily isolated) than is HMG-2.[14] We have studied four cases: the binding of HMG-E to double- and single-stranded DNA at two solvent conditions — a low-ionic-strength solvent (1 m*M* Tris-Cl/1 m*M* EDTA/1 m*M* dithiothreitol) at pH 7.5, and a higher ionic strength solvent consisting of the solvent just listed supplemented with 0.2 *M* NaCl.

Binding data for the four cases are shown in Figures 1 to 4. The data were obtained by adding small aliquots of concentrated DNA solutions to larger volumes of a protein solution and measuring fluorescence after each addition of DNA. In all cases we have studied, and as is commonly the case, the intrinsic fluorescence of the protein decreases upon adding DNA. After suitable corrections (see Section IV), the data are plotted in the form of a binding curve. We estimated binding parameters by computer simulation of binding curves over a range of reasonable values of K and n to find those values that produced a curve that best fit the experimental data.

When the binding data for such a system are plotted in the Scatchard format, there is a characteristic shape (upward convexity) if the system exhibits positive cooperativity.[15] Plots of our data in that format revealed no evidence for positive cooperativity (not shown). Other investigators have also concluded that the interactions of calf thymus and rat HMG-1s with DNA are not positively cooperative.[5,19,20] In simulating binding curves, we have therefore used McGhee and von Hippel's Equation 10, which lacks the cooperativity parameter. The best fits to our data sets that we obtained with that equation (see Section IV), are shown as solid lines in Figures 1 to 4. In Table 1, we summarize the estimates of K and n for the four cases we studied.

**B. Thermal Denaturation of Poly(dA-dT) in the Presence of HMG-1**

We have followed the denaturation of poly(dA-dT) spectrophotometrically at 260 nm in the presence and absence of chicken erythrocyte HMG-1 to determine whether that protein is capable of lowering the $T_m$ of the double-stranded polynucleotide. We chose poly(dA-dT) for these studies because its $T_m$ can be lowered by several prokaryotic single-stranded DNA binding proteins, including the gene 32 protein.[9] Recently two groups have concluded

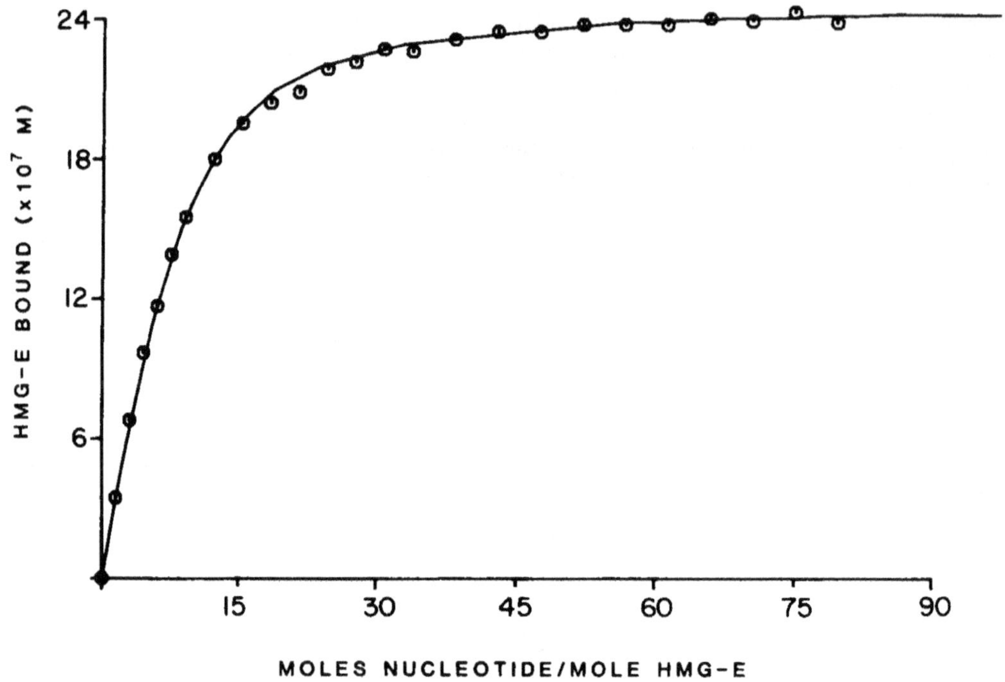


FIGURE 1. Fluorescence titration for the binding of HMG-E to double-stranded DNA in the low ionic strength solvent (1 m*M* Tris-HCl/1 m*M* EDTA/1 m*M* dithiothreitol at pH 7.5). The ratio of nucleotide to HMG-1 is given in terms of bases, not base pairs. The circles are data points. The curve is the best-fitting theoretical titration curve, for which $K = 2.5 \times 10^6 M^{-1}$ and n = 12 bases or 6 bp.

that calf thymus HMG-1[5] and HMG-1 from regenerating rat liver[7,8] can also lower the $T_m$ of poly(dA-dT).

We have examined the thermal denaturation of poly(dA-dT) at three NaCl concentrations: 10, 50, and 150 m*M*. In each case, additional components of the solvent were 1 m*M* Tris-HCl/0.25 m*M* EDTA/1 m*M* dithiothreitol and the pH was 8.1. Figures 5 to 7 show denaturation profiles in 10, 50, and 150 m*M* NaCl, respectively, at several weight ratios of HMG-1 to poly(dA-dT). To allow several curves to be shown in each figure, the curves are offset arbitrarily without reindexing the ordinate. In each figure the ordinate is labeled "Absorbance", but in several cases the curves result more from light scattering than from absorption. All curves in each figure result from heating samples of the same concentration of poly(dA-dT). If the only process contributing to the increase in apparent absorbance were the denaturation of poly(dA-dT), the increase should be nearly the same for all curves, regardless of the HMG-1-to-poly(dA-dT) ratio. (Any effect of HMG-1 on the hyperchromicity can be expected to be small.) In the curves in each figure, however, at sufficiently high HMG-1-to-poly(dA-dT) ratios, we see much larger increases in apparent absorbance than are seen in the absence of HMG-1. Thus, some process other than poly(dA-dT) denaturation is contributing to the increase in apparent absorbance. That process is the denaturation and aggregation of HMG-1, which results in light scattering.

The curves in Figure 5 were obtained at the lowest ionic strength: 10 m*M* NaCl in 1 m*M* Tris-HCl/0.25 m*M* EDTA/1 m*M* dithiothreitol. Poly(dA-dT) alone melted in a single transition centered at 40°C. Our value for this and other $T_m$s for poly(dA-dT) (in the absence of HMG-1) are in good agreement with literature values.[21] At HMG-1-to-poly(dA-dT) ratios up to 2, the increases in absorbance approximate the increase seen in the denaturation of poly(dA-dT) alone. Therefore, those increases are due essentially entirely to poly(dA-dT) denaturation. Adding HMG-1 to the poly(dA-dT) increases by about 10°C, the temperature

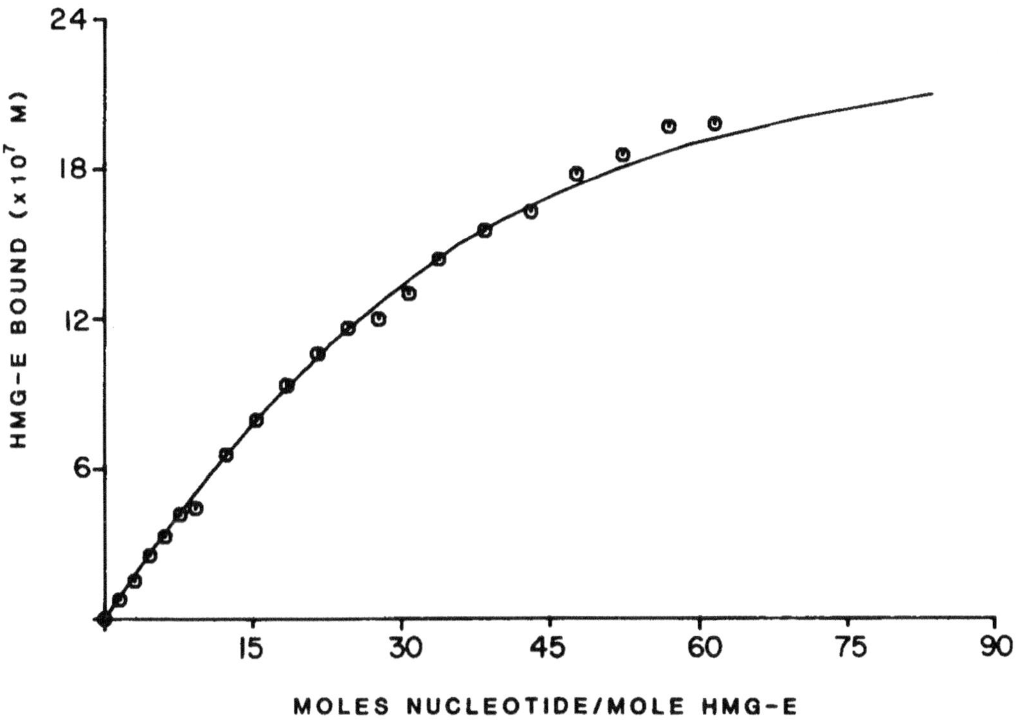


FIGURE 2. Fluorescence titration for the binding of HMG-E to double-stranded DNA in the 0.2 *M* NaCl-supplemented solvent. As in Figure 1, the ratio of nucleotide to HMG-1 is given in terms of bases. The circles are data points. The curve is the best-fitting theoretical titration curve, for which K = $4.0 \times 10^4$ $M^{-1}$ and n = 20 bases or bp.

at which the synthetic DNA is denatured. At ratios of 0.5 and 1, the denaturation curves are biphasic. We note that such curves are not, as was once believed, diagnostic of positive cooperativity in the binding of a protein to DNA.[22] At an HMG-1-to-poly(dA-dT) ratio of 6, the increase in apparent absorbance is far too great to be due solely to denaturation of poly(dA-dT). Similar increases in apparent absorbance can be observed by heating solutions of HMG-1 in the absence of polynucleotide. The increases are from light scattering that results from aggregation of heat-denatured protein. The increase in apparent absorbance at a weight ratio of 6 in Figure 5 is centered at a temperature (58°C) that is somewhat higher than the temperature at which scattering is observed from HMG-1 in the absence of poly(dA-dT). That scattering is centered at 53 to 55°C regardless of the NaCl concentration in the range from 10 to 150 m*M*. We therefore attribute the scattering at a weight ratio of 6 to denaturation of HMG-1 attached to poly(dA-dT) and to the subsequent aggregation of HMG-1 or the HMG-1/poly(dA-dT) complexes. Hyperchromicity produced by denaturation of poly(dA-dT) would be expected to account for about one fourth of the increase in apparent absorbance we observed at the weight ratio of 6. The denaturation of the polynucleotide is therefore obscured by light scattering, but we can say that at that weight ratio the $T_m$ of poly(dA-dT) was increased to at least 55°C.

At 50 m*M* NaCl (Figure 6) we observed a slight stabilization of poly(dA-dT) by HMG-1. The complications from light scattering are more serious at 50 m*M* NaCl than at 10 m*M* NaCl because at this higher ionic strength the $T_m$ of poly(dA-dT) (53°C) is close to the temperature at which the protein denatures. Even at the HMG-1-to-poly(dA-dT) ratio of 1, the increase is too great to be accounted for by the denaturation of poly(dA-dT) alone. We believe the major portion of the increases in apparent absorbance at that ratio and at a ratio of 2 are due to denaturation of HMG-1 molecules bound to poly(dA-dT) and subsequent

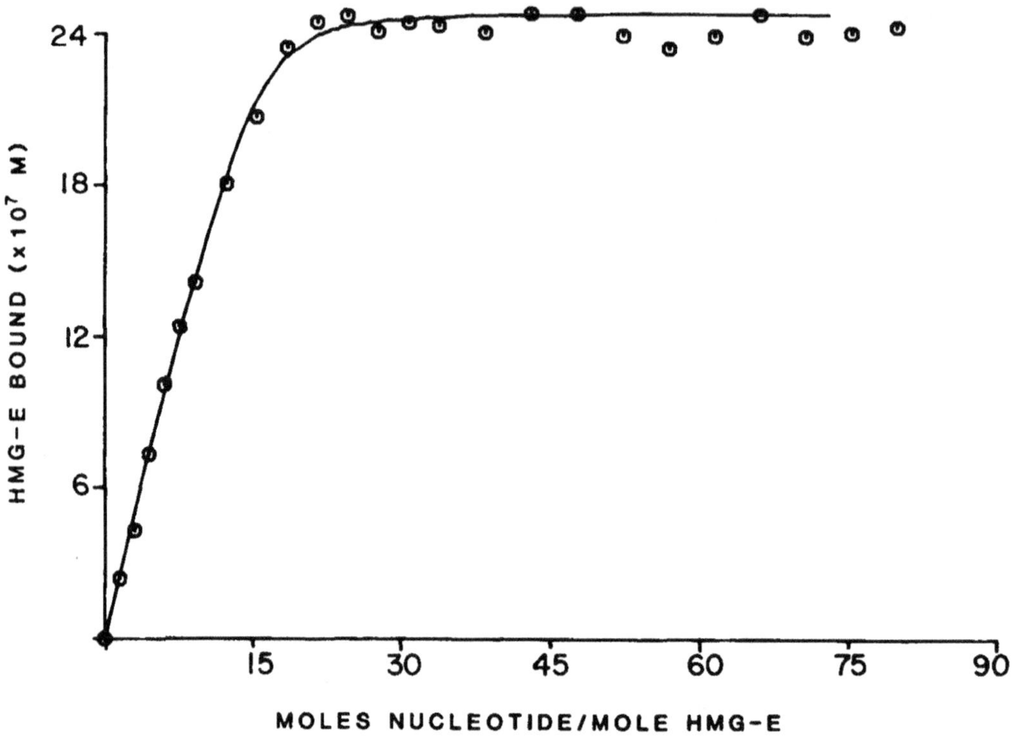


FIGURE 3. Fluorescence titration for the binding of HMG-E to single-stranded DNA in the low-ionic-strength solvent. The circles are data points. The curve is the best fitting theoretical titration curve, for which $K = 2.0 \times 10^5 M^{-1}$ and $n = 6$ bases.

aggregation. At a weight ratio of 2, a small increase is seen before 50°C. That increase accounts for no more than 10% of the increase expected from the denaturation of poly(dA-dT). No larger increase was seen in this temperature range at higher HMG-1-to-poly(dA-dT) ratios. Larger increases would have been expected if the increase at temperatures less than 50°C at the weight ratio of 2 were a reflection of a lowered $T_m$ of poly(dA-dT) due to the presence of HMG-1.

At roughly physiological ionic strengths, HMG-1 binds to columns containing immobilized single-stranded DNA but not to columns containing immobilized double-stranded DNA.[2,3] We have examined poly(dA-dT) denaturation in the presence of HMG-1 at 0.15 *M* NaCl to see if depression of the $T_m$ of the polynucleotide might be observable there even though such an effect was not seen at lower ionic strengths. The $T_m$ of free poly(dA-dT) in 0.15 *M* NaCl (65°C) leads to still greater complications from protein denaturation and light scattering. The results at a weight ratio of 2 indicate that HMG-1 attached to poly(dA-dT) denatures and aggregates at about 60°C. Free HMG-1 denatures and aggregates somewhat above 50°C and this process presumably contributes much of the increase in apparent absorbance at the weight ratio of 8. Under these conditions, then, one cannot study the effect of nondenatured HMG-1 on poly(dA-dT) denaturation.

## III. DISCUSSION

### A. Binding of HMG-E to Double- and Single-Stranded DNAs

HMG-E is well suited for studying its binding to DNA by fluorescence titration since large decreases in intrinsic fluorescence result from binding: 20 and 45% decreases for binding to double- and single-stranded DNAs, respectively. Furthermore, the overlapping

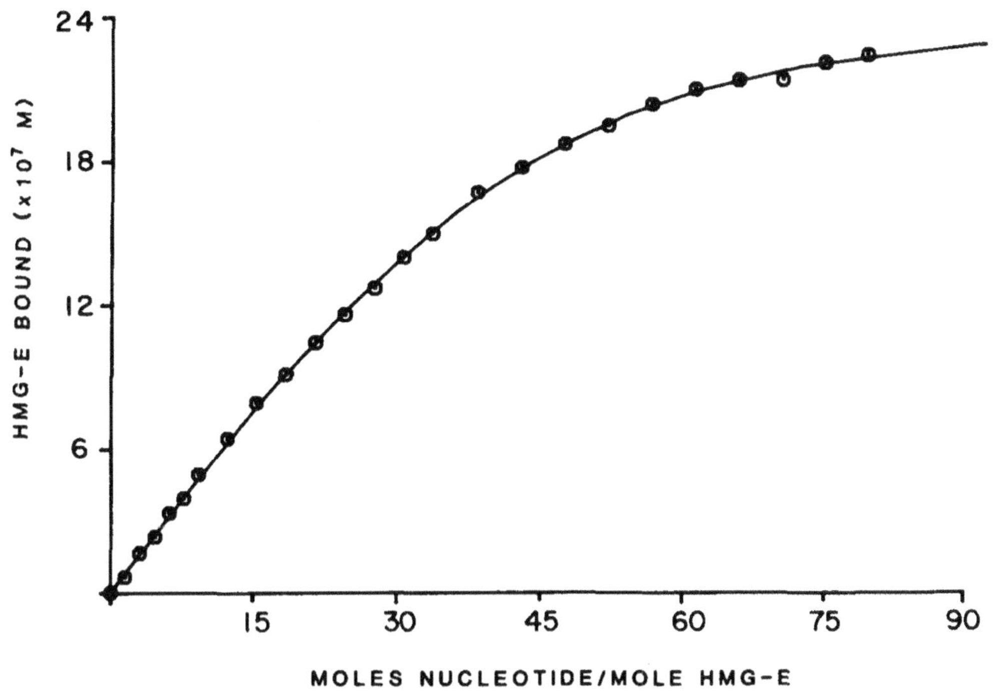


FIGURE 4. Fluorescence titration for the binding of HMG-E to single-stranded DNA in the 0.2 *M* NaCl-supplemented solvent. The circles are data points. The curve is the best-fitting theoretical titration curve, for which K = $1.1 \times 10^5$ $M^{-1}$ and n = 29 bases.

**Table 1**
**AFFINITIES AND SITE SIZES FOR THE BINDING OF HMG-E TO DOUBLE- AND SINGLE-STRANDED DNAs**

| | Double-stranded DNA | Single-stranded DNA |
|---|---|---|
| Low-ionic-strength solvent | K = $2.5 \times 10^6$ $M^{-1}$ | K = $2.0 \times 10^5$ $M^{-1}$ |
| | n = 6 bp | n = 6 bases |
| 0.2 *M* NaCl-supplemented solvent | K = $4.0 \times 10^4$ $M^{-1}$ | K = $1.1 \times 10^5$ $M^{-1}$ |
| | n = 10 bp | n = 29 bases |

binding sites model of McGhee and von Hippel[16] produced theoretical binding curves that fit our data very well. It therefore appears that the model accurately represents the interactions of HMG-E with double- and single-stranded DNAs. In none of the four cases we studied did we observe any indication of positive cooperativity in the binding of HMG-E to DNA.

Changing the solvent from our lower ionic strength to 0.2 *M* NaCl-supplemented solvent affected both K and n for the binding of HMG-E to each type of DNA. The binding site size, n, is larger for each type of DNA at the higher ionic strength. This is particularly dramatic in the case of single-stranded DNA (where n is 6 and 29 bases at low and high ionic strengths, respectively) but is also evident in the case of double-stranded DNA (where n is 6 and 10 bp at low and high ionic strengths, respectively).

The changes in n and other aspects of the interaction of HMG-E with DNA are best considered in light of the three-dimensional organization of that protein. HMG-E (as well as HMG-1 and -2) is believed to have three structural domains.[1,23] Domains A and B are DNA-binding, sequence homologous, globular domains. Domain C, which has approximately the same number of amino acid residues as does each of the other domains, is the

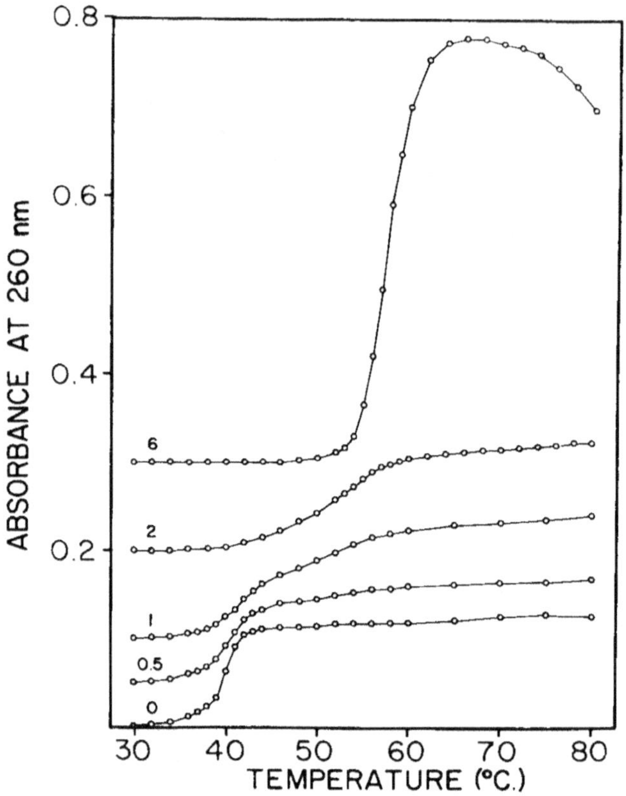


FIGURE 5. Thermal denaturation of poly(dA-dT) in the presence of HMG-1. The solvent was 10 m*M* NaCl/1 m*M* Tris-HCl/0.25 m*M* EDTA/1 m*M* dithiothreitol (pH 8.1). The poly(dA-dT) concentration in all samples was 0.011 μg/mℓ. The numbers at the leftmost portions of the curves are weight ratios of HMG-1 to poly(dA-dT).

extremely highly negatively charged COOH-terminal region of the protein. It presumably occurs as a coil, without organized secondary or tertiary structure. As a result, the volume that it occupies is much larger than the volumes of domain A and B.[1]

Domain C of HMG-E and DNA (either double- or single-stranded), as polyelectrolytes of high charge density, would both be expected to change substantially in conformation in going from low to high ionic strength conditions. Furthermore, the electrostatic forces between domain C and DNA (repulsive) and between domains A and B and DNA (attractive) would be weakened when the ionic strength is increased. Hence, the effect of salt on the system can be expected to be quite complex. With the data currently available, it is not possible to explain fully the effects of ionic strength on binding site size. One contributing factor may well be electrostatic repulsion between domain C and DNA. At the lower ionic strength, the repulsion would be very strong and it might prevent domain C from being close enough to DNA to interfere with the binding of domains A and B of an adjacent HMG-E molecule. The rather small binding site size of 6 bases (or base pairs) at the low ionic strength indicates that the DNA-binding domains of the HMG-E molecules are packed closely together. It is conceivable that at the higher ionic strength, where the repulsion would be weaker, domain C of HMG-E can approach DNA sufficiently closely to interfere with binding to DNA of domains A and B of the adjacent HMG-E molecule. That is, at the higher ionic strength, domain C may occlude bases (or base pairs) without actually binding to DNA. Regardless of the details by which the effect of increasing ionic strength is achieved, it

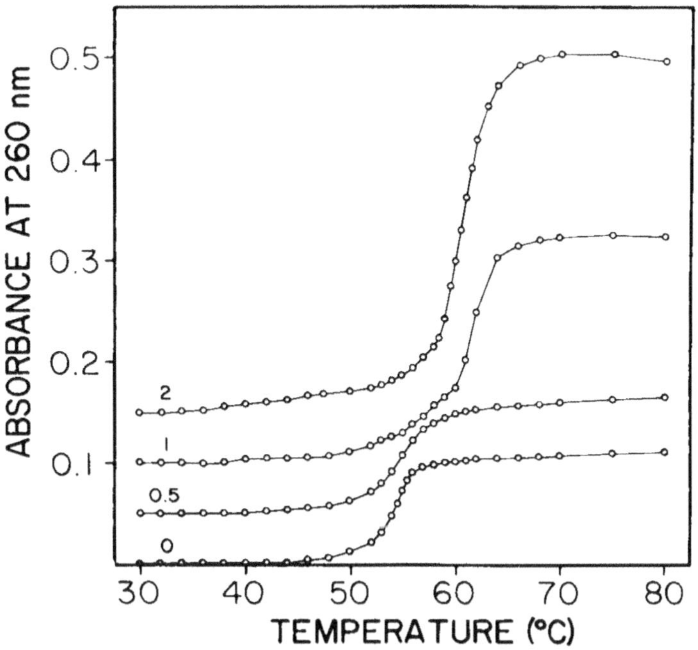


FIGURE 6. Thermal denaturation of poly(dA-dT) in the presence of HMG-1. The solvent was 50 m*M* NaCl/1 m*M* Tris-HCl/0.25 m*M* EDTA/1 m*M* dithiothreitol (pH 8.1). The poly(dA-dT) concentration in all samples was 0.011 μg/mℓ. The numbers at the leftmost portions of the curves are weight ratios of HMG-1 to poly(dA-dT).

seems likely that domain C is involved, since that region of HMG-E (and HMG-1 and -2) is distinctive in its extraordinarily high negative charge density (see Reference 1). Large effects of ionic strength on binding site size have not been found for other single-stranded DNA binding proteins,[9] none of which have regions as peculiar as domain C of the high molecular weight HMG proteins.

An increase in ionic strength also affects the affinity (K) of HMG-E for each type of DNA. Increasing the ionic strength would weaken both the repulsive forces between domain C and DNA and the attractive forces between domains A and B and DNA. The net effect on electrostatic forces will, in addition to effects due to changes in protein and DNA conformation, determine the change in K. For HMG-Es binding to double-stranded DNA, K decreases nearly two orders of magnitude upon increasing the NaCl concentration from 0 to 0.2 *M*. For binding to single-stranded DNA, K decreases only about 50% with the same increase in ionic strength. The failure to decrease more in the latter case may again be due to the influence of domain C of HMG-E; that is, the change in the strength of repulsion between domain C and DNA may be greater for single- than for double-stranded DNA. Its greater influence on changes in K for single- as compared to double-stranded DNA would be consistent with the greater postulated effect of domain C on changes in n for single- as compared to double-stranded DNA. In each case (relative sizes of effects on K and on n), the greater effects of increasing ionic strength on the interaction of HMG-E with single- as compared with double-stranded DNA may be due to a greater influence of ionic strength on the conformation of single-stranded DNA.

It is also of interest to compare the affinities of HMG-E at each ionic strength for the two types of DNA. In our lower ionic strength solvent, HMG-E binds more tightly to double- than to single-stranded DNA. At the higher ionic strength, HMG-E has a 2.8-fold greater affinity for single- than for double-stranded DNA. This is a lesser preference than might

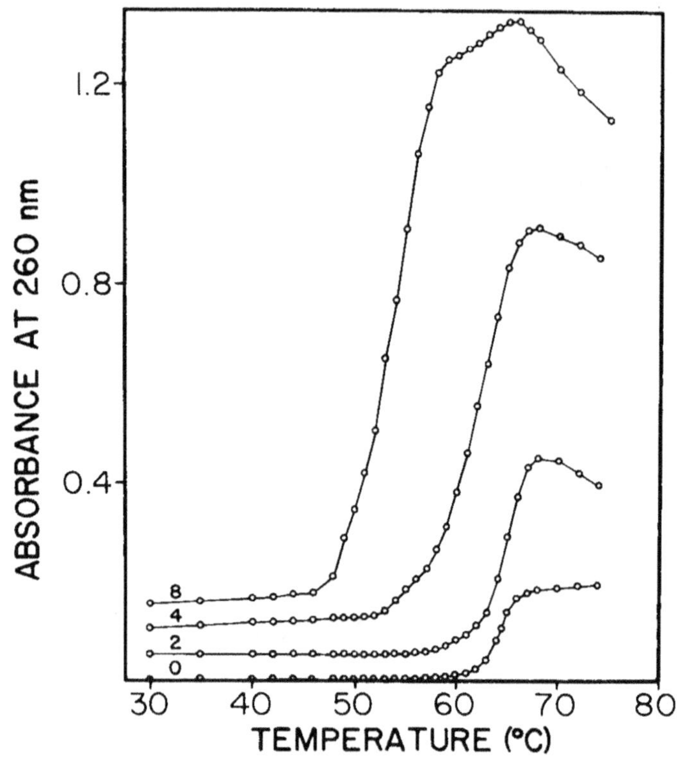


FIGURE 7. Thermal denaturation of poly(dA-dT) in the presence of HMG-1. The solvent was 150 m*M* NaCl/1 m*M* Tris-HCl/0.25 m*M* EDTA/1 m*M* dithiothreitol (pH 8.1). The poly(dA-dT) concentration in all samples was 0.017 μ/mℓ. The numbers at the leftmost portions of the curves are weight ratios of HMG-1 to poly(dA-dT).

have been expected based on DNA-cellulose chromatography of HMG proteins on columns containing immobilized single- and double-stranded DNAs.[2,3] This may indicate a difference in the binding of HMG-E to immobilized DNAs as compared to DNAs in solution. Nonetheless, the estimates of K and n from our current fluorescence titration studies indicate that HMG-E might be capable of lowering the $T_m$ of DNA, albeit weakly, at 0.2 *M* NaCl.

Finally, we note that the substantial difference in fluorescence decreases in the binding of HMG-E to double- and single-stranded DNAs suggests that the protein binds significantly differently to the two forms of DNA.

## B. Thermal Denaturation Studies

We have been unable to demonstrate a lowering of the $T_m$ of poly(dA-dT) in the presence of HMG-1 under any conditions we have tried. We observed a substantial stabilization at 10 m*M* NaCl and a slight stabilization at 50 m*M* NaCl. We have obtained readily interpretable results only at rather low HMG-1-to-poly(dA-dT) ratios because light scattering resulting from protein denaturation obscures the increase in absorption from poly(dA-dT) denaturation at even modestly high HMG-1-to-poly(dA-dT) ratios. The complications become more severe as the ionic strength is increased. While we have presented data for only chicken erythrocyte HMG-1, we have also studied the denaturation of poly(dA-dT) in the presence of HMG-1 from cultured rat hepatoma cells and of chicken erythrocyte HMG-2 and -E. We have also examined the thermal denaturation of naturally occurring DNA in the presence of HMG proteins, and we have tried variations in solvent conditions (including the addition of $MgCl_2$) that we have not described in detail in this chapter. We have also conducted the denaturations at lower rates of temperature increase. Nothing we have tried has allowed us to convincingly produce a lower polynucleotide or DNA $T_m$ in the presence of an HMG protein. With each protein we have studied and under all solvent conditions, the results have been complicated by protein denaturation and resultant light scattering.

Three groups have reported studies on the denaturation of poly(dA-dT) in the presence of a high molecular weight HMG protein. There is unfortunately little agreement in the conclusions drawn by the three groups. Before comparing results from the three laboratories, we note that HMG proteins from three different sources have been used and that the proteins were purified by different procedures. It is possible that differences obtained in studying the effects of the HMG proteins on the denaturation of poly(dA-dT) are due entirely to differences in source material and isolation procedure. Nevertheless, it is worthwhile to compare the results from the three laboratories since differences in source material and purification procedures may conceivably be inconsequential. Javaherian et al.[5] concluded that calf thymus HMG-1 is capable of dramatically lowering the $T_m$s of T7 DNA and of poly(dA-dT) in solvents containing from 25 to 75 m*M* NaCl. In a solvent that lacked NaCl, they reported that HMG-1 stabilized the polynucleotides against heat denaturation. Duguet and de Recondo have reported that HMG-1 from regenerating rat liver cytosol (their HD25) can lower the $T_m$ of poly(dA-dT) in a very low ionic strength solvent.[7,8] HMG-1 from normal rat liver cytosol (i.e., S25 in their nomenclature) was reported to stabilize poly(dA-dT) against denaturation.[8] Our results, obtained in most detail with HMG-1 from chicken erythrocytes but also with erythrocyte HMG-2 and -E[23] and with HMG-1 from cultured hepatoma cells, indicate stabilization by the HMG proteins of the heat denaturation of poly(dA-dT) at 10 and 50 m*M* NaCl. We have obtained no evidence under any of the conditions we have tried for the lowering of the $T_m$ of a double-stranded polynucleotide by a high molecular weight HMG protein.

In our hands, studying the denaturation of poly(dA-dT) in the presence of HMG proteins is greatly complicated by protein denaturation and light scattering. Javaherian et al.[5] also recognized that protein denaturation contributed to the changes in "absorbance" that they observed at temperatures well below the $T_m$ of free DNA or poly(dA-dT). We believe it is possible that light scattering was the sole contributor to those increases in "absorbance" at low temperatures. The studies of Duguet and de Recondo are more persuasive evidence for destabilization of poly(dA-dT) by HMG-1. The denaturation curves in Reference 7 are particularly interesting, and it is unfortunate that the study was not carried out at higher HMG-1 levels to denature more of the poly(dA-dT) at a lowered temperature. In another denaturation study from the same laboratory, it appears that light scattering may be contributing to the increase in apparent absorbance (Figure 2 in Reference 8), but the extent of its contribution is difficult to ascertain since the entire denaturation curve is apparently not shown. As we indicated above, we have studied the denaturation of poly(dA-dT) in the presence of HMG-1 from nuclei of cultured rat hepatoma cells (HTC cells). Those studies were complicated, in a manner similar to that described in this chapter for chicken erythrocyte HMG-1, by protein denaturation and light scattering. It would therefore not be surprising if HMG-1 from regenerating rat liver cytosol were heat denatured to produce light scattering in a poly(dA-dT) denaturation experiment. At a relatively crude level of analysis, nuclear and cytosolic HMG-1s from HTC cells are indistinguishable.[11]

We have found that light scattering from denatured protein at appropriate HMG-1-to-poly(dA-dT) ratios can contribute to a thermal denaturation curve in such a way that the curve resembles a biphasic denaturation of the polynucleotide. A convincing demonstration that a two-component denaturation curve results only from denaturation of polynucleotide would require achieving a complete conversion to the low melting state (at high HMG-1-to-poly(dA-dT) ratios). At all ratios one should observe a total increase in absorbance that is not much greater than the increase seen upon denaturation of poly(dA-dT) alone, or quantitatively account for the contribution of light scattering. At this time, these conditions have not been satisfied in any study with HMG-1 or its homologs.

Resolving the discrepancies between laboratories is important since the ability of HMG-1 to lower the $T_m$ of a double-stranded polynucleotide could have direct implications about the biological function of the protein. Also, unwinding (preferential binding to single-stranded DNA regions) is one mechanism by which HMG-1 might achieve its well documented effect of reducing the linking number of DNA that is covalently closed in its presence.[4,5,19,20,25] We believe that the available evidence, taken together, does not favor a mechanism of preferential binding to single-stranded regions. A promotion by HMG-1 of formation of single-stranded regions in a double-stranded DNA or synthetic polynucleotide has not been convincingly demonstrated. Furthermore, cytosolic HMG-1 from normal rat liver is capable of reducing the linking number of DNA under conditions similar to those in which it has been reported to stabilize poly(dA-dT) against heat denaturation.

An alternate mechanism by which HMG-1 might reduce the linking number of DNA would be to form cores around which the double-stranded DNA could wrap. Beads that might be such cores have been observed by electron microscopy of HMG-1/DNA complexes.[19,20,25] Those beads are quite fragile structures that are observed only when glutaraldehyde fixation is used in preparation for electron microscopy. Glutaraldehyde fixation is not necessary, however, for HMG-1 to reduce the linking number of DNA. We suggest that the beads observed after glutaraldehyde treatment could be artifactual products of cross-linking reversibly bound and mobile HMG-1 molecules when they happen to be juxtaposed on the DNA. The absence of positive cooperativity in the binding of HMG-1 and its homologs to DNA (this study and References 5, 19, and 20) predicts that in the absence of a cross-linking reagent HMG-1 molecules would be uniformly distributed along a DNA lattice and not clustered.

A third mechanism by which HMG-1 could reduce the linking number of DNA is inducing a small change in the DNA helix rotation upon binding of each HMG-1 molecule.[5,20] By a process of elimination, this would seem to be the most likely mechanism.

### C. Possible Functional Implications of Our Results

On several occasions[2,3,11,12,26] we have pointed out similarities between HMG-1 and the gene 32 protein of bacteriophage T4. Work we have reported in this chapter shows some differences between the proteins. The gene 32 protein is capable of lowering the $T_m$ of poly(dA-dT).[10] HMG-1, on the other hand, has not been clearly demonstrated to have that capability. Failure to lower the $T_m$ of poly(dA-dT) does not in itself rule out the possibility, however, that HMG-1 functions analogously to the gene 32 protein, since the helix-destabilizing activity of the latter protein is apparently only an artificial one. The gene 32 protein does not lower the $T_m$ of naturally occurring DNAs.[9] A more significant difference is that the gene 32 protein and other prokaryotic single-stranded DNA binding proteins known to be required for DNA replication bind cooperatively to single-stranded DNA[9] whereas HMG-1 and its homologs do not. This rather clearly indicates that HMG-1 cannot act in a manner that is analogous in all respects to the action of the gene 32 protein and other functionally similar prokaryotic proteins. This does not rule out a role of HMG-1 and its homologs in DNA replication, but we believe any such role cannot be fully analogous to that of the gene 32 protein. If HMG-1 and its homologs are involved in that process, the fact that they have as great, or somewhat greater, affinities for single- as for double-stranded DNA may be significant to the role they play.

The affinities of HMG-E and, by inference from DNA-cellulose chromatography, of HMG-1 and -2 for both forms of DNA at 0.2 *M* NaCl are not particularly high. They are, for instance, several orders of magnitude lower than the overall affinity ($K\omega$) of the gene 32 protein for single-stranded DNA at the same ionic strength.[27] In chromatin, however, HMG-1 and its homologs might interact with proteins as well as with DNA. Domain C, with its high negative charge density, is well suited to interact with histones. Interactions of HMG-1 and -2 with histone H1 have in fact been demonstrated,[28,29] and we have recently obtained evidence for interaction of HMG-1 and its homologs with core histones (L. Wen and G. R. Reeck, unpublished observations). An interaction of HMG-1 or of its homologs with DNA and with histones would probably be stronger than an interaction with either alone.

An intriguing possibility that our current data do not address is that HMG-1 and its homologs might possess nucleosome assembly activity. Polyanions, including polyglutamic acid, have been shown to facilitate the assembly of nucleosomes from core histones and DNA.[30] Since the COOH-terminal domains of HMG-1 and its homologs are polyanions,[31] it seems entirely possible that these HMG proteins could act as nucleosome assembly factors. An assembly factor with affinity for DNA on domains distinct from the domain with assembly activity could be imagined to be part of DNA replication or of transcription in eukaryotes. It is in such a context, admittedly hypothetical at this point, that we believe the somewhat greater affinity of HMG-1 and its homologs for single- than for double-stranded DNA may be important.

## IV. METHODS

### A. Binding of HMG-E to DNA

HMG-E was purified under nondenaturing conditions as previously described.[14] Salmon sperm DNA (Sigma Type III) was purified by repeated chloroform/isoamyl alcohol (24:1) extraction. Single-stranded DNA was obtained from double-stranded salmon sperm DNA by heat denaturation (15 min in a boiling-water bath) and rapid cooling in an ice water mixture. DNA concentrations were determined spectrophotometrically using $A_{260}^{0.1\%}$ values of 20 and 30 for double- and single-stranded DNA, respectively. In calculating concentrations of bases, we used a value of 330 daltons for the molecular mass of a nucleotide (as a residue in DNA). For calculations involving HMG-E we used an extinction coefficient of 0.8 ($A_{280}^{0.1\%}$) and a molecular weight of 26,000.

Fluorescence measurements were made with a Perkin Elmer MPF-44A Spectrofluorimeter (we are grateful to Dr. Frederick W. Oehme, College of Veterinary Medicine, for use of that instrument). Excitation and emission wavelengths were 280 and 335 nm, respectively, and 10-nm slits were used at both wavelengths. DNA and protein solutions in the appropriate solvents were deoxygenated by aspiration and subsequent bubbling of nitrogen gas through the samples. The protein solution ($2.5 \times 10^{-6}$ *M*) was placed in a cuvette and overlaid with mineral oil. Aliquots (5 $\mu\ell$) of a DNA solution (0.5 mg/m$\ell$) were introduced under the layer of mineral oil with a Hamilton syringe. After stirring 10 sec with a tiny magnetic stir bar in the cuvette, the fluorescence was measured. The above approach was followed to reduce the oxygen level in the samples, since oxygen quenches strongly. The sensitivity of the spectrofluorimeter was set so that the fluorescence of the protein solution, before adding DNA, was about 95. The fluorescence of a buffer blank was from 0.5 to 1. We corrected for dilution (which was 12% at maximum) and absorption of exciting radiation by DNA by carrying out parallel fluorescence measurements with the protein and DNA solutions in 1 *M* NaCl. Under those conditions we would expect no interaction of HMG-E and DNA. For any given amount of DNA added, the fluorescence in 1 *M* NaCl (at that level of addition) was divided into the fluorescence of the protein solution (no added DNA) at 1 *M* NaCl to obtain a correction factor. The fluorescence in the low ionic strength buffer (or in that buffer supplemented with 0.2 *M* NaCl) at the same amount of added DNA was multiplied by the correction factor from the measurements in 1 *M* NaCl, after having subtracted the fluorescence of the buffer itself. The corrected fluorescence was subtracted from the fluorescence of the protein solution (without DNA) to obtain a change in fluorescence. Titrations in all three solvents (added NaCl concentrations of 0, 0.2, and 1*M*) were all carried out at least twice. The data we report are the averages of values obtained in replicate titrations. The approach we have taken in handling data is similar to that used by von Hippel and his colleagues.[15,27]

**B. Simulating Binding Curves**

The DNA concentration, D, is measured in *bases*. The total protein concentration, $L_T$, is distributed between bound, $L_b$, and free, $L_f$, protein. The titration curve is a plot of $L_b$ against the ratio of DNA to total protein, $D/L_T$.

The working equation for uncooperative binding to overlapping sites n bases long is Equation 10 of McGhee and von Hippel:[16]

$$\frac{\nu}{L_f} = K(1 - n\nu)\left(\frac{1 - n\nu}{1 - (n - 1)\nu}\right)^{n-1} \quad (1)$$

where $\upsilon = L_b/D$ and K is the intrinsic association constant. This equation is equivalent to

$$\frac{1}{KL_f} = \frac{Q}{\nu}\left(\frac{Q}{Q + \nu}\right)^{n-1} \quad (2)$$

where $Q = 1 - n\upsilon$.

The construction of a simulated binding curve begins by specifying the site size, n, the intrinsic equilibrium constant, K, and the total ligand concentration, $L_T$. A series of values of $L_f$ is selected, ranging from a little more than zero to a little less than $L_T$. The calculation is done initially for 24 values of $L_f$ equally spaced between $L_T/25$ and 24 $L_T/25$. To define the titration curve over an extended range of $D/L_T$, it is often necessary to further subdivide the lowest and highest intervals in $L_f$. For each value of $L_f$, $\nu$ is found using Equation 2 and a binary search over the range of possible values $0 < \nu < 1/n$. In programming the binary search, it is helpful to note that, as $\nu$ is adjusted upward, the right-hand side of Equation 2 decreases.

Once $\nu$ is known for the entire range of $L_f$, the corresponding values of $L_b$ and $D/L_T$ are found and the titration curve is constructed:

$$L_b = L_T - L_f$$

$$\frac{D}{L_T} = \frac{L_b}{\nu L_T}$$

**C. Fitting Simulated Data to Experiments**

The maximum possible fluorescence change ($\Delta F_{max}$) will be obtained if sufficient DNA is added to adsorb all of the protein present. If binding is tight, then $\Delta F_{max}$ will in fact be reached experimentally. For looser binding, not all of the protein will be bound, even at the highest DNA concentration tested.

For the particular set of data treated here, the problem is alleviated by the fact that the experiments were done in pairs. Titration of the protein (HMG-E) with single- and double-stranded DNA was done at low salt, where the binding was tight, and at higher salt, where the binding was less tight. The tight binding at low salt defined $\Delta F_{max}$ with little ambiguity for each pair (19.8 and 45.0% decreases for double- and single-stranded DNAs, respectively), and the same values of $\Delta F_{max}$ were then used to fit the matching high salt data.

The data, converted initially to fluorescence changes ($\Delta F$) at a series of DNA-to-protein ratios (in units of DNA bases per mole of protein), were further converted to molar concentration of bound ligand by the following equation: $L_b = L_T (\Delta F/\Delta F_{max})$.

The first step in the fitting procedure was to ask the computer to plot $L_b$ vs. $D/L_T$ for each of a series of values of $\Delta F_{max}$ covering the entire range of plausible values. Simulated

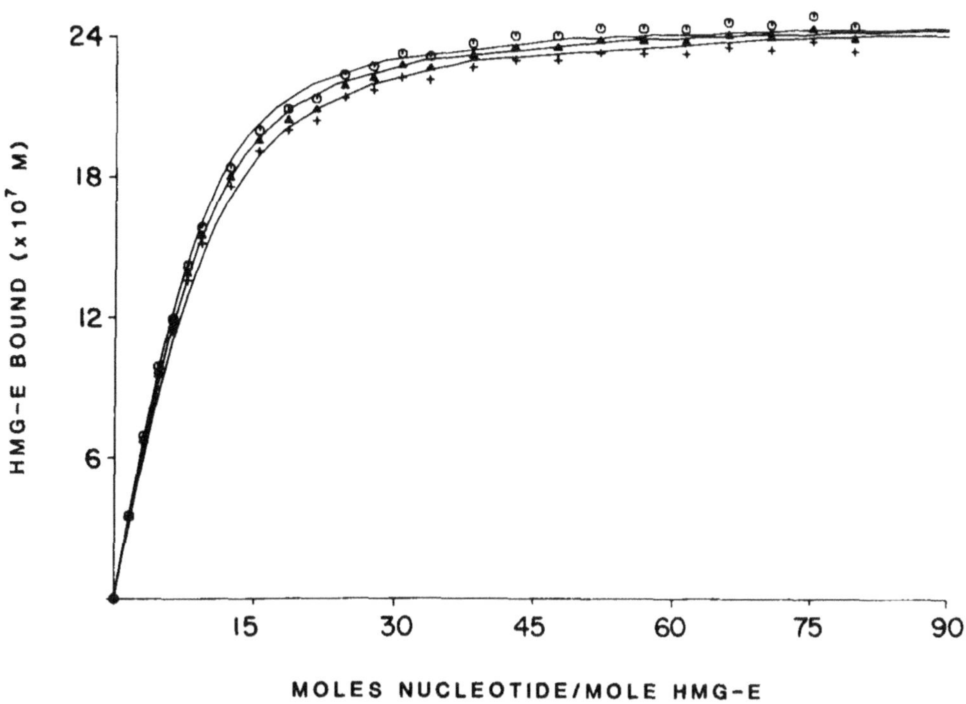


FIGURE 8. Determination of $\Delta F_{max}$ and K for the binding of HMG-E to single-stranded DNA in the low-ionic-strength solvent. Points shown as crosses, triangles, and circles are experimental data calculated for three choices of $\Delta F_{max}$: 46.0, 45.0, and 44.0%, respectively. The curves are theoretical titration curves for three different values of K. For the top, middle, and lowest curves the values of K were $2.4 \times 10^5 M^{-1}$, $2.0 \times 10^5 M^{-1}$, and $1.6 \times 10^5 M^{-1}$, respectively. In each case n was 6 bases. The range in $\Delta F_{max}$ spans the possible values of $\Delta F_{max}$ since, for all K, data points shown as circles are too high in the plateau region and data points shown as crosses are too low in the plateau region. The intermediate value of $\Delta F_{max}$ was selected. The middle titration curve was judged to be the best fit to the data points calculated with the selected value of $\Delta F_{max}$ (triangles).

data plots were then constructed for several values of n. For each site size, a range of equilibrium constants was selected, and the simulations were plotted together (e.g., as in Figure 8). Several such collections of plots were assembled, each showing simulated data for a range of equilibrium constants at one site size. These plots were then overlaid in turn on the reduced experimental data plots and inspected for fit.

For every value of n, it was possible to identify one combination of equilibrium constant and $\Delta F_{max}$ that gave the best match between simulation and experiment. For tightly binding cases, the correct $\Delta F_{max}$ was particularly easy to identify by agreement with the simulations in the nearly flat region of the titration curve at high DNA-to-$L_T$ ratios.

The selection of the best overall match between simulation and experiment was then a matter of comparing the optimal fits for the various site sizes with each other. In each case, there was clearly a best choice of n. In general, when n was too low, the simulated titration curve lay to the left of the data at low DNA to $L_T$ ratios, crossed the experimental points at intermediate ratios, and was flatter than the data as the binding approached saturation. Conversely, too large a site size produced a simulation lying to the right of the data at low saturation, crossing the data at moderate saturation, and sloping too steeply at high saturation. This procedure defined n to the nearest integer for site sizes less than 10 and within $\pm 2$ for n = 30.

In attempting to simulate the binding curve for the interaction of HMG-E with double-stranded DNA at 0.2 *M* NaCl, we released our constraint that $\Delta F_{max}$ for this case be the same as $\Delta F_{max}$ for the binding of HMG-E to double-stranded DNA at low ionic strength.

This led, however, to no improvement in the quality of the fit. We therefore reimposed the constraint. Failure to fit the data over the entire range of points for this case probably results from systematic error in the data at the highest nucleotide-to-HMG-E ratios. That such error should appear in this case, and not the others we studied, is not surprising since the association constant is the lowest of the four cases and $\Delta F_{max}$ for binding to double-stranded DNA is about half $\Delta F_{max}$ for binding to single-stranded DNA.

### D. Thermal Denaturation Experiments

Poly(dA-dT), by which we mean the double-stranded, alternating copolymer of dA and dT, was used as supplied by Sigma (P 0883). HMG-1 was purified under nondenaturing conditions as previously described.[14] Concentrations were determined spectrophotometrically, using $A_{260}^{0.1\%} = 20$ for poly(dA-dT) and $A_{280}^{0.1\%} = 0.8$ for HMG-1. Protein and poly(dA-dT) were combined in 0.2-mℓ samples with either 10, 50, or 150 m*M* NaCl in 1 m*M* dithiothreitol/0.25 m*M* EDTA/1 m*M* Tris-HCl (pH 8.1). Denaturations were performed in a Gilford Model 250 spectrophotometer equipped with a thermo-programmer and Gilford chart recorder. Cuvettes of pathlength 1 cm and volume 0.3 mℓ were used. The absorbance at 260 nm was monitored as the temperature was increased at 1°C/min.

## ACKNOWLEDGMENTS

This work was supported by NIH grants GM-29203 (to Gerald R. Reeck and David J. Cox), CA-17782 (to Gerald R. Reeck), GM-22243 (to David J. Cox), and by the Kansas Agricultural Experiment Station. Gerald R. Reeck is the recipient of NIH Research Career Development Award CA-00425. This is publication number 83-28-B, Kansas Agricultural Experiment Station.

## REFERENCES

1. **Reeck, G. R. and Teller, D. C.,** High Mobility Group proteins: purification, properties, and amino acid sequence comparisons, in *Progress in Nonhistone Protein Research,* Vol. 2, Bekhor, I., Ed., CRC Press, Boca Raton, Fla., 1984, chap. 1.
2. **Bidney, D. L. and Reeck, G. R.,** Purification from cultured hepatoma cells of two nonhistone chromatin proteins with preferential affinity for single-stranded DNA: apparent analogy with calf thymus HMG proteins, *Biochem. Biophys. Res. Commun.,* 85, 1211, 1978.
3. **Isackson, P. J., Fishback, J. L., Bidney, D. L., and Reeck, G. R.,** Preferential affinity of high molecular weight High Mobility Group non-histone chromatin proteins for single-stranded DNA, *J. Biol. Chem.,* 254, 5569, 1979.
4. **Javaherian, K., Liu, L. F., and Wang, J. C.,** Nonhistone proteins $HMG_1$ and $HMG_2$ change the DNA helical structure, *Science,* 199, 1345, 1978.
5. **Javaherian, K., Sadeghi, M., and Liu, L. F.,** Nonhistone proteins $HMG_1$ and $HMG_2$ unwind DNA double helix, *Nucleic Acids Res.,* 6, 3569, 1979.
6. **Bonne, C., Sautiere, P., Duguet, M., and de Recondo, A.-M.,** Identification of a single-stranded DNA binding protein from rat liver with High Mobility Group protein 1, *J. Biol. Chem.,* 257, 2722, 1982.
7. **Duguet, M. and de Recondo, A.-M.,** A deoxyribonucleic acid unwinding protein isolated from regenerating rat liver. Physical and functional properties, *J. Biol. Chem.,* 253, 1660, 1978.
8. **Bonne, C., Duguet, M., and de Recondo, A.-M.,** Rat liver DNA binding proteins: physiological variations, *FEBS Lett.,* 6, 292, 1979.
9. **Kowalczykowski, S. C., Bear, D. G., and von Hippel, P. H.,** Single-stranded DNA binding proteins, in *The Enzymes,* Vol. 14, Part A, 3rd ed., Boyer, P. D., Ed., Academic Press, New York, 1981, 373.
10. **Alberts, B. M. and Frey, L.,** T4 bacteriophage gene 32: a structural protein in the replication and recombination of DNA, *Nature (London),* 227, 1313, 1970.
11. **Isackson, P. J., Bidney, D. L., Reeck, G. R., Neihart, N. K., and Bustin, M.,** High Mobility Group chromosomal proteins isolated from nuclei and cytosol of cultured hepatoma cells are similar, *Biochemistry,* 19, 4466, 1980.

12. **Shastri, K., Isackson, P. J., Fishback, J. L., Land, M. D., and Reeck, G. R.,** Influence of nonhistone chromatin protein HMG-1 on the enzymatic digestion of purified DNA, *Nucleic Acids Res.*, 10, 5059, 1982.
13. **Tyrell, D., Isackson, P. J., and Reeck, G. R.,** Two-dimensional gel electrophoresis of nonhistone chromatin proteins with nonequilibrium pH gradient electrophoresis as the first dimension, *Anal. Biochem.*, 119, 433, 1982.
14. **Isackson, P. J., Debold, W. A., and Reeck, G. R.,** Isolation and separation of chicken erythrocyte High Mobility Group non-histone chromatin proteins by chromatography on phosphocellulose, *FEBS Lett.*, 119, 337, 1980.
15. **Kelly, R. C., Jensen, D. E., and von Hippel, P. H.,** DNA "melting" proteins. IV. Fluorescence measurements of binding parameters for bacteriophage T4 gene 32-protein to mono- , oligo- , and poly-nucleotides, *J. Biol. Chem.*, 251, 7240, 1976.
16. **McGhee, J. D. and von Hippel, P. H.,** Theoretical aspects of DNA-protein interactions: co-operative and non-co-operative binding of large ligands to a one-dimensional homogeneous lattice, *J. Mol. Biol.*, 86, 469, 1974.
17. **Jensen, D. E., Kelly, R. C., and von Hippel, P. H.,** DNA "melting" proteins. II. Effects of bacteriophage T4 gene 32-protein on the conformation and stability of nucleic acid structures, *J. Biol. Chem.*, 251, 7215, 1976.
18. **Isackson, P. J., Clow, L. G., and Reeck, G. R.,** Comparison of the salt dissociations of high molecular weight HMG non-histone chromatin proteins from double-stranded DNA and from chromatin, *FEBS Lett.*, 125, 30, 1981.
19. **Bonne, C., Duguet, M., and de Recondo, A.-M.,** Single-strand DNA binding protein from rat liver: interactions with supercoiled DNA, *Nucleic Acids Res.*, 8, 4955, 1980.
20. **Duguet, M., Bonne, C., and de Recondo, A.-M.,** Single-strand deoxyribonucleic acid binding protein from rat liver changes the helical structure of deoxyribonucleic acid, *Biochemistry*, 20, 3598, 1981.
21. **Sober, H. A., Ed.,** *Handbook of Biochemistry*, 2nd ed., Chemical Rubber Company, Cleveland, 1970, H-22.
22. **McGhee, J. D.,** Theoretical calculations of the helix-coil transition of DNA in the presence of large, cooperatively binding ligands, *Biopolymers*, 15, 1345, 1976.
23. **Reeck, G. R., Isackson, P. J., and Teller, D. C.,** Domain structure in high molecular weight high mobility group nonhistone chromatin proteins, *Nature (London)*, 300, 76, 1982.
24. **Isackson, P.,** Isolation of High Mobility Group Nonhistone Chromatin Proteins and Studies of Their DNA Binding Properties, Ph.D. dissertation, Kansas State University, Manhattan, 1980.
25. **Mathis, D. J., Kindelis, A., and Spadafora, C.,** HMG proteins (1 + 2) form beaded structures when complexed with closed circular DNA, *Nucleic Acids Res.*, 8, 2577, 1980.
26. **Isackson, P. J. and Reeck, G. R.,** Removal of degradation products from calf thymus High Mobility Group non-histone chromatin proteins by chromatography on immobilized double-stranded DNA, *Biochim. Biophys. Acta*, 697, 378, 1982.
27. **Newport, J. W., Lonberg, N., Kowalczykowski, S. C., and von Hippel, P. H.,** Interactions of bacteriophage T4-coded gene 32 protein with nucleic acids. II. Specificity of binding to DNA and RNA, *J. Mol. Biol.*, 145, 105, 1981.
28. **Smerdon, M. J. and Isenberg, I.,** Interactions between the subfractions of calf thymus H1 and nonhistone chromosomal proteins HMG1 and HMG2, *Biochemistry*, 15, 4242, 1976.
29. **Cary, P. D., Shooter, K. V., Goodwin, G. H., Johns, E. W., Olayemi, J. Y., Hartman, P. G., and Bradbury, E. M.,** Does High-Mobility-Group non-histone protein HMG 1 interact specifically with histone H1 subfractions?, *Biochem. J.*, 183, 657, 1979.
30. **Stein, A., Whitlock, J. P., and Bina, M.,** Acidic polypeptides can assemble both histones and chromatin *in vitro* at physiological ionic strength, *Proc. Natl. Acad. Sci. U.S.A.*, 76, 5000, 1979.
31. **Walker, J. M., Gooderham, K., Hastings, J. R. B., Mayes, E., and Johns, E. W.,** The primary structures of non-histone chromosomal proteins HMG 1 and 2, *FEBS Lett.*, 122, 264, 1980.

Chapter 3

# IMMUNOBIOCHEMICAL STUDIES OF THE CHANGE IN HMG PROTEINS IN RESPONSE TO ESTROGENIC EFFECT

**Ching Sung Teng and Christina T. Teng**

## TABLE OF CONTENTS

## I. INTRODUCTION

High mobility group (HMG) proteins are a group of chromosomal nonhistone proteins which associate with the active (or expressed) genes in the eukaryotic cells.[1-3] They exist universally in the eukaryotic kingdom, from yeast to mammalian species.[4-6] HMG proteins are closely associated with the nucleosomes which are the basic structural units of chromosome.[7,8] The importance of HMG proteins in correlation with gene regulation has been postulated, yet their functional role in gene expression is still unknown. The amount of HMG protein increase is proportional with the proliferational state of the tissue.[9,10] The possibility of HMG protein serving as the initiation signal for chromosomal replication has been suggested.[10]

In our laboratory, we are interested in the studies of genomic activity in correlation with the quantitative changes of nonhistone proteins during hormone stimulation. We have selected chick oviduct as a model system to investigate whether the increase of HMG proteins is in parallel with cell differentiation and the activation of gene expression. The advantage of using this system is that chick oviduct has a drastic response to estrogenic hormones for epithelial differentiation and ovalbumin gene expression. In chick oviduct after maturation the major cellular population (50 to 60%) in the epithelial layer is tubular gland cells. The differentiation of tubular gland cell can be precociously induced in immature chick oviduct by an extensive stimulation with estrogen.[11] Consequently the change of HMG proteins during the processes of cellular differentiation can be monitored.

## II. PREPARATION OF HMG PROTEINS FROM ESTROGEN-INDUCED CHICK OVIDUCT

### A. Isolation and Purification

Initially, we have established the procedure for isolation of HMG proteins from chick oviducts which have been extensively stimulated by in vivo estrogen treatment for 3 weeks. Applying the perchloric acid extraction of purified nuclei procedures,[12] we have obtained the total HMG proteins which range from 55 to 60 mg of protein per kilogram of oviduct (approximately 65 mg/kg of oviduct from active laying hen).[13] Total HMG proteins have been separated into several major protein fractions by CM-Sephadex® column chromatography. HMG-1 and -2 were eluted with NaCl at concentration ranging from 0.05 to 0.1 *M* (see Figures 1 and 2). Due to the fact that these two proteins tend to aggregate with each other, a further separation by reducing agent (dithiothreitol) and re-fractionation from CM-Sephadex® becomes necessary. Figures 3 and 4 indicate that in the presence of the reducing agent the separation of HMG-1 and -2 can be accomplished.

HMG-14 and -17 were eluted from CM-Sephadex® column with NaCl at the concentration of 0.25 *M*. These two proteins tend to be eluted at a close range of salt concentration; therefore, a selective extraction method with 18% trichloroacetic acid (TCA) has been applied for their separation (see Figure 5).

Electrophoretic separation by acrylamide gel indicated that in the estrogen-stimulated chick oviduct, there are four major HMG proteins, namely HMG-1, -2, -14, and -17 (see Figure 6). In addition, two distinct protein bands with higher electrophoretic mobility than HMG-17 were present (see Figure 6). They are tentatively designated as HMG-18 and -19. The presence of these two avian tissue-specific HMG proteins has been reported in chick erythrocyte and oviduct as HMG-Y proteins that were associated with the active nucleosomes.[14]

### B. Characterization

The amino acid composition of proteins HMG-1, -2, -14, and -17 was analyzed and is presented in Table 1. The amounts of acidic (or basic) amino acids represent about one quarter of the total amino acid composition. The concentration of cysteine, methionine,

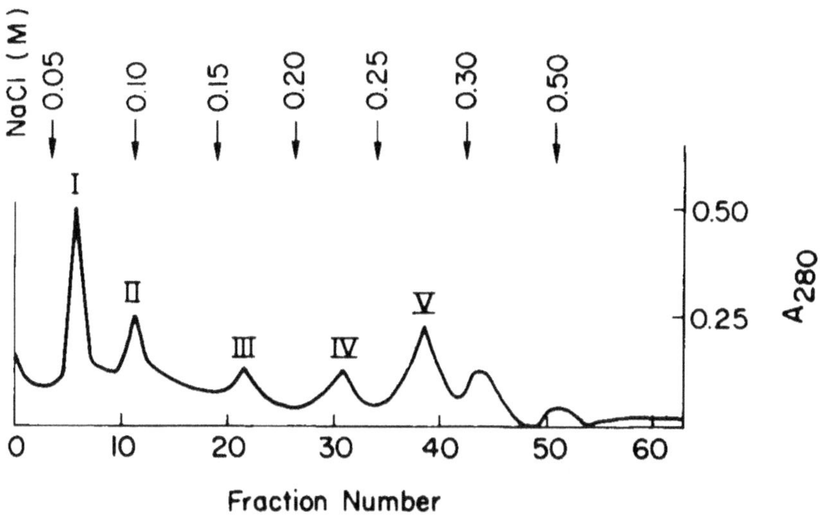


FIGURE 1. Separation of HMG proteins by CM-Sephadex® column chromatography. Approximately 30 mg of total HMG protein was dissolved in 7.5 m*M* borate buffer with 0.05 *M* NaCl and applied to a CM-Sephadex® column (1 × 8 cm). The column was eluted stepwise with various concentrations of NaCl (as indicated above the arrows).

isoleucine, leucine, threonine, and phenylalanine is relatively low. The amounts of aspartic, glutamic, proline, lysine, and arginine are generally present in high concentration. This characteristic amino acid composition is not distinguishable from that of other HMG proteins reported in yeast, wheat, avian erythrocyte, and calf tissue.[5,6,15,16] There is no drastic difference between the HMG-1 and -2 proteins of chick oviduct in terms of amino acid composition and molecular weight (28,000 and 27,000 daltons) as determined by the procedure of Weber and Osborn[17] (see Figure 7).

## III. INDUCTION OF ANTIBODY AGAINST HMG-1 PROTEIN

Purified HMG-1 protein has been injected into rabbits for producing serum containing antibody against this protein. The IgG fraction of antisera was tested against the pure HMG-1 protein by using double-immunodiffusion plates. There is only one precipitation line found in the plate (see Figure 8). This antibody has been used for the immunofluorescence and immunomicrocomplement-fixation studies.

## IV. DETECTION OF HMG-1 PROTEIN IN THE SUBCELLULAR FRACTIONS OF CHICK OVIDUCT CELL

### A. Localization by Immunofluorescence

The antisera have been used for the studies on cellular localization of HMG-1 protein. The fibroblasts from immature chick oviduct were cultured on glass slides and HMG-1 distribution in the cell was viewed by immunofluorescence microscopy with anti-(HMG-1) serum. We have discovered that both nucleus and cytoplasm are fluorescent labeled. The intensity of the fluorescence is higher in the nucleus than that in the cytoplasm (see Figure 9). A control experiment has been demonstrated by applying the same technique except the anti-tubulin serum was used. We found that the intensity of the fluorescence is concentrated in the cytoplasm instead of the nucleus (see Figure 10).

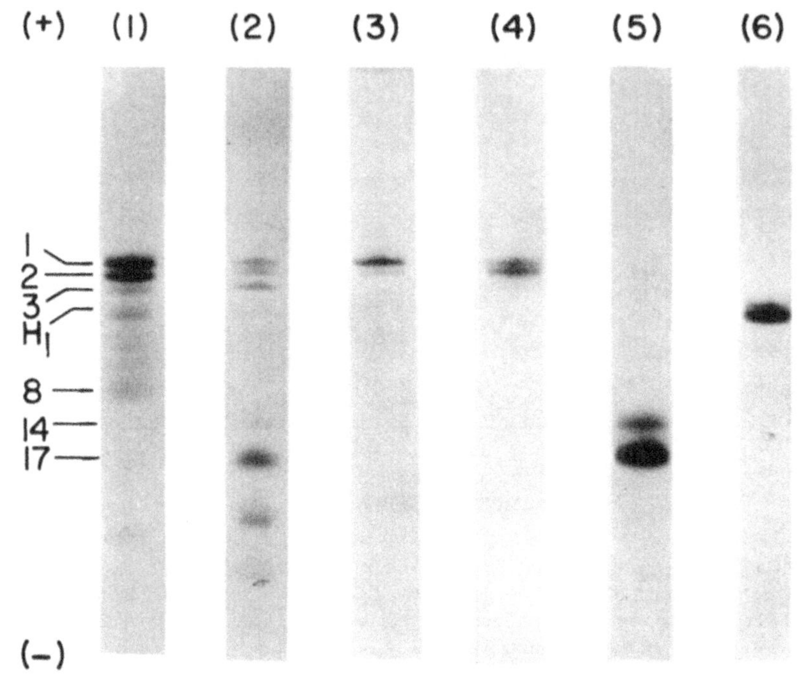


FIGURE 2. Acetic acid/urea/PAGE of HMG proteins. Tube gels (5 × 100 mm) consisting of 15% acrylamide, 0.1% NN′-methylenebisacrylamide, 0.9 *M* acetic acid, 2.5 *M* urea, 0.125% NNN′N′-tetramethylphenylenediamine and 0.125% ammonium persulfate were polymerized overnight at 23°C. Gels were pre-run at 2 mA per gel for 5 hr at 23°C. At the end of the pre-run, voltage was increased from 90 to 165 V, after which it remained constant. Running buffer consisted of 0.9 *M* acetic acid. Protein samples were dissolved directly in 0.9 *M* acetic acid containing 15% sucrose. Appropriate samples (40 to 60 μg of protein) were adjusted to 100 μℓ final volume with 0.9 *M* acetic acid/15% sucrose, then reduced with 5% β-mercaptoethanol. Electrophoresis was at 1 mA per gel (95 to 100 V) for 5.5 hr at 23°C. Gels were stained for 30 min with 0.5% Amido Black 10B in a solution containing 60% water, 33% methanol, and 6.5% acetic acid and destained in the same solution overnight. Gels (1) and (2) were calf thymus and hen oviduct HMG protein, respectively. Protein (peaks I, II, and IV) eluted from the CM-Sephadex® column (Figure 1) was individually resolved in gels (3), (4), and (5), respectively. Histone H1 isolated from the first 3X acetone precipitation of $HClO_4$ extract was in gel (6).

## B. Probe by Nuclease Digestion

In order to understand where the HMG-1 protein preferentially locates in the nucleus of the chick oviduct cell, nuclease digestion experiments have been performed. After limited DNAase I digestion of estrogen-treated chick oviduct nuclei, a very small amount of proteins HMG-1 and -2 was released which had been resolved by SDS-PAGE (polyacrylamide gel electrophoresis) (see Figure 11-gel 3, 4). However, when the same supernatant sample was run in acetic acid/urea/polyacrylamide gel, no HMG protein was visible (see Figure 12-gel 3, 4). A similar observation in parallel to this finding has been demonstrated in calf thymus and liver nuclei.[18]

On the other hand, after limited digestion of oviduct nuclei by micrococcal nuclease which release more proteins HMG-1 and -2 (see Figure 11-gel 5, 6). Clear bands of protein HMG-1 and -2 are visible in the acetic acid/urea/polyacrylamide gel (see Figure 12-gel 6, 7). Both of these two nucleases preferentially digest the transcriptionally active region of the genome in various tissues.[19-22] Our experiments indicated that a portion of proteins HMG-1 and -2 is associated with this active region and can be solubilized by this treatment. This association is in agreement with what has been observed in calf thymus, brain, and avian erythrocytes.[21,23,24]

It has been reported that the effect of the two nucleases tested on chromatin is different,

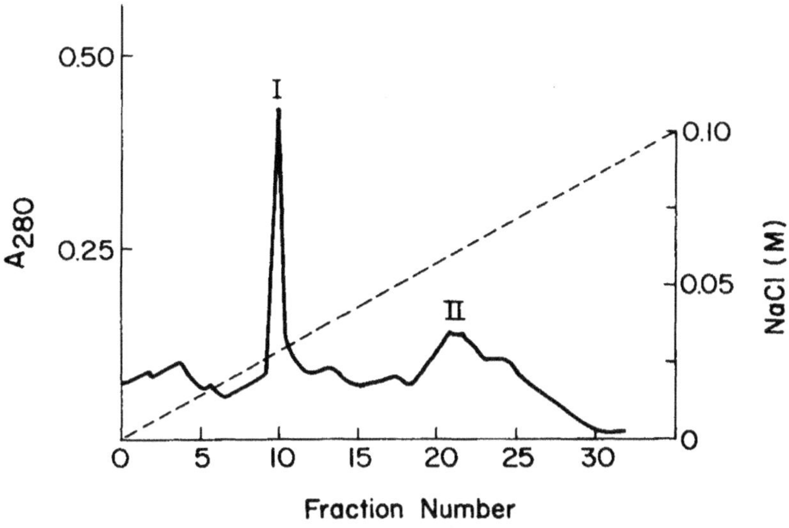


FIGURE 3. CM-Sephadex®-chromatographic separation of HMG-1 and HMG-2 proteins. The samples from peaks I and II of Figure 1 containing proteins HMG-1 and -2 were combined and reduced with dithiothreitol by the procedure of Ruiz-Carillo and Allfrey.[39] Approximately 15 mg of the reduced sample was rechromatographed on another CM-Sephadex® column by using a linear NaCl gradient (0 to 0.1 *M*) in 7.5 m*M* borate buffer with 0.5 m*M* PMSF. Protein HMG-1 was eluted first, followed by protein HMG-2.

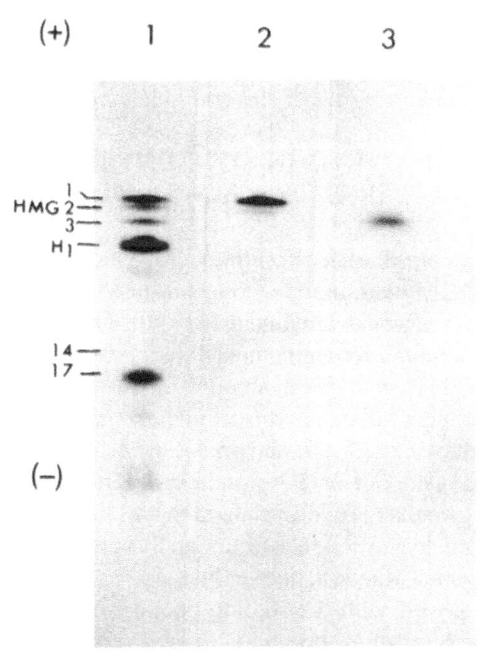


FIGURE 4. Identification of HMG-1 and -2 proteins by acetic acid/urea/gel electrophoresis. Gel (1) was hen oviduct total HMG protein. The individual proteins HMG-1 and -2 obtained from linear NaCl elution of CM-Sephadex® (Figure 3) were resolved in gel (2) and (3), respectively.

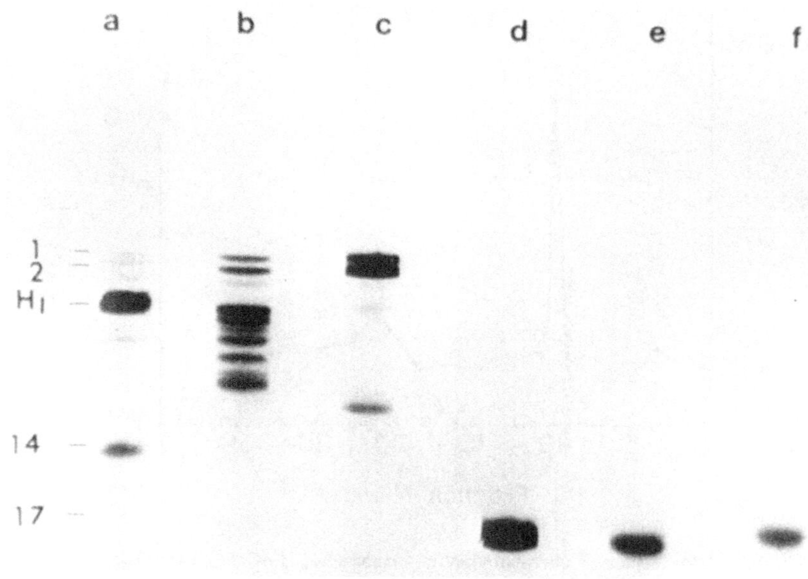


FIGURE 5. Purification of HMG-17 protein. Columns (a) and (b) are total HMG proteins from chick oviduct and calf thymus, respectively. Column (c) is the crude HMG-1 and -2 preparation. Columns (d) through (f) are different preparations of HMG-17 protein. ("H-1" = histone H-1.)

with the DNAase I preferentially digesting the DNA strands within the nucleosomes and the DNA in the spacer region, and with the micrococcal nuclease digesting the DNA in the internucleosomal spacer region.[25-27] In chick oviduct nuclei, the preferential release of proteins HMG-1 and -2 after micrococcal nuclease treatment may indicate that these proteins are associated with the linker region of the core nucleosomes.

## V. CHANGE OF HMG-PROTEIN CONTENT IN CELLULAR FRACTIONS AFTER ESTROGEN STIMULATION

### A. Detection by Microcomplement-Fixation

The amount of HMG-1 protein in the cytoplasm, microsomal, or nuclear fraction of chick oviduct (with or without estrogen treatment in vivo) was extracted by 5% perchloric acid and determined by immunomicrocomplement-fixation technique.[28,29]

Figure 13 indicates that, at a fixed amount of the antigen (200 ng of $HClO_4$-soluble protein), the percent of reactivities in cytosol, microsomal, and nuclear fractions are 22, 48, and 85 for control oviduct and 28, 65, and 96 for the estrogen-treated oviduct, respectively. This indicates that the content of HMG-1 protein in the $HClO_4$-extractable protein is generally higher in the estrogen stimulated oviduct than in that of the control. Furthermore, the content of HMG-1 protein in nuclear fraction is four- and twofold higher than that extracted from the cytosol and microsomal fraction, respectively.

This result correlates well with the finding obtained by immunofluorescence study suggesting a large amount of HMG-1 protein is indeed concentrated in the nucleus.

### B. Measurement by Radioimmunoassay

#### *1. Standardization of the Radioimmunoassay Technique*

In order to obtain more specific quantities of HMG-1 in the $HClO_4$-soluble fraction, it was further measured by radioimmunoassay. Purified oviduct HMG-1 proteins were labeled

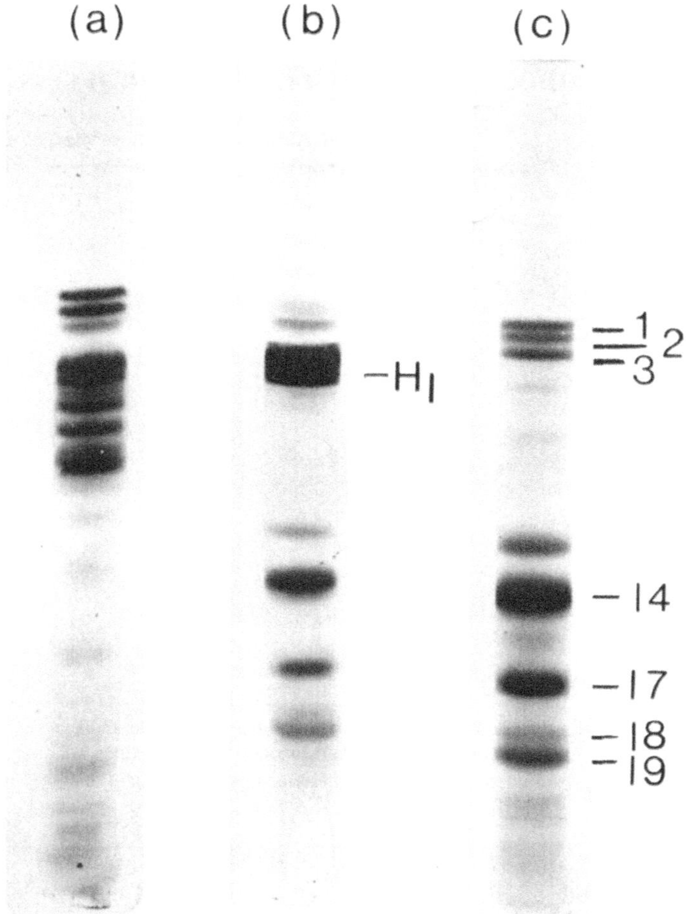


FIGURE 6. Acetic acid/urea/PAGE of HMG proteins isolated from chick oviduct receiving estrogen stimulation for 3 weeks. (a) Total HMG proteins extracted from calf thymus nuclei; (b) $HClO_4$-soluble protein (from chick oviduct which received estrogen treatment daily for 3 weeks) precipitated with 6 volumes of acetone (H-1 = histone H1); (c) total HMG protein from oviduct nuclei after removal of histone H1. Gel electrophoresis with 15% (w/v)-polyacrylamide gel. The details for electrophoresis have been described in the legend of Figure 2. The numbers on the right side of gel (c) represent HMG-1, -2, -3, -14, and -17. The protein bands -18 and -19 are the newly discovered HMG proteins.

with $^{125}I$ by the chloramine-T method originally developed by Hunter and Greenwood.[30] $^{125}I$-labeled HMG proteins were used for radioimmunoassay.

All dilutions for the radioimmunoassay were made in 10 m*M*-Tris/HCl buffer, pH 7.5, containing 1% bovine serum albumin (BSA). In a final volume of 0.5 mℓ the following were added: 0.1 mℓ of rabbit anti-(HMG-1) serum (1:500 initial dilution), 0.1 mℓ of $^{125}I$-labeled HMG-1 (60 to 80 μCi/μg, about 60,000 cpm) and 0.1 mℓ of standard HMG-1 from oviduct or 20 to 50 μℓ of unknown sample (in 10 m*M*-Tris/HCl buffer) and adjusted to 0.5 mℓ with the dilution buffer. The mixture was incubated for 2 hr at 37°C, followed by 18 hr at 4°C. Then 0.1 mℓ of goat anti-(rabbit IgG) antibody (0.2 mg) was added and the mixture was incubated for 6 hr at 4°C. Finally, 1.5 mℓ of m*M*-Tris/HCl buffer, pH 7.5 was added to each tube and the immunoprecipitate was isolated by centrifugation at 1200 × *g* for 20 min at 4°C. The supernatant was decanted and the radioactivity of the pellet counted in a Beckman gamma counter.

**Table 1**
**AMINO ACID COMPOSITION OF CHICK OVIDUCT HMG-1, -2, -14, AND -17 PROTEINS**

| | Content (mol/100 mol of total amino acid) | | | |
|---|---|---|---|---|
| Amino acid | HMG-1 | HMG-2 | HMG-14 | HMG-17 |
| Asp | 10.6 | 9.7 | 9.6 | 9.4 |
| Thr | 4.5 | 3.7 | 3.0 | 2.2 |
| Ser | 7.4 | 7.2 | 4.5 | 2.5 |
| Glu | 15.8 | 15.1 | 12.8 | 10.5 |
| Pro | 8.4 | 8.8 | 12.2 | 12.9 |
| Gly | 7.0 | 6.3 | 10.7 | 11.2 |
| Ala | 8.3 | 7.6 | 16.3 | 18.4 |
| Val | 4.6 | 6.7 | 3.0 | 2.5 |
| Cys | Trace | Trace | — | — |
| Met | 1.1 | 1.5 | 0.1 | — |
| Ile | 1.8 | 1.5 | 0.2 | — |
| Leu | 3.9 | 2.8 | 1.5 | — |
| Tyr | 2.8 | 1.4 | 0.1 | — |
| Phe | 2.9 | 2.4 | 0.1 | Trace |
| Lys | 14.3 | 18.2 | 20.9 | 24.3 |
| His | 1.7 | 2.4 | 0.2 | — |
| Arg | 4.8 | 4.6 | 4.8 | 4.1 |

*Note:* The protein (150 μg) was hydrolyzed with acid and analyzed for amino acid content on a Beckman model 121 amino acid analyzer as described by Spackman et al.[36]

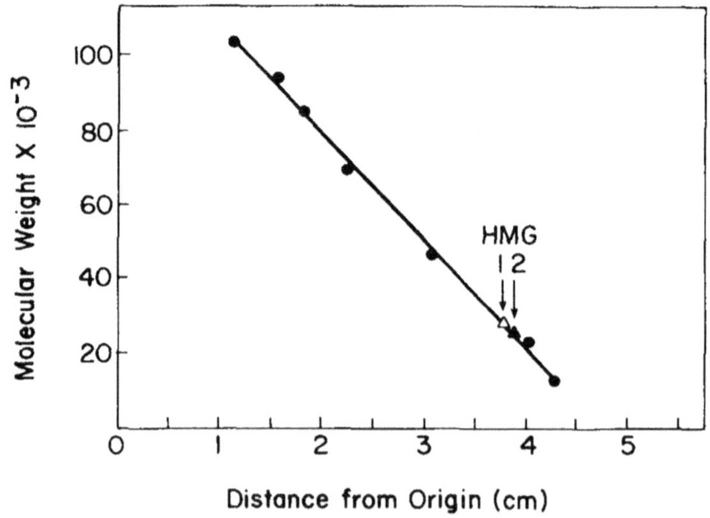


FIGURE 7. Immunological reaction between HMG-1 protein and its antiserum by immunodiffusion. Wells (1) and (2) contain antiserum from a rabbit immunized with HMG-1 protein; wells (3) and (4) contain control serum from this rabbit. Center well contains 25 μℓ of pure HMG-1 protein (1 mg/mℓ).

Figure 14 indicates that in the radioimmunoassay mixture, increasing amounts of HMG-1 from oviduct were added as nonradioactive competitors, and the decrease of radioactivity in the immunoprecipitate was expressed as a percentage of the "noncompetitive" (no HMG-1 protein) control values. A standard curve for the radioimmunoassay of HMG-1 protein

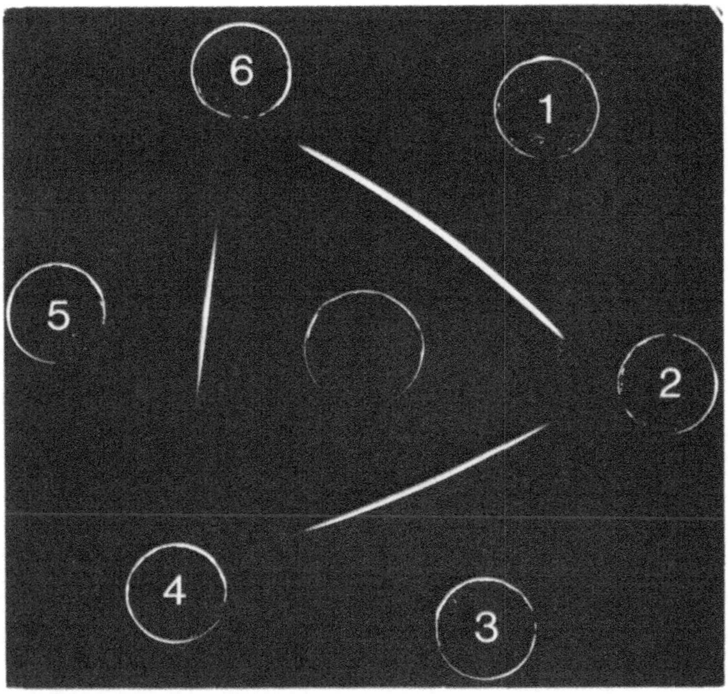


FIGURE 8. Immunodiffusion study with purified HMG-1 protein. Wells (1), (3), and (5) contain purified antiserum from three individual rabbits which received the immunization of HMG-1 protein. Wells (2), (4), and (6) contain control serum obtained from these rabbits before receiving HMG protein. The center well contains 25 μg of HMG-1.

was obtained. Based on this standard, the amount of HMG-1 protein in the $HClO_4$-soluble fraction can be measured with radioimmunoassay and expressed as micrograms per milligram of DNA.

*2. Stimulation of HMG-1 Protein by Estrogen*

After estrogen stimulation, the amount of HMG-1 in cytosol, microsomal, and nuclear $HClO_4$-soluble fraction increased from 1.05 to 4.2, 0.16 to 0.56, and 2.1 to 3.4, respectively. The increase due to estrogen is 4-, 3.5-, and 1.6-fold in these three respective fractions (Table 2). After estrogenic action, a drastic change was observed in the cytoplasm. In contrast, a relatively lower induction of HMG-1 protein was found in the nucleus. One could argue that the immunodetectable HMG-1 in the cytoplasm is due to the leakage of this protein from the nucleus during the process of nuclei isolation. However, the measurement of HMG-1 protein from oviduct nuclei obtained by nonaqueous medium did not support this argument (Table 2).

From this observation it may be interpreted that HMG-1 can shuttle between cytoplasm and nucleus in chick oviduct cell. After estrogen stimulation HMG-1 is actively synthesized and stored in the cytoplasm. It is likely that only a proportion of it will be transported into the nucleus. In eukaryotic cells, HMG proteins probably follow the same synthetic pattern as histones which are synthesized in the cytoplasm and then transported into the nucleus.[31,32]

## VI. FUTURE PROSPECTS

The immunobiochemical assay technique used to detect the localization and quantity

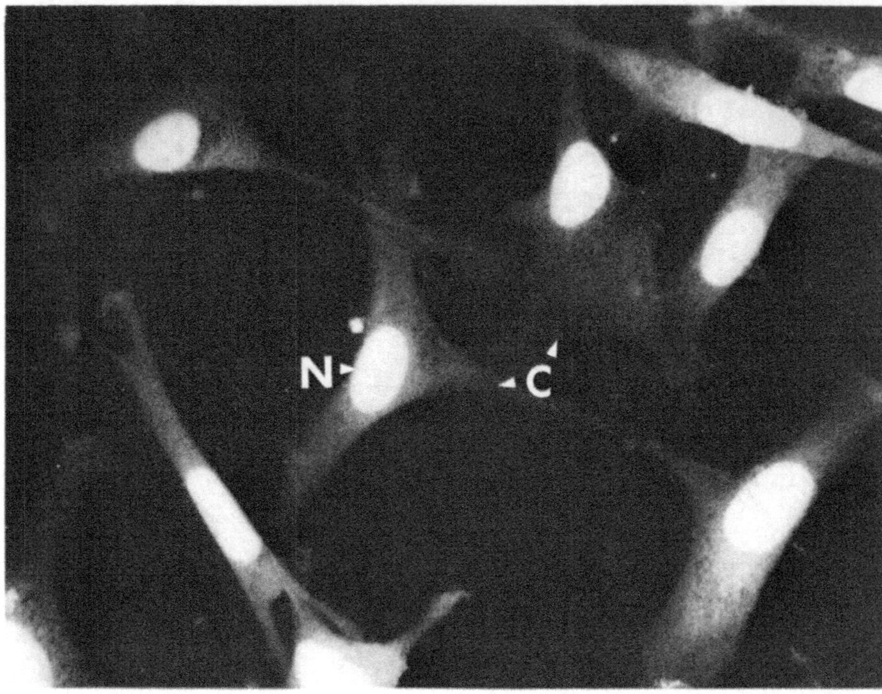


FIGURE 9. Localization of HMG-1 protein in chick oviduct fibroblast by immunofluorescence. The slides containing the fibroblasts from immature chick oviduct were fixed in 3% (v/v) formaldehyde for 20 min and then in acetone for 7 min. The solutions were precooled at − 20°C. The slides were incubated with 50 μℓ of anti-(HMG-1) serum (the optimal concentration has been found to be 0.1 mg/mℓ) at 37°C for 30 min. The cells were then incubated with 50 μℓ of anti-(rabbit IgG) goat antiserum with fluorescein label at 37°C for 30 min. Excell serum was removed from the slides and a coverslip was mounted in 99% (v/v) glycerol with Kronig wax. The fluorescence photomicrograph was taken with Kodak Ektachrome® film (ASA 200).[40] The arrows point to nucleus (N) and cytoplasm (C). (Magnification × 1040.)

change of HMG protein in chick oviduct cell is dependent on the antibody produced against HMG protein in rabbit. This technique was hindered by the limited specificity of the antibody so induced. Therefore we have adopted the monoclonal technique for inducing monospecific antibody against HMG protein.[33]

Total chromosomal HMG proteins (see Figure 6-gel c) from estrogen-stimulated chick oviduct were injected into mice for antibody induction. The spleen cells from the mice were fused with myeloma cells to form hybridomas. Subsequently the purified HMG protein 17 (see Figure 5) was used to select for the hybridomas that produce antibody against HMG protein 17. The hybridomas which produced antibody against HMG-17 were detected by an enzyme-linked immunosorbent assay (ELISA).[34]

The positive hybridoma clones were cultured and injected into mice to produce large quantities of ascites and solid tumors. The antibody in the ascites fluid and tumors was extracted and its immunoglobulin G fraction purified by Protein-A-Sepharose® column chromatography.[35]

A solid-phase radioimmunoassay technique has been developed to detect and characterize this antibody as follows: The assay was performed in 12 × 75 mm polystyrene test tubes that had been coated with antigen (10 μg of HMG-17 protein in 0.6 mℓ of phosphate buffered saline [PBS] containing 0.1 % gelatin) overnight. The antibody IgG fraction (1:20 dilution in PBS/gelatin buffer) was added to each tube and the mixture was incubated for 2 hr at 37°C. The tubes were washed three times with 0.6 mℓ of PBS/gelatin buffer followed by

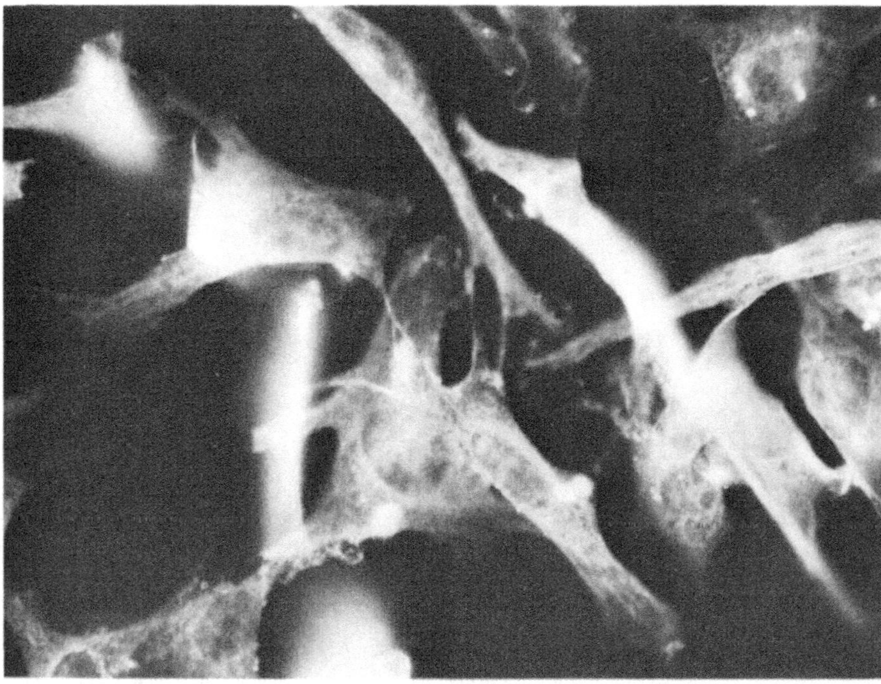

FIGURE 10. Localization of tubulin in chick oviduct fibroblast by immunofluorescence. The same technique described in the legend of Figure 9 was used except the slides were incubated with anti-tubulin serum.

the addition of 50 μℓ $^{125}$I-labeled rabbit anti-mouse IgG containing 150,000 cpm of radioactivity. After a 2-hr incubation at room temperature, the tubes were washed three times in PBS/gelatin buffer, and the radioactivity in the tube was measured by a Beckman gamma counter. This anti-(HMG protein 17) IgG interacted only with HMG protein 17 but not with other chromosomal proteins, e.g., histone H1, 95 K protein (a chick oviduct-specific chromosomal protein)[29] and HMG protein 1, 2, and 14 (see Table 3).

For future studies, it will be necessary to induce monospecific antibody against each individual HMG protein, e.g., HMG-1, -2, and -14. Applying this monoclonal antibody to the study of chromosomal components in oviduct is more sensitive, accurate, and specific than the conventional serological methods. We anticipate that this new approach will facilitate future studies on chromatin and chromosome structure and function.

## VII. CONCLUSION AND SUMMARY

Nuclear HMG proteins were isolated from estrogen-stimulated chick oviduct. These were proteins HMG-1, -2, -14 and -17, which are equivalent to the classification of calf thymus HMG proteins. In addition two extra HMG proteins which have lower electrophoretic mobility than HMG-17 were discovered in the chick oviduct. The HMG-1, -2, -14 and -17 proteins have been extracted and purified by $HClO_4$ and CM-Sephadex® column chromatography. The proteins have a high content of acidic and basic amino acids.

Antiserum against HMG-1 protein has been induced in the rabbit. With this antiserum and the limited nuclease digestion technique, the localization of HMG protein in cellular fractions of chick oviduct cell has been identified.

The quantitative change of HMG-1 in the cellular fractions of chick oviduct before or after estrogen stimulation has been detected by radioimmunoassay. The content of HMG-1

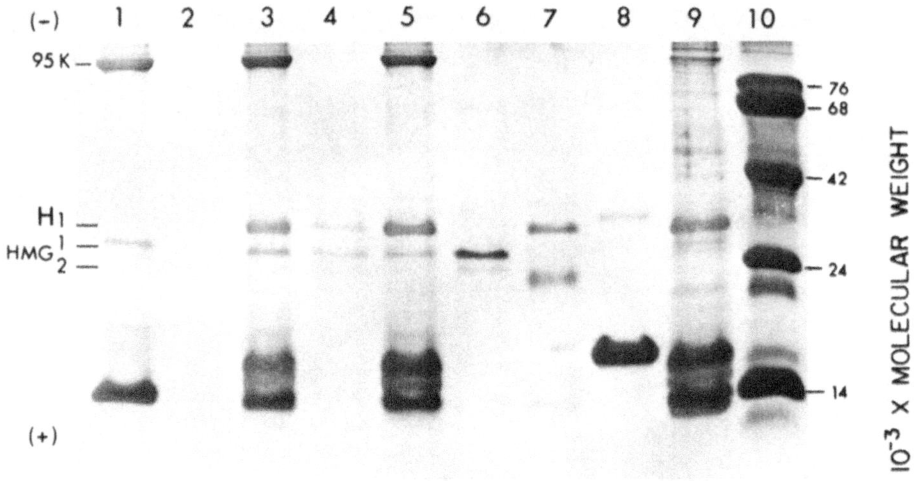


FIGURE 11. Separation of nuclear protein after digestion with nucleases. Chick oviduct nuclei were suspended in RSB buffer (10 m*M* Tris/HCl, pH 7.4, 10 m*M* NaCl, 3 m*M* $MgCl_2$, 0.5 m*M* PMSF) at a DNA concentration of 2.3 mg/mℓ. Then 10% digestion of the nuclear suspension was accomplished by the addition of DNAse I (40 μg/mℓ for 10 min at 37°C) or micrococcal nuclease (7 μg/mℓ for 10 min at 37°C). Control sample was incubated without the nuclease. Protein samples were prepared and separated by gel electrophoresis. Gel: (1) control pellet proteins (60 μg); (2) control supernatant proteins (10 μg, representing 25% of the total supernatant proteins); (3) DNAase-I-digested nuclear pellet protein (60 μg); (4) DNAase-I-digested supernatant protein (20 μg, representing 25% of the total supernatant proteins); (5) micrococcal-nuclease-digested nuclear pellet protein (60 μg); (6) micrococcal-nuclease-digested supernatant protein (30 μg, representing 25% of the total supernatant proteins); (7) DNAase I enzyme alone (20 μg); (8) micrococcal nuclease enzyme alone (20 μg); (9) hen oviduct nuclear acid-soluble (0.2 *M* $H_2SO_4$) proteins; (10) standards: lysosome, chymotrypsinogen, ovalbumin, BSA, and conalbumin. The separation was in an 18% acrylamide slab gel with 2% (w/v) SDS.

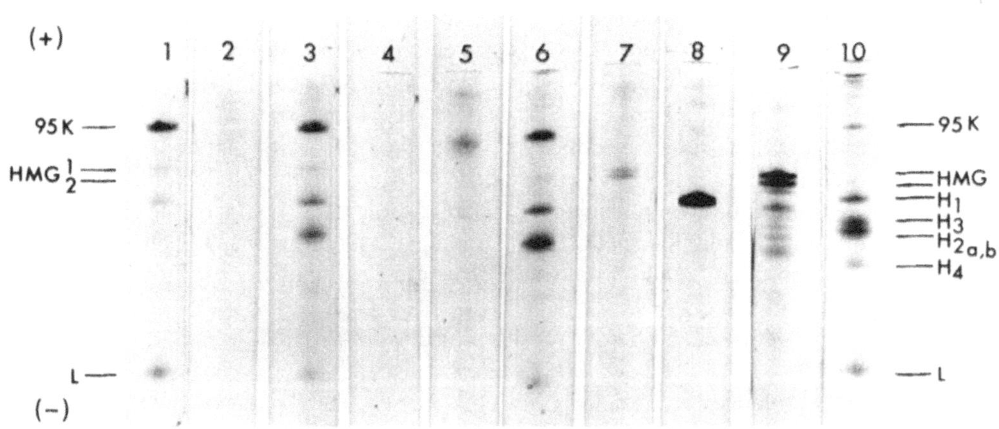


FIGURE 12. Separation of nuclear protein after nuclease digestion in an acetic acid/urea/gel. Protein samples from previous preparation (Figure 11) were resolved in acetic acid/urea/gel preparation. Gels: (1) control pellet proteins (30 μg); (2) control supernatant proteins (10 μg); (3) DNAase-I-digested pellet proteins (30 μg); (4) DNAase-I-digested supernatant proteins (20 μg); (5) DNAase-I- alone (20 μg); (6) micrococcal-nuclease-digested pellet proteins (30 μg); (7) micrococcal-nuclease-digested supernatant proteins (30 μg); (8) micrococcal nuclease alone (20 μg); (9) total calf thymus HMG proteins (40 μg); (10) total acid-soluble (in 0.2 *M* $H_2SO_4$) hen oviduct nuclear proteins. ("H" = histone.)

is higher in the nuclear fraction than in the cytosol and microsomal fractions. After administration of estrogen to young chick in vivo for 5 days, the quantity of HMG-1 is increased

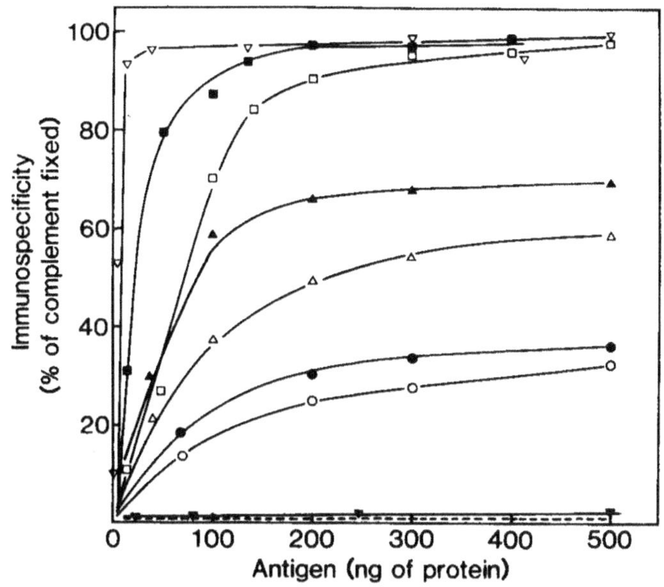


FIGURE 13. Microcomplement fixation of HMG-1 in cellular fractions of control or estrogen-induced chick oviduct. HMG-1 protein or $HClO_4$-extractable protein was fixed with rabbit antiserum against hen oviduct HMG-1 protein. The procedure for microcomplement-fixation was performed by using the following antigens: (▽) HMG-1; (▼), histone H1; (○,●) cytosol protein; (△,▲) microsomal protein; (□,■) nuclear $HClO_4$-soluble protein. The open symbols represent the control chicks and the closed symbols represent the estrogen-treated chicks; (----) represents control serum incubated with HMG-1.

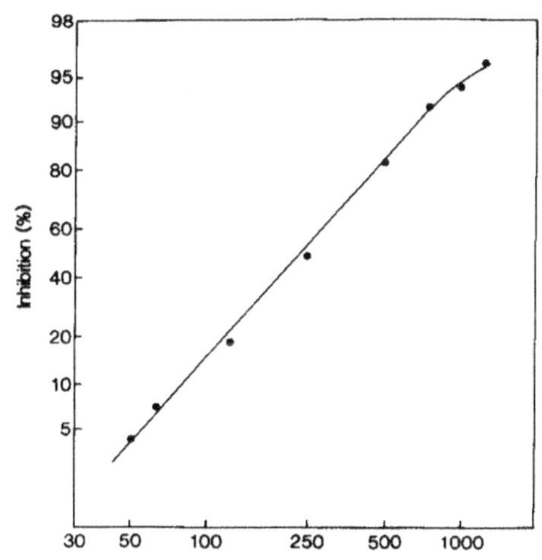


FIGURE 14. Standard curve for the radioimmunoassay of HMG-1 protein. $^{125}I$-labeled HMG-1 was added to the assay medium containing rabbit anti-(HMG-1) serum at a final dilution of 1:2500. Increasing amounts of HMG-1 from hen oviduct were added as nonradioactive competitors, and the decrease of radioactivity in the immunoprecipitate was expressed as a percentage of the "noncompetitive" (no HMG-1 protein) control value. Each point represents the mean of six different determinations.

**Table 2**
**DETERMINATION OF HMG-1 IN SUBCELLULAR FRACTIONS**

| | Protein (μg/mg of DNA) | | | | | | |
|---|---|---|---|---|---|---|---|
| | Control | | | Estrogen treated | | | |
| Cellular fraction | $HClO_4$-soluble | HMG-1 | % | $HClO_4$-soluble | HMG-1 | % | HMG-1 stimulation ( -fold) |
| Cytosol | 265 ± 14.3 | 1.05 ± 0.13 | 0.4 | 1087 ± 56.8 | 4.20 ± 0.55 | 0.4 | 4 |
| Microsomal | 26 ± 1.35 | 0.16 ± 0.02 | 0.6 | 78 ± 3.86 | 0.56 ± 0.07 | 0.7 | 3.5 |
| Nuclear | 117 ± 5.73 | 2.11 ± 0.26 | 1.8 | 180 ± 9.30 | 3.36 ± 0.43 | 1.9 | 1.58 |
| Nuclear (nonaqueous medium) | 121 ± 6.29 | 2.30 ± 0.32 | 1.9 | 183 ± 9.12 | 3.63 ± 0.44 | 2.0 | 1.57 |

*Note:* Oviducts of newborn chicks, 5 days to 1 week old, with or without estrogen treatment, were fractionated into three subcellular fractions. Each fraction was extracted with 5% $HClO_4$ and the soluble fraction was precipitated with acetone. The total protein content in $HClO_4$-soluble fractions was determined by the method of Lowry et al.[37] The DNA content in the oviduct was determined by the diphenylamine method as modified for this system by Teng and Teng.[38] The content of HMG-1 in $HClO_4$-soluble fractions was determined by radioimmunoassay technique. In a separate experiment, nuclei were isolated with a nonaqueous medium. After obtaining the nuclei, the $HClO_4$-soluble fraction was extracted and protein determined as described before. The values presented are means ±SD for three to five determinations.

**Table 3**
**DETECTION OF HMG-17 ANTIBODY MONOSPECIFICITY**

| Protein serving as antigen | $^{125}$I-labeled anti-mouse IgG bound (cpm) |
|---|---|
| Buffer only | 523 ± 19 |
| Buffer + antibody (IgG) | 808 ± 25 |
| HMG-17 | 2450 ± 45 |
| Histone H1 | 862 ± 23 |
| "95K protein" (chick oviduct) | 841 ± 24 |
| HMG-1 | 539 ± 20 |
| HMG-2 | 536 ± 16 |
| HMG-14 | 760 ± 22 |

*Note:* A solid-phase radioimmunoassay was used to detect the monospecificity of the IgG fraction of HMG-17 antibody obtained from the Protein A-Sepharose® column. The values presented are means ±SD of three to four determinations.

4-fold in the cytosol, 3.5-fold in the microsomal fraction, and 1.6-fold in the nuclear fraction. The finding of large amounts of HMG-1 in cytoplasm of oviduct cell after receiving estrogen stimulation is not due to its leakage from the nucleus. We anticipate that HMG-1 is synthesized in the cytoplasm and then transported into the nucleus. The synthesis and transportation of HMG protein is probably regulated by estrogen.

The induction of a monospecific antibody against HMG-17 has been described. Finally, we have stressed the importance of using monospecific antibody for the future studies on the change of chromosomal components during estrogen stimulation.

## REFERENCES


1. **Weisbrod, S. and Weintraub, H.,** Isolation of a subclass of nuclear proteins responsible for conferring a DNase I-sensitive structure on globin chromatin, *Proc. Natl. Acad. Sci. U.S.A.*, 76, 630, 1979.
2. **Weisbrod, S., Groudine, M., and Weintraub, H.,** Interaction of HMG 14 and 17 with actively transcribed genes, *Cell*, 19, 289, 1980.
3. **Albanese, I. and Weintraub, H.,** Electrophoretic separation of a class of nucleosomes enriched in HMG 14 and 17 and actively transcribed globin genes, *Nucleic Acids Res.*, 8, 2787, 1980.
4. **Weber, S. and Isenberg, I.,** High mobility group proteins of Saccharomyces cerevisiae, *Biochemistry*, 19, 2236, 1980.
5. **Hamana, K. and Iwai, K.,** High mobility group nonhistone chromosomal proteins also exist in Tetrahymena, *J. Biochem.*, 86, 789, 1979.
6. **Spiker, S., Mardian, J. K. W., and Isenberg, I.,** Chromosomal HMG proteins occur in three eukaryotic kingdoms, *Biochem. Biophys. Res. Commun.*, 82, 129, 1978.
7. **Sandeen, G., Wood, W. I., and Felsenfeld, G.,** The interaction of high mobility proteins HMG 14 and 17 with nucleosomes, *Nucleic Acid Res.*, 17, 3757, 1980.
8. **Mardian, J. K. W., Paton, A. E., Bunick, G. J., and Olins, D. E.,** Nucleosome cores have two specific binding sites for nonhistone chromosomal proteins HMG 14 and HMG 17, *Science*, 209, 1537, 1980.
9. **Kuehl, L.,** Synthesis of high mobility group proteins in regenerating rat liver, *J. Biol. Chem.*, 254, 7276, 1979.
10. **Seyedin, S. M. and Kistler, W. S.,** Levels of chromosomal protein high mobility group 2 parallel the proliferative activity of testis, skeletal muscle, and other organs, *J. Biol. Chem.*, 254, 11264, 1979.
11. **Andrews, G. K. and Teng, C. S.,** Studies on sex-organ development, prenatal effect of oestrogenic-hormone on tubular gland cell morphogenesis and ovalbumin gene expression in the chick Mullerian duct, *Biochem. J.*, 182, 271, 1979.
12. **Rabbani, A., Goodwin, G. H., and Johns, E. W.,** Studies on the tissue specificity of the high-mobility-group non-histone chromosomal proteins from calf, *Biochem. J.*, 173, 497, 1978.
13. **Teng, C. S., Andrews, G. K., and Teng, C. T.,** Studies on the high-mobility-group non-histone proteins from hen oviduct, *Biochem. J.*, 181, 585, 1979.
14. **Goodwin, G. H., Wright, C. A., and Johns, E. W.,** The characterisation of ISF monomer nucleosomes from hen oviduct and the partial characterisation of a third HMG 14/17 like protein in such nucleosomes, *Nucleic Acids Res.*, 12, 2761, 1981.
15. **Rabbani, A., Goodwin, G. H., and Johns, E. W.,** High mobility group non-histone chromosomal proteins from chicken erythrocytes, *Biochem. Biophys. Res. Commun.*, 81, 351, 1978.
16. **Goodwin, G. H., Sanders, C., and Johns, E. W.,** A new group of chromatin-associated proteins with a high content of acidic and basic amino acids, *Eur. J. Biochem.*, 38, 14, 1973.
17. **Weber, K. and Osborn, M.,** The reliability of molecular weight determinations by dodecyl sulfate-polyacrylamide gel electrophoresis, *J. Biol. Chem.*, 244, 4406, 1969.
18. **Goodwin, G. H. and Johns, E. W.,** Are the high-mobility-group non-histone chromosomal proteins associated with "active" chromatin? *Biochem. Biophys. Acta*, 519, 279, 1978.
19. **Weintraub, H. and Groudine, M.,** Chromosomal subunits in active genes have an altered conformation. Globin genes are digested by deoxyribonuclease I in red blood cell nuclei but not in fibroblast nuclei, *Science*, 193, 848, 1976.
20. **Garel, A. and Axel, R.,** Selective digestion of transcriptionally active ovalbumin genes from oviduct nuclei, *Proc. Natl. Acad. Sci. U.S.A.*, 73, 3966, 1976.
21. **Levy, W. B. and Dixon, G. H.,** Changes in the sequence diversity of polyadenylated cytoplasmic RNA during testis differentiation in rainbow trout (salmo gairdnerii), *Eur. J. Biochem.*, 74, 61, 1977.
22. **Bloom, K. S. and Anderson, J. N.,** Fractionation of hen oviduct chromatin into transcriptionally active and inactive regions after selective micrococcal nuclease digestion, *Cell*, 15, 141, 1978.
23. **Levy, W. B., Wong, N. C. W., and Dixon, G. H.,** Selective association of the trout-specific H6 protein with chromatin regions susceptible to DNase I and DNase II: possible location of HMG-T in the spacer region between core nucleosomes, *Proc. Natl. Acad. Sci. U.S.A.*, 74, 2810, 1977.
24. **Vidali, G., Boffa, L. C., and Allfrey, V. G.,** Selective release of chromosomal proteins during limited DNase I digestion of avian erythrocyte chromatin, *Cell*, 12, 409, 1977.
25. **Noll, M.,** Subunit structure of chromatin, *Nature (London)*, 251, 249, 1974.
26. **Axel, R., Cedar, H., and Felsenfeld, G.,** The structure of the globin genes in chromatin, *Biochemistry*, 14, 2489, 1975.
27. **Burgoyne, L. A., Mobbs, J. D., and Marshall, A. J.,** Chromatin structure: a property of the higher structure of chromatin and the time course of its formation during chromatin replication, *Nucleic Acids Res.*, 3, 3293, 1976.
28. **Wasserman, E. and Levine, L.,** Quantitative micro-complement fixation and its use in the study of antigenic structure by specific antigen-antibody inhibition, *J. Immunol.*, 87, 290, 1961.

29. **Teng, C. T. and Teng, C. S.,** Immuno-biochemical studies of a non-histone chromosomal protein in embryonic and mature chick oviduct. *Biochem. J.*, 185, 169, 1980.
30. **Hunter, W. M. and Greenwood, F. C.,** Preparation of iodine-131 labelled human growth hormone of high specific activity. *Nature (London)*, 194, 495, 1962.
31. **Block, D. P. and Brack, S. D.,** Evidence for cytoplasmic synthesis of nuclear histone during spermiogenesis in the grasshopper chortophaga viridifasciata (De Geer). *J. Cell Biol.*, 22, 327, 1964.
32. **Robbins, E. and Borun, T. W.,** The cytoplasmic synthesis of histones in Hela cells and its temporal relationship to DNA replication, *Proc. Natl. Acad. Sci. U.S.A.*, 57, 409, 1967.
33. **Teng, C. S., Teng, C. T., and Chan, T. S.,** Detection of monoclonal antibody to high-mobility-group protein 17 from chick oviduct, *Biochem. J.*, 203, 471, 1982.
34. **Kennett, R. H.,** Enzyme-linked antibody assay with cells attached to polyvinyl chloride plates, in *Monoclonal Antibodies*, Kennett, R. H. and McKearn, T. J., Eds., Plenum Press, New York, 1980, 376.
35. **Tucker, D. F., Begent, R. H. J., and Hogg, N. M.,** Characterization of immune complexes in serum by absorption in staphylococcal protein A: model studies and application to sera of rats bearing a gross virus-induced lymphoma, *J. Immunol.*, 121, 1644, 1978.
36. **Spackman, D. H., Stein, W. H., and Moore, S.,** Automatic recording apparatus for use in the chromatography of amino acids, *Anal. Chem.*, 30, 1190, 1958.
37. **Lowry, O. H., Rosebrough, M. J., Farr, A. L., and Randall, R. J.,** Protein measurement with the folin phenol reagent, *J. Biol. Chem.*, 193, 265, 1951.
38. **Teng, C. S. and Teng, C. T.,** Changes in chemical composition and oestradiol-binding capacity in chromatin during the differentiation of chick Mullerian ducts, *Biochem. J.*, 172, 361, 1978.
39. **Ruiz-Carillo, A. and Allfrey, V. G.,** A method for the purification of histone fraction $F_3$ by affinity chromatography, *Arch. Biochem. Biophys.*, 154, 185, 1973.
40. **Teng, C. T. and Teng, C. S.,** Changes in quantities of high-mobility-group protein 1 in oviduct cellular fractions after oestrogen stimulation, *Biochem. J.*, 198, 85, 1981.

Chapter 4

# PHOSPHORYLATION OF THE HIGH MOBILITY GROUP NONHISTONE PROTEINS

**Jaswant S. Bhorjee**

## TABLE OF CONTENTS

# I. INTRODUCTION

One of the most challenging problems in eukaryotic genome biology is to elucidate the relationship between chromatin structure and gene activity. The metazoan genome is a complex structure in which DNA and its associated proteins, in particular the histones, form the nucleosomes — the fundamental repeat subunit of chromatin.[1,2] Although core histones are essential for the structural integrity of the nucleosomes, it is becoming increasingly clear that chromatin-associated nonhistone proteins may critically affect the structure-function relationships of nucleosomes in chromatin. For recent reviews see Mathis et al.[3] and Cartwright et al.[4]

The long-standing interest in the nonhistone chromosomal (NHC) proteins as potential regulators or modulators of gene activity among the students of chromosome biology (including our laboratory), is based upon several studies with native and reconstituted chromatins demonstrating tissue specificity of the bulk nonhistones in the synthesis of specific mRNAs.[5,6] However, no chromosomal protein(s) has been identified so far that establishes a direct causal relationship between its presence and transcription per se. The problem lies in the vast heterogeneity of the total NHC proteins, most of which may indeed be structural in nature, and the lack of precise methodology that may assist in purifying a possible regulatory protein(s) which may be present in only a few molecules per cell nulceus. This is not to suggest that productive effort or progress in studies on chromosomal proteins has been lacking. Actually, two groups — the high mobility group (HMG) and the DNA-binding class of the NHC proteins — have been subjects of intensive investigations in this regard, as reported in this volume.

## A. High-Mobility Group Proteins

The HMG class of the nuclear nonhistone proteins is among the most studied.[7] The HMG proteins are so defined because of their relative mobilities in acid-urea polyacrylamide gels. Historically, HMG proteins were first observed as impurities in histone H1 fractions from calf thymus.[8] Twenty peptides were identified initially. Only four — HMG-1, -2, -14, and -17 — are now characterized as authentic HMGs (see Figure 1). They seem to be ubiquitous in distribution and like the histones, are a highly conserved group of nuclear proteins.[7,9,10] HMG proteins can be characteristically extracted by 5% perchloric acid (PCA)[11] or 0.35 *M* NaCl[12] from whole cells, purified nuclei, or chromatin. The four HMG proteins can be subdivided into two groups based on their structural and functional similarities: HMG-1 and -2 have similar molecular weights (estimated to be 26,500 and 26,000 daltons, respectively),[13,14] exhibit a high degree of amino acid sequence homology,[7] and cross-react immunologically.[15] Likewise, HMG-14 and -17 are smaller in size (with approximate molecular weights of 10,700 and 9200 daltons, respectively),[7,12] and are closely related in amino acid sequence.[16,17] Furthermore, the acid solubility properties of these two subgroups also seem to be different. For example, HMG-1 and -2 are soluble in 2% trichloroacetic acid (TCA) and insoluble in 10% TCA, and HMG-14 and -17 are soluble in 10% TCA. The structural studies so far do indicate, however, that the four HMGs are clearly distinct proteins. HMG proteins have a high content of both basic (~25%) and acidic (~30%) amino acid residues and partial sequence analyses show that the charge distribution in the molecules is polarized.[7,18] The HMG proteins are present in high concentrations (~$10^6$ molecules per cell nucleus) in most animal cells,[19,20] approximating about 5% of the chromatin mass, and equal to about 10 to 20% of the nucleosome content, indicating thereby predominantly a structural role for these nuclear proteins.

### 1. *Nucleosomic Distribution*

All four HMG proteins are present in chromatin subunits.[21-23] This result has generated a great upsurge in interest in this group of the NHC proteins. Use of several endonucleases

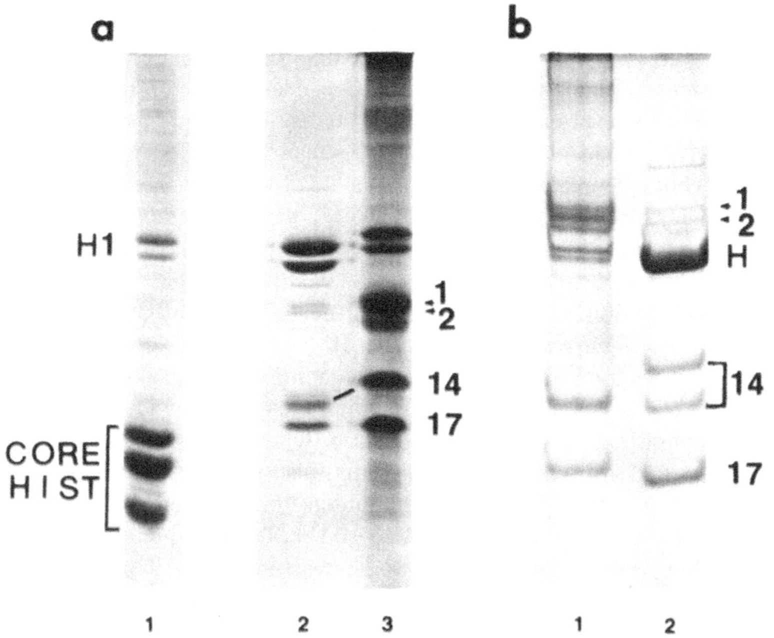


FIGURE 1. Electrophoretic analysis of total HeLa nuclear proteins extracted by 5% PCA. (a) 17% Polyacrylamide SDS-slab gel:[53] lane 1, total HeLa histones; lane 2, HeLa HMG proteins; lane 3, erythrocyte HMG proteins. (b) 20% Polyacrylamide acid-urea slab gel (Panyim, S. and Chalkley, R., *Arch. Biochem. Biophys.*, 130, 337, 1969): lane 1 erythrocyte HMG proteins, lane 2 HeLa HMG proteins. Protein load, 50 μg per lane. Stained with Coomassie brilliant blue R-250. Erythrocyte HMG proteins were a gift from Gerald R. Reeck, and were obtained as 2% TCA soluble material from the 0.35 *M* NaCl extract of chicken erythrocyte chromatin.

as probes for chromatin structure has revealed that HMG-1 and -2 (or the trout HMG-T analogue),[21,23-25] and HMG-14 and -17 (or the trout H6 analogue)[21-26] are enriched in the micrococcal nuclease sensitive mononucleosome fractions containing transcribed DNA sequences. It is not the purpose of this article to review the exhaustive literature on the relationship of nuclease sensitivity of chromatin regions to the transcriptionally active genes. This has been a subject of several recent reviews.[3,4,27] Work is cited here which in the author's judgment may help amplify the significance of the HMG proteins in regard to their post-synthetic modifications.

The best evidence for a specific role for HMG proteins in the alteration of chromatin structure around actively transcribed genes comes from the reconstitution studies of Weintraub and colleagues.[28,29] Actively transcribed globin DNA sequences in chick erythrocyte chromatin are preferentially sensitive to pancreatic DNase I.[30] Based on this landmark finding, it has been clearly demonstrated that the DNase-I sensitivity of the chick erythrocyte globin gene in the 0.35 *M* NaCl stripped chromatin can be reconstituted by the 0.35 *M* NaCl extract or even by purified HMG 14/17 when mixed at a ratio of 1 mol HMG per 10 mol of nucleosomes.[29] Furthermore, Stalder et al.[31] have shown that DNA cleavages in the globin chromatin domain are tissue-specific; for example, whereas both the adult and embryonic β-globin genes were found to be very sensitive to DNase I in embryonic red cells, only the adult gene remained hypersensitive in adult erythroid cells. This result suggests that genes which are "potentially" active (i.e., in the preactivation state) also exhibit hypersensitivity to DNase I. It was later suggested that the developmental stage-specific switch in the DNase I-sensitivity of the adult β gene may result, among other alterations, from the association of HMG-14 and -17 with the newly activated adult gene.[32] The special chromatin conformation, i.e., preferentially sensitive to DNase I, of actively transcribed gene sequences seems to be a general property of all transcriptionally competent genes regardless of the rate of their expression,[33-35] and was found to be dependent on the presence of HMG-14 and -17.[35] In contrast, however, there are several studies that argue against the exclusive enrichment of HMG-14 and -17 in the nuclease-sensitive fraction of transcribed DNA sequences.[21,25,26,36] For example, though HMG-14 and -17 were enriched in micrococcal nuclease generated chick erythrocyte mononucleosomes,[21,26] the DNA associated with such nucleosomes was not enriched in globin sequences,[26] nor were they found to have been preferentially released by the DNase-I treatment of thymus and liver nuclei.[36] Moreover, using two-dimensional hybridization method for mapping nucleosomes by gel electrophoresis, Varshavsky and colleagues have demonstrated that HMG-14 and -17 are associated with highly repetitive, presumably transcriptionally inactive, regions of HeLa chromatin.[37] Furthermore, HMG proteins were not found to be preferentially associated with the mononucleosomes enriched in the amplified dihydrofolate reductase (DHF-reductase) gene.[38] The association of some HMG-14 and -17 with satellite DNA has also been described.[39] In connection, an important observation is that given certain conditions of ionic strength and HMG:nucleosome ratios HMG-14 and -17 can bind to bulk nucleosomes,[40] though preferential binding of these HMGs to β-globin gene-containing nucleosomes was indicated. The reason(s) for this discrepancy in the findings on the exclusive association of HMG-14 and -17 with active nucleosomes is not clear. It may simply be due to the fact that for the most part the studies emphasizing exclusive presence of HMG-14 and -17 in actively transcribed DNA sequences have examined genes transcribed essentially at high rates, for example, globin in chick erythroid tissue. Such genes may well differ in some subtle manner from the normally expressed (housekeeping) genes. Cedar's group[35,41] has addressed this question by examining the DNase-I sensitivity of the nick translated chicken erythrocyte bulk chromatin, under conditions where the radiolabel is concentrated in the expressed DNA sequences. They found, in agreement with Weisbrod et al.,[29] that all the actively transcribed DNA sequences were preferentially degraded by DNase I and that this sensitivity could be reconstituted by HMG-14 and -17.[41] These HMG proteins, however, were not necessary for the conformation which makes the actively transcribed genes preferentially accessible to micrococcal nuclease.[35,36] Regardless, it is now a tenet of faith that actively transcribed genes or the genes committed to transcription possess a conformation that is more accessible to nucleases, and most probably the low molecular weight HMGs (14/17), in conjunction with some yet undefined factor(s)[29] are involved in determining the transcriptional potential of chromatin domains. How do HMG proteins recognize transcriptionally competent nucleosomes? How do they generate or maintain the DNase-I sensitive conformation of such nucleosomes? More urgently, how do they even interact within the basic nucleosome structure? These are all unanswered questions. Analyses of the DNA and the core histones by gel electrophoresis,[42] and sedimentation properties,[40] of the HMG-14 and -17 bound and unbound nucleosomes have revealed no significant differences. The above discussion reinforces the concept that some yet unidentified chromosomal elements (e.g., histones or other NHC proteins), or even post-synthetic modifications of the HMG proteins, may facilitate the HMG-induced transcriptionally competent features in chromatin.

*2. Post-Synthetic Modification*

Post-synthetic modifications (e.g., phosphorylation, acetylation, methylation) of the chromosomal proteins have been the subject of several reviews.[43,44] Acetylation of histones[45] in particular, and phosphorylation of the nonhistones,[46] have been correlated to transcriptional activation of chromatin, possibly by relaxation of constraints on the associated DNA sequences. Similarly, in several tissues and cell types examined, the HMG proteins have been shown to undergo covalent modifications by acetylation,[47] poly(ADP)-ribosylation,[48,49] methylation,[50] glycosylation,[49] and phosphorylation.[51-53] The possibility that such modifications may, at least in part, explain how the HMG proteins induce transcriptionally competent conformation of chromatin regions, albeit nonspecifically (i.e., in all the genes that are transcribed), is of potential value in this regard.

***a. Phosphorylation in Cell Cultures***

It is only recently that phosphorylation of the HMG proteins has been studied. Beginning with Saffer and Glazer's[51] report on [$^{32}$P] phosphate incorporation into HMG-14 and -17 in Ehrlich ascites and L1210 mouse tumor cells, it has now been documented that these low molecular weight HMG proteins are also phosphorylated in HeLa,[52,53] Chinese hamster ovary (CHO),[54,55] and several other mouse and human transformed cells.[55] Except for the CHO cell study[54] where all four HMG proteins were found to incorporate $^{32}$P, HMG-14 and -17 were the only HMG proteins found to be phosphorylated. HMG-1 and -2 seem not to be phosphorylated. In almost all of these studies the isolated HMG proteins were separated by electrophoresis in acid-urea or SDS-polyacrylamide gels, the proteins stained, and the labeled phosphoproteins identified by autoradiography. In none of these studies were the phosphorylated sites identified. In regard to the low levels of $^{32}$P incorporation into HMG-1 and -2 noted in CHO cells,[54] the radioactivity was monitored in the peptide bands excised from SDS-polyacrylamide gels. This, however, can lead to spurious data. For example, minor peptides in a mixture that may not be visible by standard staining procedures, and which may be phosphorylated, may comigrate in SDS-polyacrylamide gels with mobilities similar to the proteins of interest, thereby resulting in contamination of the sample. Alternatively, it is possible that some $^{32}$P-labeling of HMG-1 and -2 could occur as a result of ADP-ribosylation.[48,49] In vitro poly(ADP-ribose) modification of mouse Friend cell HMGs[49] and trout H6[56] has been reported. In contrast, however, in the HMG proteins isolated from HeLa cells labeled in vivo with [$^{32}$P] orthophosphate, no HMG-associated ADP-ribose residues were detected[52,57] by the criteria of hydrolysis in dilute alkali or by snake-venom phosphodiesterase. An interesting result in regard to phosphorylation of HMG-14 and -17 is the cell cycle dependence of phosphorylation of these proteins in synchronized HeLa cells[53] (see Figure 2). It was shown by Bhorjee[53] that there is a sevenfold increase in $^{32}$P incorporation into HMG-14 in the $G_2$-phase relative to $G_1$ and it was indicated that phosphorylation of HMG-14 may correlate to the events that follow $G_2$, i.e., chromatin condensation and cytokinesis. It was also shown that the differential phosphorylation of HMG-14 and -17 in synchronized HeLa cells was independent of the net synthesis of these proteins as the cells progressed through the cell cycle (see Figures 2 and 6). In another study, no difference was reported in the incorporation of $^{32}$P-phosphate into HMG-14 and -17 in synchronized CHO cells.[55] In this latter study, however, only $G_1$ and early-S phase cells were monitored, and no significant differences in phosphorylation of HMG-14 and -17 would have been observed since major cell cycle stage-specific differential phosphorylation of these HMGs was noted beginning in the late-S phase and in $G_2$ in synchronized HeLa cells (see Figure 2 and Reference 53). Arfmann et al.[54] have reported phosphorylation of all four HMG proteins in interphase and metaphase CHO cells. They observed that HMG-17 was the predominant phosphate acceptor during interphase. However, the pattern of $^{32}$P incorporation changed during metaphase when the phosphorylation of HMG-17 decreased to half the interphase value while that of other HMG proteins increased by a factor of three. This result seems to be in agreement with the HeLa cell cycle study[53] where phosphorylation of HMG-14 had increased dramatically in $G_2$, a stage prior to mitosis. In both HeLa[53] and CHO[54] cells the stage-specific $^{32}$P incorporation into HMG-14 seems to parallel the phosphorylation of histone H1. This suggests that the phosphorylated molecules of HMG-14 may replace histone H1 in the actively transcribed regions of chromatin. In no study was any relationship between HMG protein phosphorylation and DNA synthesis apparent. Although no causal relationship

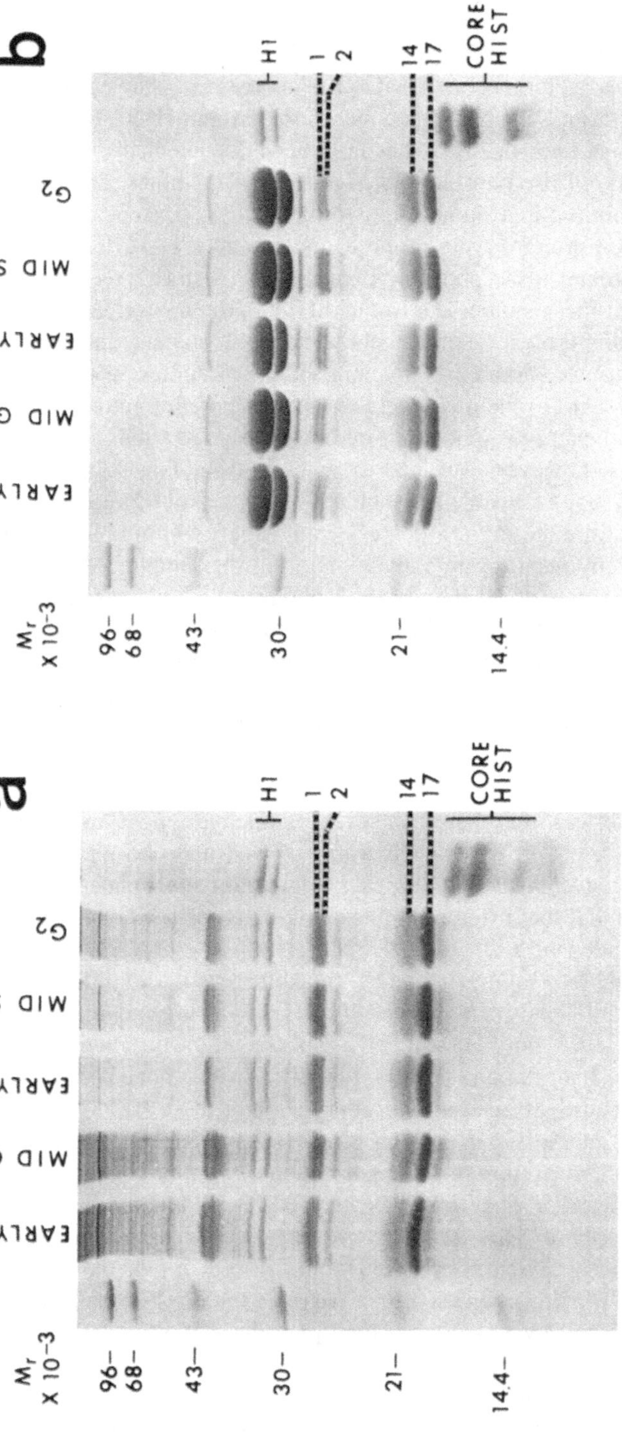
b
a
EARLY G1
MID G1
EARLY S
MID S
G2
H1
1
2
14
17
CORE HIST
Mr X 10-3
96-
68-
43-
30-
21-
14.4-

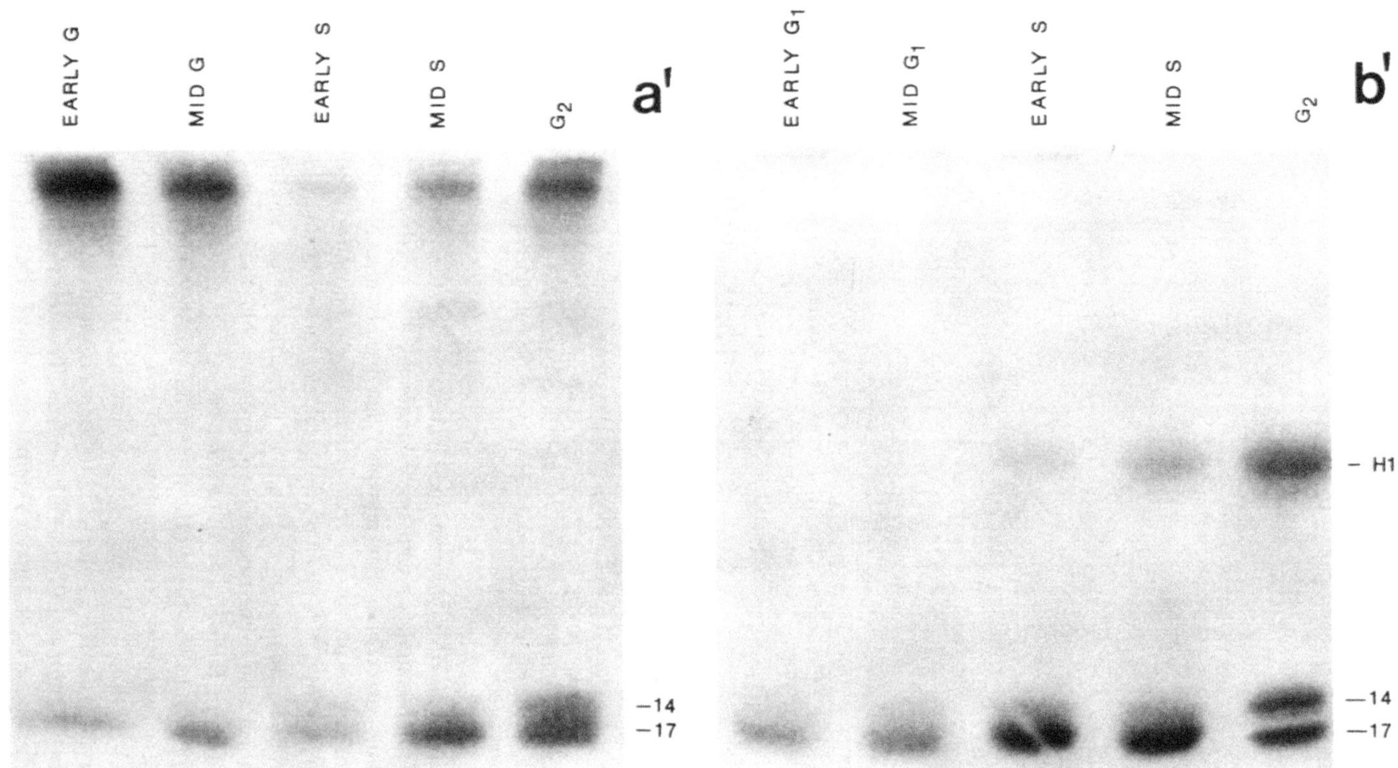


FIGURE 2. 17% Polyacrylamide SDS-gel electrophoretic analysis of the cell cycle stage-specific [$^{32}$P]phosphate labeled HMG proteins extracted by 0.35 *M* NaCl (a) or 5% PCA (b) from nuclei of HeLa cells synchronized by the double-thymidine block technique of Puck (*Science*, 144, 565, 1964). The proteins were processed for electrophoresis as described previously.[53] Protein load, 50 μg per lane. In both (a) and (b) the far right lanes show total HeLa histones; the far left lanes show molecular weight marker proteins. (a) and (b) Coomassie blue stained gel patterns; (a′) and (b′) corresponding autoradiogram patterns. (a, a′, and b from Bhorjee, J. S., *Proc. Natl. Acad. Sci. U.S.A.*, 78, 6944, 1981. With permission.)

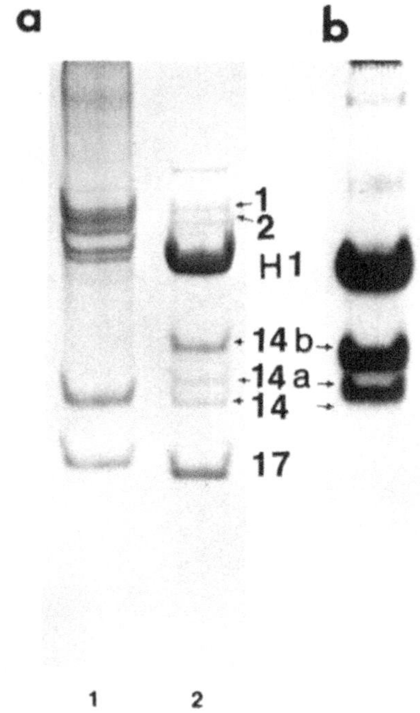


FIGURE 3. Electrophoretic analysis of the HMG proteins prepared as 5% PCA extracts of nuclei isolated from [$^{32}$P]phosphate-labeled HeLa cells. Proteins were separated on 20% polyacrylamide acid-urea slab gel. Protein load, 50 μg. (a) Coomassie blue stained gel: lane 1, unlabeled chicken erythrocyte HMG proteins (see Figure 1); lane 2, HeLa HMG proteins. (b) Autoradiogram of gel in (a) showing $^{32}$P incorporation into HeLa HMG proteins. Note that HeLa HMG-17, comigrating with erythrocyte HMG-17 (as marker), is not phosphorylated. Also note that HeLa HMG-14 exhibits three electrophoretic forms (14, 14a, 14b) by Coomassie stain. Two of these (14a, 14b) substantially incorporate $^{32}$P, whereas the third (14), seems to be very weakly phosphorylated and hence the label is visible only after overexposure of the X-ray film.

between phosphorylation of HMG proteins and generation of the active chromatin conformation has been demonstrated, a recent study[55] indicates that micrococcal nuclease generated mononucleosomes are enriched in phosphorylated molecules of HMG-14 and -17. Furthermore, contradictory results on butyrate-induced hyperphosphorylation of HMG-14 and -17 have also been reported.[52,55] Since phosphorlyation causes an alteration in the charge properties, it may result in changes in the binding properties of the modified proteins to DNA. Hence, phosphorylation in the case of HMG proteins, like the histones[43] and the bulk nonhistones,[44] could also be biologically significant.

One productive approach to identifying the role of the phosphorylated HMG proteins in generation of the DNase-I sensitive conformation of actively transcribed nucleosomes, is to study the nuclease sensitivity of chromatins reconstituted with purified native and enzymatically dephosphorylated HMG-14 and -17. In their efforts to purify HMG-14 and -17 for such an experiment, Mellon and Bhorjee[57] have recently discovered that when displayed on acid-urea polyacrylamide gels, the major Coomassie positive HMG-17 band comigrating with erythrocyte HMG-17 as marker (gift from Gerald Reeck) is not phosphorylated (see Figure 3). When reexamined by electrophoresis on both SDS-polyacrylamide gels and acid-urea polyacrylamide gels, the results indicated the following: (1) the $^{32}$P radioactivity comigrating with HMG-17 in standard-size SDS-polyacrylamide gels migrates with a slightly

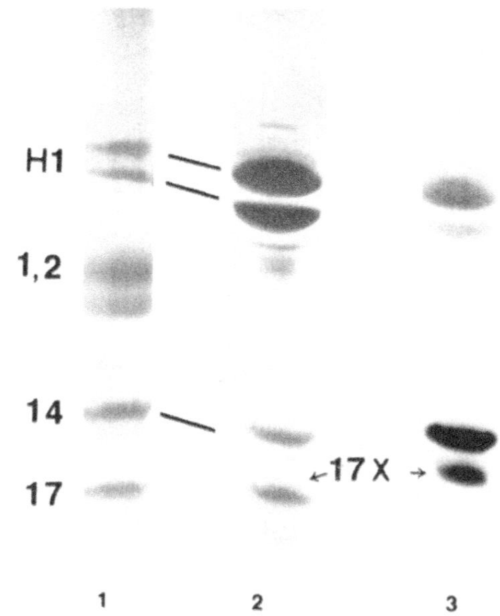


FIGURE 4. Electrophoretic separation of protein 17X from bulk HMG-17 using long (16 cm) 17% polyacrylamide SDS-slab gel. HeLa HMG proteins were extracted by 5% PCA from purified nuclei obtained from [$^{32}$P]phosphate labeled cells. Lane 1, unlabeled chicken erythrocyte HMG proteins (see Figure 1); lane 2, 70 μg of HeLa HMG proteins; lane 3, autoradiogram of proteins shown in lane 2. Lane 1 and 2 stained with Coomassie brilliant blue. Note that HeLa HMG-17, comigrating with erythrocyte HMG-17 (as marker), is not phosphorylated — a result similar to that found using the acid-urea polyacrylamide electrophoresis system (Figure 3). On the other hand, protein 17X, which is masked by the bulk HMG-17 in standard-size SDS-gels (see Figures 1 and 2), is heavily phosphorylated.

slower mobility as a minor peptide (labeled 17X) relative to the major Coomassie positive HMG-17 in long (16-cm) SDS-polyacrylamide gels, and it is this 17X peptide which is heavily phosphorylated (see Figure 4); and (2) the 17X peptide migrates with the apparent mobility of histone H1 (or slightly slower) in acid-urea gels[57] (see acid-urea to SDS two-dimensional gel electrophoretic analysis of $^{32}$P-labeled HeLa HMGs in Figure 5). The significance of protein 17X alone, or in conjunction with HMG-17 (since it coextracts with HMG proteins) in the nucleosome hierarchy is not known. It appears as of now that 17X may constitute the highly phosphorylated molecular species of HMG-17. It also is conceivable that protein 17X is an entirely different, though HMG-like, polypeptide and as such may be involved in some manner in directing the binding of HMGs to the nucleosomes. Clarification on any such relationships between protein 17X, other HMGs, and nucleosomes, will have to wait until the amino acid sequence information on protein 17X is available.

It must also be noted that HeLa HMG-14 appears as two to three well defined protein species in acid-urea polyacrylamide gels (see Figures 1 and 3). That these indeed are electrophoretic forms of HMG-14 is indicated by their identical migration as single HMG-14 species in the second dimension in SDS-polyacrylamide gels.[57,58] Multiple electrophoretic forms of HMG-14 have been seen in chicken erythrocyte chromatin.[59] In view of the above findings on the differences in indication of $^{32}$P incorporation into HMG-17 and the appearance of multiple forms of HMG-14 in acid-urea gels, the cell cycle-dependent phosphorylation of HMG proteins[53] was reexamined. When HMG proteins isolated from [$^{32}$P] orthophosphate labeled synchronized HeLa cells were analyzed using the acid-urea polyacrylamide gel

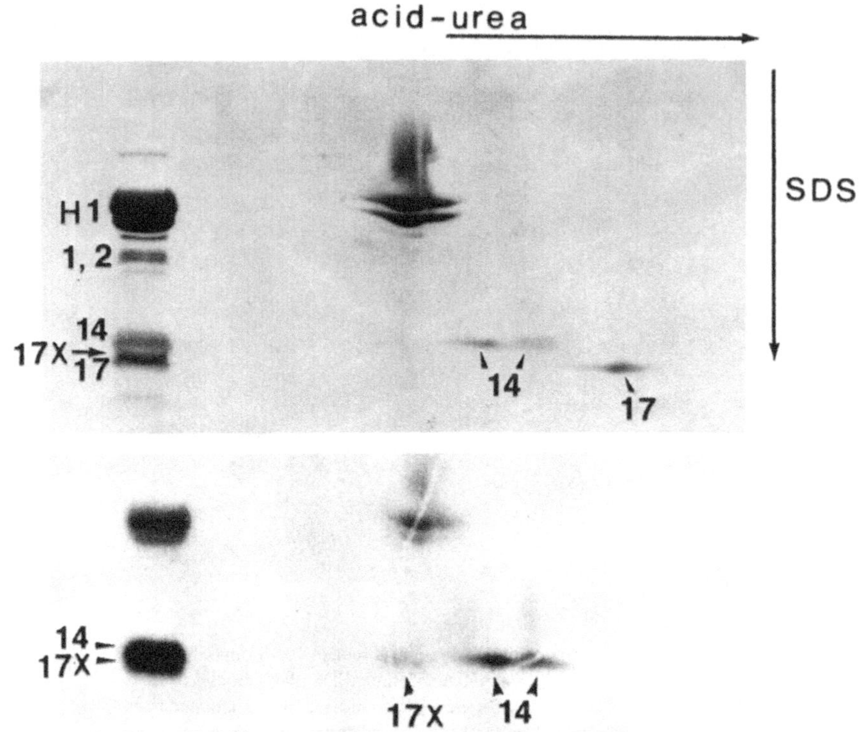


FIGURE 5. Two-dimensional electrophoretic analysis of [$^{32}$P]phosphate-labeled HeLa HMG proteins obtained as 5% PCA soluble nuclear extracts. Top panel: 70 μg of total HMG protein was electrophoresed in 20% polyacrylamide acid-urea slab gel in the horizontal direction, the lane cut out, equilibrated in 2% SDS-1% β-mercaptoethanol at 60°C for 5 min, and placed on 17% polyacrylamide SDS-slab gel and electrophoresed in the second dimension. As shown in the left-most lane, total HMG proteins electrophoresed in the second dimension only. (Stained with Coomassie blue.) Bottom panel: Autoradiogram of the top panel. Note that the bulk Coomassie positive HMG-17 (top panel) is not phosphorylated. Highly phosphorylated protein 17X (bottom panel) migrates with the apparent mobility of histone H1 in acid-urea electrophoresis system (horizontal direction), but comigrates with bulk HMG-17 in the SDS-electrophoresis system. Protein 17X is not visible by Coomassie stain (top panel) because of the extremely small amounts in which it is present in the total HMG protein extracts.

electrophoresis (PAGE) system, HMG-17 did not show any $^{32}$P radioactivity[60] (see discussion above on 17X protein). HMG-14 and its various electrophoretic forms exhibited the previously reported[53] cell cycle-dependent phosphorylation in synchronized HeLa cells.[60] In regard to the reports of Saffer and Glazer[51,55] on the appearance of $^{32}$P radioactivity in HMG-14 and -17 electrophoretic bands in acid-urea polyacrylamide gels, a careful examination of the figures shown (in particular, Figure 1 in Reference 55) reveals that there are three major protein bands in the area of migration for HMG-14 and -17. The two slower moving bands have been labeled as HMG-14 and -17 which are shown to be phosphoproteins on the autoradiogram. The third and the fastest moving band is labeled as "ubiquitin" based upon the migration properties on HMG-20.[61] Although the originally labeled HMG-20 by Walker et al.[61] was subsequently shown to be ubiquitin by amino acid sequence comparison,[61,62] HMG-20 migrates much ahead of HMG-17 in a weak triplet with HMG-18 and -19.[61] Furthermore, the acid electrophoretic gel conditions employed by Walker et al.[61] were different and did not contain urea. Moreover, in two different studies, HeLa[52] and trout,[63] ubiquitins have been shown to migrate between histones H2B and H4 in the acid-urea polyacrylamide gel system similar to that used by Saffer and Glazer.[55] It, therefore, is not clear whether the protein band labeled ubiquitin[55] is indeed ubiquitin, or as discussed earlier (see also Figure 3), the unphosphorylated bulk HMG-17. We have examined the migration behavior of calf ubiquitin (gift from LeRoy Kuehl) in the acid-urea polyacrylamide gel system and found it to migrate much faster than HeLa HMG-17, probably at the position of HMG-20 indicated by Walker et al.[61] In view of the results discussed here on the different migration properties of highly phosphorylated protein 17X in two different gel electrophoresis systems (acid-urea and SDS), and the appearance of multiple electrophoretic forms of HMG-14, we suggest caution in identification of the HMG proteins by apparent migration in a single electrophoretic system alone.

***b. In Vitro Phosphorylation by Protein Kinases***

The cyclic nucleotide-independent kinases NI and NII,[64] and the cyclic nucleotide-dependent kinases,[65,66] have been shown to phosphorylate HMG proteins in vitro. Protein kinase NII from rat liver, and not NI, was found to phosphorylate HMG-17 and probably HMG-14.[64] Similarly, Glioma cell nuclear protein kinase NII phosphorylated C6 Glioma cell HMG proteins 14 and 17.[67] NI kinase showed little activity. In both studies HMG-1 and -2 were not phosphorylated. Likewise, the major HMG protein to be phosphorylated by the bovine cGMP-dependent protein kinase was HMG-14.[66] HMG-17 showed a minor site for phosphorylation, and HMG-1 and -2 were not phosphorylated.[66] In contrast, Sun et al.[65] demonstrated that purified cAMP-dependent protein kinase from calf thymus will phosphorylate calf HMG-1. This latter study[65] also describes a useful method for recovery of the thiophosphorylated proteins by affinity chromatography on organomercurial Sepharose® columns.

If phosphorylation represents the modification of serine residues in HMG-14 and -17 as has been suggested,[66,67] then it may be on the serine site(s) in a common sequence (-Arg-Arg-Ser-Ala-Arg-Leu-Ser-Ala-) shared by calf thymus HMG-14[16] and -17,[17] and trout testis H6.[68] In addition, a major phosphorylation site for cGMP-dependent kinase in calf thymus HMG-14 has been identified to be between residues 4 and 13 in a sequence: -Lys-Val-Ser (P)-Ser-Ala-Glu-Gly-Ala-Ala-Lys-.[66] Nuclear protein kinases are a highly complex group of enzymes.[69] In vitro phosphorylation, therefore, may not reflect the in vivo situation. Nonetheless, these studies are valuable for identifying the specific site(s) on the HMG proteins that can potentially be modified by phosphorylation, and hence, provide an important avenue for exploring their function in vivo.

*3. Cellular Distribution of the HMG Proteins and their Phosphorylation*

The presence of the four major HMG proteins has been reported from all the tissues and cell types examined, and in general these proteins seem to be conserved in amino acid sequence.[7] The comparison of HMG proteins from various sources has been based on structural characteristics since no functional assays are yet known. The only species-specific HMG proteins described are HMG-T (analogous to HMG-1 and -2) and H6 (analogous to HMG-14 and -17) from trout testis,[68,70] and HMG-E from birds.[71] HMG-like protein D1 has also been reported from *Drosophila* chromatin.[72] In regard to tissue specificity, quantitative differences in the recovery of HMG-2 have been recorded in several rat[73] and chicken[74] tissues. By contrast, HMG-1 levels did not change. On the other hand, Rabbani et al.[75] found no significant differences in the quantitative appearance of HMG-1 and -2 from calf thymus, spleen, liver, and kidney. In a cell cycle study,[53] though a slight increase (approximately 20%) in the amounts of the HMG proteins in the $G_1$ and S phases of synchronized HeLa cells have been reported, the overall quantitative distribution of all four HMG proteins was remarkably similar (see Figures 2 and 6). Seyedin and Kistler[73] have suggested a striking correlation between the levels of HMG proteins and the proliferative activity, which is reciprocal to the H1° levels in terminally differentiated tissues.[77] Similarly, quantitative tissue-specific differences in HMG-2 subfractions have been shown to occur in differentiating chick skeletal muscle.[74] However, in agreement with Kuehl's[76] analysis of HMG protein

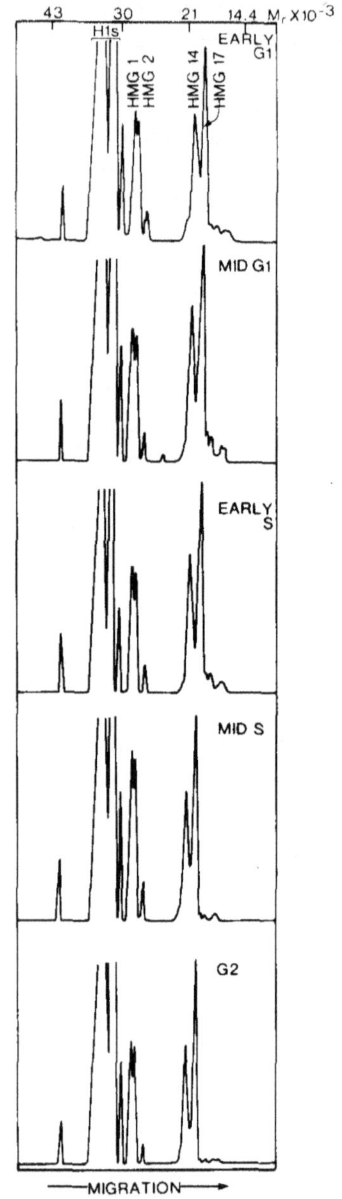


FIGURE 6. Soft-laser densitometer scans of the cell cycle stage-specific HMG proteins obtained as 5% PCA extracts of nuclei purified from synchronized HeLa cells (see Figure 2). A 50-μg quantity of the stage-specific HMG proteins was electrophoresed on 17% polyacrylamide SDS-slab gel, stained with Coomassie blue, and densitometry performed using a Zeineh soft-laser scanning densitometer equipped with an electronic integrator (Biomed).

synthesis in regenerating rat liver, no relationship was apparent between the cell cycle stage-specific distribution of HMG proteins and DNA synthesis.[53]

HMG-2 is known to exist in multiple electrophoretic forms.[59,78] The microheterogeneity of HMGs and the tissue-specific variations in their distribution may, at least in part, be a reflection of their post-synthetic modifications. It is also likely that the discrepancy in the tissue- or species-specific distribution of HMG proteins may simply be due to the different

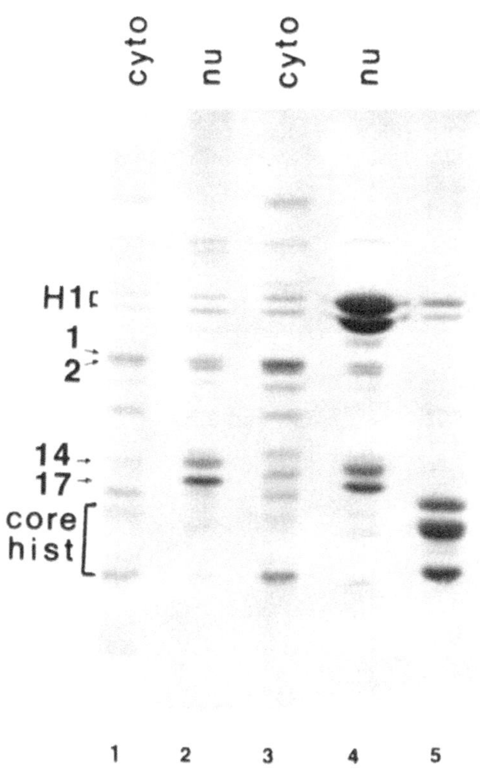


FIGURE 7. Electrophoretic distribution of the nuclear and cytoplasmic HMG proteins. HeLa cells were fractionated into nuclei and cytoplasm (post-nuclear supernatant) as described previously (Bhorjee, J. S. and Pederson, T., *Biochemistry*, 12, 2766, 1973). HMG proteins were extracted by 0.35 *M* NaCl or 5% PCA from both nuclei and the cytoplasm (less than 1% contamination by nuclei as determined by phase-contrast microscopy or [$^3$H]thymidine incorporation), and prepared for electrophoresis on 17% polyacrylamide SDS-slab gel. Protein load, 40 μg in each lane. Lanes 1 and 2, 0.35 *M* NaCl extracted HMG proteins from the cytoplasm and nuclei, respectively. Lanes 3 and 4, 5% PCA extracted HMG proteins from the cytoplasm and nuclei, respectively. Lane 5, total HeLa histones. Note that HMG-1 is enriched by almost 2.5-fold in the cytoplasm relative to the nuclei. The cytoplasmic proteins that appear to be comigrating in the region of the nuclear HMG-14 and -17 are probably not proteins 14 and 17.

source materials used for extraction of these proteins; for example, the heterogeneity of the embyronic chick muscle,[74,79] the use of whole tissues or cells,[73,74] or purified nuclei.[53,75] This may be an important consideration. *In situ* immunofluorescence staining of cultured cells with anti-HMG-1 antibody[80,81] shows a cytoplasmic location for HMG-1. Although leakiness of nuclei during HMG protein fractionation may cause the appearance of some HMG-1 in the cytoplasm,[82] a dynamic equilibrium of HMG-1 and -2 between the cytoplasm and the nucleus is indicated by the intracellular movement and localization of the microinjected $^{125}$I-labeled HMG-1 in fused cultured cells.[83] Seyedin and Kistler[73] have demonstrated the cytoplasmic presence of HMG-1 and -2 in several rat tissues, and though possible nuclear contamination of the cytoplasmic fraction has been suggested, all the cytoplasmic HMG protein cannot be attributed to nuclei. Results presented here (see Figure 7) on the intracellular distribution of the HMG proteins in HeLa cells show that approximately twofold more (as determined by densitometry) HMG-1 is present in the cytoplasm relative to the nucleus.

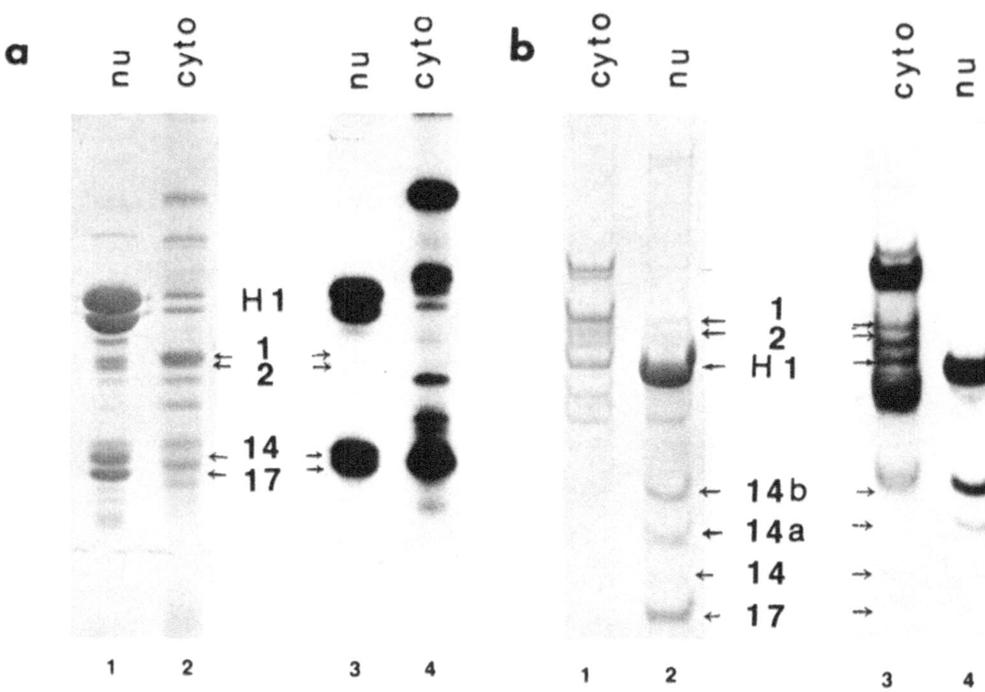


FIGURE 8. Electrophoretic analysis of [$^{32}$P]phosphate incorporation into the nuclear and cytoplasmic HMG proteins. Nuclear and cytoplasmic HMG proteins were extracted with 5% PCA from $^{32}$P-labeled HeLa cells as described under Figure 7, and electrophoresed on 17% polyacrylamide SDS-slab gel (a) or 20% polyacrylamide acid-urea slab gel (b). Protein load, 50 μg per lane. In each panel lanes 1 and 2 Coomassie stained pattern; lanes 3 and 4 corresponding autoradiograms. Note that neither the cytoplasmic nor the nuclear HMG proteins 1 and 2 exhibit any $^{32}$P incorporation. The phosphoproteins corresponding to the cytoplasmic HMG-1 and -2 (panel b, lane 3) in acid-urea gel are not HMG-1 and -2. This is clearly evident from the SDS-gel separation (8a). The phosphoprotein corresponding to HMG-17 in the cytoplasm (panel a, lane 4) separated by SDS-gel is not HMG-17.

HMG-2, however, is equally distributed between the two compartments. No HMG-14 and -17 seems to be present in the cytoplasm. Furthermore, no selective loss or alteration in the distribution of any of the HMG proteins was observed when HeLa cells were Dounce-homogenized in the presence of NonidetP-40®. This result is significant since the outer nuclear membrane is apparently disrupted by the detergent, and if nuclear leakiness were to cause a major selective loss of specific HMG proteins,[82] then this should have been apparent under these conditions. This observation also reinforces the concept of the in vivo existence of a dynamic equilibrium in the intracellular appearance of HMG-1 and -2 as documented by the *in situ* immunofluorescence studies.[30,31] Moreover, Mathew et al.[23] have shown that up to 50% of HMG-1 and -2 can be released by washing nuclei with 0.14 *M* NaCl. This suggests the presence of two populations of the high molecular weight HMG proteins based upon their affinity of binding to chromatin. Such subpopulations of HMG proteins may reflect their different modified forms. In several experiments, however, no $^{32}$P incorporation into cytoplasmic or nuclear HMG-1 and -2 was observed (see Figure 8).

## II. CONCLUSIONS

Because of the limited number of studies available, it would be premature at this stage to make any generalized comments on the significance of the HMG protein post-synthetic modifications in gene activity. It is, however, already clear that such modifications, in particular their phosphorylation and acetylation, may be highly important in assessing the role(s) of these nuclear nonhistone proteins in the structure-function organization of chromatin. For example, covalent modifications can allow for a vast array of structural heterogeneity in otherwise conserved molecules, such as the HMG proteins, and hence, provide for an equivalent degree of functional flexibility.

All four HMG proteins bind to DNA.[13,14,84,85] This, however, may not be their primary mode of interaction in chromatin since the relative binding affinity of HMG-14 and -17 is severalfold higher for core-particle nucleosomes than for naked DNA.[29,40] Since phosphorylation modifies the binding of histones to DNA[43] and of the nonhistone proteins in chromatin,[44] it would be a reasonable assumption that phosphorylation of the HMG proteins causes some such similar effect on the conformation of the transcriptional potential of chromatin domains. One such positive correlation between transcription and the phosphorylation of HMG-17 has been suggested to occur during the early induction in nerve growth factor stimulated rat PC12 nerve-like cells in culture.[86] Furthermore, enrichment of phosphorylated HMG-14 and -17 protein molecules has been demonstrated in the micrococcal nuclease generated mononucleosomes from HT-29 transformed human cells.[55] Whether the binding of the HMG proteins to nucleosomes, or the presence of their modified forms, affects transcription *per se*, is an open question. Post-synthetic modifications of proteins are dynamic events, and these might be affected further in vivo by the interplay of several factors, for example other regulatory enzymes, histones, and other nonhistone proteins.

The studies of post-synthetic modifications of HMG proteins are still in infancy. The early results from published and unpublished studies on the phosphorylation of the HMG proteins presented here open an important new avenue for further investigations on the nature and the mechanism of the HMG protein interactions in the dynamics of chromatin architecture.

## REFERENCES


1. **Kornberg, R. D.,** Structure of chromatin, *Ann. Rev. Biochem.*, 46, 931, 1977.
2. **McGhee, J. D. and Felsenfeld, G.,** Nucleosome structure, *Ann. Rev. Biochem.*, 49, 1115, 1980.
3. **Mathis, D., Oudet, P., and Chambon, P.,** in *Prog. Nucleic Acid Res. Mol. Biol.*, 24, 1, 1980.
4. **Cartwright, I. L., Abmayr, S. M., Fleischmann, G., Lowenhaupt, K., Elgin, S. C. R., Keene, M. A., and Howard, G. C.,** Chromatin structure and gene activity: the role of nonhistone chromosomal proteins, *Crit. Rev. Biochem.*, 13, 1, 1982.
5. **Axel, R., Cedar, H., and Felsenfeld, G.,** Synthesis of globin ribonucleic acid from duck-reticulocyte chromatin *in vitro*, *Proc. Natl. Acad. Sci. U.S.A.*, 70, 2029, 1973.
6. **Gilmour, R. D. and Paul, J.,** Tissue-specific transcription of the globin gene in isolated chromatin, *Proc. Natl. Acad. Sci. U.S.A.*, 70, 3440, 1973.
7. **Goodwin, G. H., Walker, F. M., and Johns, E. W.,** The high mobility group (HMG) nonhistone chromosomal proteins, in *The Cell Nucleus*, Vol. 6 (Part C), Busch, H., Ed., Academic Press, New York, 1978, 181.
8. **Johns, E. W.,** Preparative methods for histone fractions from calf thymus, *Biochem. J.*, 92, 55, 1964.
9. **Levy-Wilson, B. and Dixon, G. H.,** A study of localization of high mobility group proteins in chromatin, *Can. J. Biochem.*, 56, 480, 1978.
10. **Sterner, R., Boffa, L. C., and Vidali, G.,** Comparative structural analysis of high mobility group proteins from a variety of sources, *J. Biol. Chem.*, 253, 3830, 1978.
11. **Sanders, C. and Johns, E. W.,** A method for the large-scale preparations of two chromatin proteins, *Biochem. Soc. Trans.*, 2, 547, 1974.
12. **Goodwin, G. H., Sanders, C., and Johns, E. W.,** A new group of chromatin associated proteins with a high content of acidic and basic amino acids, *Eur. J. Biochem.*, 38, 14, 1973.
13. **Goodwin, G. H., Shooter, K. V., and Johns, E. W.,** Interaction of a non-histone chromatin protein (high mobility group protein 2) with DNA, *Eur. J. Biochem.*, 54, 427, 1975.
14. **Shooter, K. V., Goodwin, G. H., and Johns, E. W.,** Interactions of a purified non-histone chromosomal protein with DNA and histone, *Eur. J. Biochem.*, 47, 263, 1974.
15. **Bustin, M., Hopkins, R. B., and Isenberg, I.,** Immunological relatedness of high mobility group chromosomal proteins from calf thymus, *J. Biol. Chem.*, 253, 1694, 1978.

16. **Walker, J. M., Goodwin, G. H., and Johns, E. W.,** The primary structure of the nucleosome-associated chromosomal protein HMG 14, *FEBS Lett.*, 100, 394, 1979.
17. **Walker, J. M., Hastings, J. R. B., and Johns, E. W.,** The primary structure of a nonhistone chromosomal protein, *Eur. J. Biochem.*, 76, 461, 1977.
18. **Walker, J. M., Goodwin, G. H., and Johns, E. W.,** The similarity between the primary structures of two non-histone chromosomal proteins, *Eur. J. Biochem.*, 62, 461, 1976.
19. **Johns, E. W., Goodwin, G. H., Walker, J. M., and Sanders, C.,** Chromosomal proteins related to histones, in *The Structure and Function of Chromatin*, Vol. 28, Ciba Foundation Symp. (New Series), Elsevier, Amsterdam, 1975, 95.
20. **Goodwin, G. H., Walker, J. M., and Johns, E. W.,** Studies on the degradation of high mobility group nonhistone chromosomal proteins, *Biochem. Biophys. Acta*, 519, 233, 1978.
21. **Goodwin, G. H., Mathew, C. G. P., Wright, C. A., Venkov, C. D., and Johns, E. W.,** Analysis of the high mobility group proteins associated with salt-soluble nucleosomes, *Nucleic Acids Res.*, 7, 1815, 1979.
22. **Levy-Wilson, B., Wong, N. C. W., and Dixon, G. H.,** Selective association of the trout-specific H6 protein with chromatin regions susceptible to DNase I and DNase II: possible location of HMG-T in the spacer region between core nucleosomes, *Proc. Natl. Acad. Sci. U.S.A.*, 74, 2810, 1977.
23. **Mathew, C. G. P., Goodwin, G. H., and Johns, E. W.,** Studies on the association of the high mobility group non-histone chromatin proteins with isolated nucleosomes, *Nucleic Acids Res.*, 6, 167, 1979.
24. **Levy-Wilson, B. and Dixon, G. H.,** Partial purification of transcriptionally active nucleosomes from trout testis cells, *Nucleic Acids Res.*, 5, 4155, 1978.
25. **Levy-Wilson, B., Connor, W., and Dixon, G. H.,** A subset of trout testis nucleosomes enriched in transcribed DNA sequences contains high mobility group proteins as major structural components, *J. Biol. Chem.*, 254, 609, 1979.
26. **Mathew, C. G. P., Goodwin, G. H., Wright, C. A., and Johns, E. W.,** The high mobility group proteins and transcribed nucleosomes, *Cell Biol. Int. Rep.*, 5, 37, 1982.
27. **Weisbrod, S.,** Active chromatin, *Nature (London)*, 297, 289, 1982.
28. **Weisbrod, S. and Weintraub, H.,** Isolation of a subclass of nuclear proteins responsible for conferring a DNase I-sensitive structure on globin chromatin, *Proc. Natl. Acad. Sci. U.S.A.*, 76, 630, 1979.
29. **Weisbrod, S., Groudine, M., and Weintraub, H.,** Interaction of HMG 14 and 17 with actively transcribed genes, *Cell*, 19, 289, 1980.
30. **Weintraub, S. and Groudine, M.,** Transcriptionally active and inactive conformations of chromosomal subunits, *Science*, 193, 848, 1976.
31. **Stalder, J., Groudine, M., Dodgson, J. B., Engel, J. D., and Weintraub, H.,** Hb switching in chickens, *Cell*, 19, 973, 1980.
32. **Staider, J., Larsen, A., Engel, J. D., Dolan, M., Groudine, J., and Weintraub, H.,** Tissue-specific DNA cleavages in the globin chromatin domain induced by DNase I, *Cell*, 20, 451, 1980.
33. **Garel, A., Zolan, M., and Axel, R.,** Genes transcribed at diverse rates have a similar conformation in chromatin, *Proc. Natl. Acad. Sci. U.S.A.*, 73, 3966, 1976.
34. **Levitt, A., Axel, R., and Cedar, H.,** Nick translation of active genes in intact nuclei, *Dev. Biol.*, 96, 496, 1979.
35. **Gazit, B. and Cedar, H.,** Nuclease sensitivity of active chromatin, *Nucleic Acids Res.*, 8, 5143, 1980.
36. **Goodwin, G. H. and Johns, E. W.,** Are the high mobility group proteins associated with "active" chromatin? *Biochem. Biophys. Acta*, 519, 279, 1978.
37. **Levinger, L., Barsoum, J., and Varshavsky, A.,** Two-dimensional hybridization mapping of nucleosomes. Comparison of DNA and protein patterns, *J. Mol. Biol.*, 146, 287, 1981.
38. **Barsoum, J., Levinger, L., and Varshavsky, A.,** On the chromatin structure of the amplified, transcriptionally active gene for dihydrofolate reductase in mouse cells, *J. Biol. Chem.*, 257, 5274, 1982.
39. **Mathew, C. G. P., Goodwin, G. H., Igokemenes, T., and Jones, E. W.,** The protein composition of rat satellite chromatin, *FEBS Lett.*, 125, 25, 1981.
40. **Sandeen, G., Wood, W. I., and Felsenfeld, G.,** The interaction of high mobility proteins HMG 14 and 17 with nucleosomes, *Nucleic Acids Res.*, 8, 3757, 1980.
41. **Gazit, B., Panet, A., and Cedar, H.,** Reconstitution of a deoxyribonuclease I-sensitive structure on active genes, *Proc. Natl. Acad. Sci. U.S.A.*, 77, 1787, 1980.
42. **Weisbrod, A. and Weintraub, H.,** Isolation of actively transcribed nucleosomes using immobilized HMG 14 and 17 and an analysis of α-globin chromatin, *Cell*, 23, 391, 1981.
43. **Johnson, E. M. and Allfrey, V. G.,** Post-synthetic modifications of histone primary structure: phosphorylation and acetylation as related to chromatin conformation and function, in *Biochemical Actions of Hormones*, Vol. 5, Litwack, G., Ed., Academic Press, New York, 1978, 1.
44. **Kleinsmith, L. J.,** Phosphorylation of nonhistone proteins, in *The Cell Nucleus*, Vol. 6, Busch, H., Ed., Academic Press, New York, 1978, 221.

45. **Vidali, G., Boffa, L. C., Bradbury, E. M., and Allfrey, V. G.,** Butyrate supression of histone deacetylation leads to accumulation of multiacetylated forms of histones H3 and H4 and increased DNase I sensitivity of the associated DNA sequences, *Proc. Natl. Acad. Sci. U.S.A.*, 74, 2239, 1978.
46. **Gershey, E. L. and Kleinsmith, L. J.,** Phosphorylation of nuclear proteins in avian erythrocytes, *Biochem. Biophys. Acta*, 194, 519, 1969.
47. **Sterner, R., Vidali, G., Heinrikson, R. L., and Allfrey, V. G.,** Postsynthetic modification of high mobility group proteins, *J. Biol. Chem.*, 253, 7601, 1978.
48. **Giri, C. P., West, M. H. P., and Smulson, M. E.,** Nuclear protein modification and chromatin substructure. I. Differential poly (adenosine diphosphate) ribosylation of chromosomal proteins in nuclei versus isolated nucleosomes, *Biochemistry*, 17, 3495, 1978.
49. **Reeves, R., Chang, D., and Chung, S. H.,** Carbohydrate modifications of the high mobility group proteins, *Proc. Natl. Acad. Sci. U.S.A.*, 78, 6704, 1981.
50. **Boffa, L. C., Sterner, R., Vidali, G., and Allfrey, V. G.,** Post-synthetic modifications of nuclear proteins. High mobility group proteins are methylated, *Biochem. Biophys. Res. Commun.*, 89, 1322, 1979.
51. **Saffer, J. D. and Glazer, R. D.,** The phosphorylation of high mobility group proteins 14 and 17 from Ehrlich ascites and L1210 *in vitro*, *Biochem. Biophys. Res. Commun.*, 93, 1280, 1980.
52. **Levy-Wilson, B.,** Enhanced phosphorylation of high-mobility-group proteins in nuclease-sensitive mononucleosomes from butyrate-treated HeLa cells, *Proc. Natl. Acad. Sci. U.S.A.*, 78, 2189, 1981.
53. **Bhorjee, J. S.,** Differential phosphorylation of nuclear high mobility group proteins HMG 14 and HMG 17 during the cell cycle, *Proc. Natl. Acad. Sci. U.S.A.*, 78, 6944, 1981.
54. **Arfmann, H. A., Haase, E., and Schroter, H.,** High mobility group proteins from CHO cells and their modifications during cell cycle, *Biochem. Biophys. Res. Commun.*, 101, 137, 1981.
55. **Saffer, J. D. and Glazer, R. D.,** The phosphorylation of high mobility group proteins 14 and 17 and their distribution in chromatin, *J. Biol. Chem.*, 257, 4655, 1982.
56. **Wong, N. C. W., Poirier, G. G., and Dixon, G. H.,** Adenosine diphosphoribosylation of certain basic chromosomal proteins in isolated trout testis nuclei, *Eur. J. Biochem.*, 77, 11, 1977.
57. **Mellon, I. and Bhorjee, J. S.,** Two-dimensional gel electrophoretic analysis of the high mobility group (HMG) nonhistone chromosomal proteins, *J. Cell Biol.*, 95, 69a, 1982.
58. **Mellon, I. and Bhorjee, J. S.,** Unpublished data.
59. **Tyrell, D., Isackson, P. J., and Reeck, G. R.,** Two-dimensional gel electrophoresis of nonhistone chromatin proteins with nonequilibrium pH gradient electrophoresis as the first dimension, *Anal. Biochem.*, 119, 433, 1982.
60. **Bhorjee, J. S.,** Unpublished data.
61. **Walker, J. M., Goodwin, G. H., and Johns, E. W.,** The isolation and identification of ubiquitin from the high mobility group (HMG) non-histone fraction, *FEBS Lett.*, 90, 327, 1978.
62. **Schlesinger, D. H., Goldstein, G., and Niall, H. D.,** The complete amino acid sequence of ubiquitin, an adenylate cyclase stimulating polypeptide probably universal in living cells, *Biochemistry*, 14, 2214, 1975.
63. **Kuehl, L., Lyness, T., Dixon, G. H., and Levy-Wilson, B.,** Distribution of high mobility group proteins among domains of trout testis chromatin differing in their susceptibility to micrococcal nuclease, *J. Biol. Chem.*, 255, 1090, 1980.
64. **Inoue, A., Tei, Y., Hasuma, T., Yukioka, M., and Morisawa, S.,** Phosphorylation of HMG 17 by protein kinase NII from rat liver cell nuclei, *FEBS Lett.*, 117, 68, 1980.
65. **Sun, I. Y. C., Johnson, E. M., and Allfrey, V. G.,** Affiity purification of newly phosphorylated protein molecules, *J. Biol. Chem.*, 255, 742, 1980.
66. **Walton, G. M., Spiess, J., and Gill, G. N.,** Phosphorylation of high mobility group 14 protein by cyclic nucleotide-dependent protein kinases, *J. Biol. Chem.*, 257, 4661, 1982.
67. **Harrison, J. J. and Jungmann, R. A.,** Phosphorylation of high mobility group proteins 14 and 17 by nuclear protein kinase NII in rat C6 glioma cells, *Biochem. Biophys. Res. Commun.*, 108, 1204, 1982.
68. **Watson, D. C., Wong, N. C. W., and Dixon, G. H.,** The complete amino-acid sequence of a trout testis non-histone protein, H6, localized in a subset of nucleosomes and its similarity to calf-thymus non-histone proteins HMG-14 and HMG-17, *Eur. J. Biochem.*, 95, 193, 1979.
69. **Kish, V. M. and Kleinsmith, L. J.,** Purification and assay of nuclear protein kinases, in *Methods in Cell Biology*, Vol. 19, Stein, G., Stein, J., and Kleinsmith, L. J., Eds., Academic Press, New York, 1978, 101.
70. **Watson, D. C., Peters, E. H., and Dixon, G. H.,** The purification, characterization and partial sequence determination of a trout testis non-histone protein, HMG-T, *Eur. J. Biochem.*, 74, 53, 1977.
71. **Vidali, G., Boffa, L. C., and Allfrey, V. G.,** Selective release of chromosomal proteins during limited DNAase I digestion of avian erythrocyte chromatin, *Cell*, 12, 409, 1977.
72. **Alfageme, C. R., Rudkin, G. T., and Cohen, L. H.,** Locations of chromosomal proteins in polytene chromosomes, *Proc. Natl. Acad. Sci. U.S.A.*, 73, 2038, 1976.

73. **Seyedin, S. M. and Kistler, W. S.,** Levels of chromosomal protein high mobility group 2 parallel the proliferative activity of testis, skeletal muscle, and other organs, *J. Biol. Chem.*, 254, 11264, 1979.
74. **Gordon, J. S., Rosenfeld, B. I., Kaufman, R., and Williams, D. L.,** Evidence for a quantitative tissue-specific distribution of the high mobility group chromosomal protein, *Biochemistry*, 19, 4395, 1980.
75. **Rabbani, A., Goodwin, G. H., and Johns, E. W.,** Studies on the tissue specificity of the high-mobility-group non-histone chromosomal proteins from calf, *Biochem. J.*, 173, 497, 1078.
76. **Kuehl, L.,** Synthesis of high mobility group proteins in regenerating rat liver, *J. Biol. Chem.*, 254, 7276, 1979.
77. **Pehrson, J. and Cole, R. D.,** Histone H1° accumulates in growth-inhibited cultured cells, *Nature (London)*, 285, 43, 1980.
78. **Mathew, C. G. P., Goodwin, G. H., Gooderham, K., Walker, J. M., and Johns, E. W.,** A comparison of the high mobility group non-histone chromatin protein HMG2 in chicken thymus and erythrocytes, *Biochem. Biophys. Res. Commun.*, 87, 1243, 1979.
79. **Gordon, J. S., Kaufman, R., and Rosenfeld, B. I.,** Independent control during myogenesis of histone and high-mobility-group (HMG) chromosomal protein heterogeneity, *Arch. Biochem. Biophys.*, 211, 709, 1981.
80. **Bustin, M. and Neihart, N. K.,** Antibodies against chromosomal HMG proteins stain the cytoplasm of mammalian cells, *Cell*, 16, 181, 1979.
81. **Isackson, P. J., Bidney, D. L., Reeck, G. R., Neihart, N. K., and Bustin, M.,** High mobility group chromosomal proteins isolated from nuclei and cytosol of cultured hepatoma cells are similar, *Biochemistry*, 19, 4466, 1980.
82. **Gordon, J. S., Bruno, J., and Lucas, J. J.,** Heterogeneous binding of high mobility group chromosomal proteins to nuclei, *J. Cell Biol.*, 88, 373, 1981.
83. **Rechsteiner, M. and Kuehl, L.,** Microinjection of the nonhistone chromosomal protein HMG 1 into bovine fibroblasts and HeLa cells, *Cell*, 16, 901, 1979.
84. **Isackson, P. J., Fishback, J. L., Bidney, D. L., and Reeck, G. R.,** Preferential affinity of high molecular weight high mobility group non-histone chromatin proteins for single-stranded DNA, *J. Biol. Chem.*, 254, 5569, 1979.
85. **Isackson, P. J. and Reeck, G. R.,** Nonhistone chromatin proteins HMG-14 and HMG-17 bind preferentially to single-stranded DNA, *Nucl. Acids Res.*, 9, 3779, 1981.
86. **Halegoua, S. and Patrick, J.,** Nerve growth factor mediates phosphorylation of specific proteins, *Cell*, 22, 571, 1980.

Chapter 5

# IMMUNOLOGICAL STUDIES ON THE STRUCTURE AND FUNCTION OF HMG PROTEINS

**Michael Bustin**

## TABLE OF CONTENTS

## I. INTRODUCTION

One of the difficulties in studying the role of chomosomal proteins in maintaining the structure and regulating the function of chromatin and chromosomes is the lack of specific biological assays for their detection and quantitation. This is particularly true for the non-histone chromosomal (NHC) proteins which are frequently identified by mobilities in polyacrylamide gels. Since this group is composed of many proteins, some of which are posttranslationally modified, identification of a component by its mobility in gels is, at best, tenuous.

Studies on the chromosomal organization of specific proteins are further complicated by the fact that the chromatin fiber is a very dynamic structure involved in a variety of complex cellular events such as cell division, replication, and transcription. Such events may affect the manner in which proteins bind to DNA or the manner in which they interact with each other.

Antibodies of defined specificity can be used as versatile reagents to identify a protein and study its cellular localization. Furthermore, microinjection of functional antibodies into living cells can be used to study the cellular function of an antigen under in vivo conditions.[1-6] Previous studies have shown that it is feasible to use immunological approaches to study the specificity, chromosomal location, and cellular function of histones.[7]

Meaningful immunological studies on chromosomal proteins require well defined, specific antibodies. When polyclonal sera are used, the production of specific antisera necessitates rigorous purification of the antigen. The use of monoclonal antibodies circumvents the problem of antigen purification for antibody production. However, pure antigen is still necessary for definition of specificity. Furthermore, in some cases the narrow specificity of monoclonal antibodies leads to unexpected nonspecific reactions.[8]

The high mobility group (HMG) chromosomal proteins are among the few nonhistone proteins which have been purified to homogeneity and whose primary sequence is known. As such, it is possible to elicit specific, well characterized sera against these proteins and use the antisera for studies aimed at an understanding of their function.[9] Such studies may be of particular importance since the four major HMG proteins HMG-1, -2, -14, and -17 are relatively abundant in the cell and are present in several eukaryotic kingdoms.[10] Experimental evidence suggests that HMG-14 and -17 may be associated with transcribable regions of chromatin.[11] The cellular functions of proteins HMG-1 and -2 are not known.

The purpose of this chapter is to summarize the information presently available on the structure and function of HMG proteins which was obtained by immunological techniques. These volumes contain several other chapters dealing with the application of immunological techniques to specific studies involving HMG proteins. This review will not cover those studies.

## II. PRODUCTION OF ANTISERA AND ASSAY FOR ANTIGEN SPECIFICITY

Antibodies to HMG proteins can be elicited by injecting purified proteins into rabbits at multiple intradermal sites.[9] In contrast to histones, the HMG are good immunogens and most rabbits injected give sera containing between 0.05 to 0.1 mg/mℓ antibody. Pure antibodies can be obtained from such sera by affinity chromatography.[12]

We encountered difficulties in our attempts to produce monoclonal antibodies to HMG proteins. Several clones which originally were positive proved to be unstable. Nevertheless, we succeeded in obtaining two stable clones, both of which produced IgM type antibodies.[13]

Another potential source of antibodies to HMG proteins is sera from patients suffering from certain autoimmune diseases. Screening of a limited number of such sera revealed that a high proportion of the patients contained detectable levels of anti-HMG-17 antibodies.[14]

**Table 1**
**IMMUNOLOGICAL ASSAYS USED TO STUDY HMG PROTEINS**

| | Antigen needed | | Antisera needed[a] | |
|---|---|---|---|---|
| Assay | Concentration | Amount | Dilutions | Amount |
| Immunodiffussion | 0.1—5.0 mg/mℓ | (10—50 μℓ) | 0—1:10 | (10—50 μℓ) |
| Precipitin | 0.05—1.0 mg/mℓ | (100 μℓ) | 0—1:10 | (200—400 μℓ) |
| Microcomplement fixation | 0.05—1.0 mg/mℓ | (100 μℓ) | 1:1000—1:5000 | (1000 μℓ) |
| Radioimmunoassay | 0.0001—0.01 mg/mℓ | (100 μℓ) | 1:1000—1:5000 | (100 μℓ) |
| ELISA | 0.00001—0.01 mg/mℓ | (50—100 μℓ) | 1:1000—1:50,000 | (100 μℓ) |
| "Western" blots | 0.1—10 mg/mℓ | (10—20 μℓ) | 1:200—1:100 | (10,000 μℓ) |

[a] All experiments done with the same antiserum preparation.

Fewer patients contained antibodies to HMG-1 and -2, and the titer of these was significantly lower than the anti-HMG-17 titer. Whether this is a general phenomenon or a chance finding in the limited number of samples examined remains to be seen. It should be noted that lymphocytes from these patients potentially can be a source of human monoclonal antibodies for HMG proteins.

A variety of immunological assays have been used to study the interaction between HMG and anti-HMG antibodies. Ouchterlony immunodiffusion,[9] precipitin,[9] microcomplement fixation,[9] solid phase radioimmunoassay,[15] and enzyme-linked immunoassay[14] have been used to study the specificity of the antisera. Table 1 compares the sensitivity of the various assays used for the study of HMG-anti-HMG interaction. While all the assays verified that the antisera contained antibodies reacting with the antigen, the most convincing method is the "western" blots,[16] in which proteins transferred from polyacrylamide gels to diazobenzyloxymethyl (DBM) or nitrocellulose paper are reacted with antibodies and $^{125}$I-protein A and the location of the reacting bands visualized by autoradiography. Figure 1 depicts such blots. It can be seen that antihistone antibodies do not react with HMG and vice versa. The anti-HMG-1 sera does not distinguish between HMG-1 and -2. The anti-HMG-1, -2 did not react with HMG-17 and anti-HMG-17 did not cross react with HMG-1 or -2.

Initial studies on the specificity of the antisera were designed to see whether immunological methods could be used for assaying the presence of such proteins and identifying the type of HMG present in a mixture of proteins. The microcomplement fixation technique is sensitive to minor antigenic variations between related proteins. In some cases it can distinguish between proteins which differ by as little as one residue, rendering it useful in assessing the sequence relationship among similar proteins.[18] Obviously such studies do not have the precision nor provide the information available from determination of the entire primary sequence. On the other hand, the microcomplement fixation technique is equally or more reliable than peptide mapping or determination of terminal amino acids. Furthermore, it provides information as to whether antibodies elicited against HMG protein present in one type of tissue or organism can be used to study HMG proteins in a heterologous system. The use of heterologous sera in the study of chromosomal proteins is advantageous in that it minimizes possible complications due to the presence of antibodies against contaminating, immunogenic, species-specific proteins.[9]

As judged by microcomplement fixation, HMG-1 and -2 purified from calf thymus are similar proteins with a sequence difference of approximately 6%.[9] Studies involving inhibition in solid phase radioimmunoassay suggested that the difference may be somewhat larger.[15] Indeed, anti-HMG-2 sera readily distinguishes between the two proteins.[13] Comparison of the amino acid sequence of the two proteins, purified from calf thymus, shows that they differ in about 15% of their sequence.[19] Most of the sequence differences involve conservative substitution of amino acids. The fact that antisera to HMG-1 recognized both

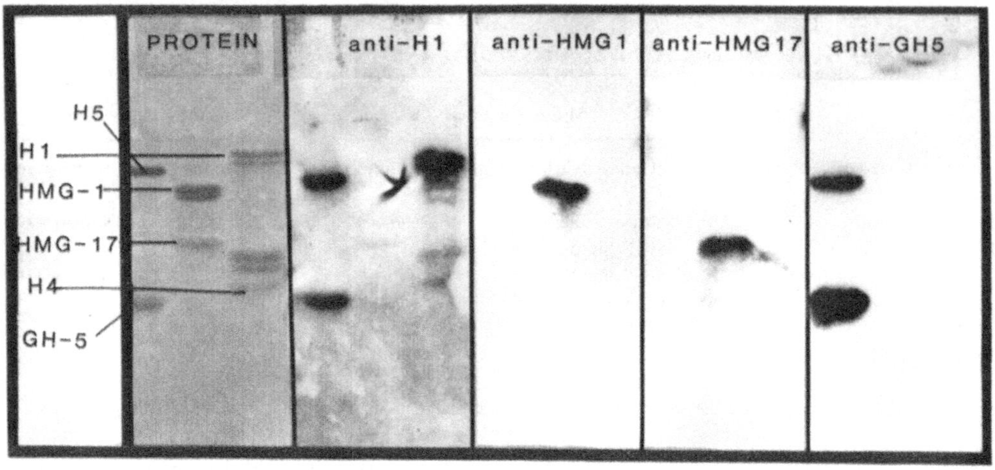


FIGURE 1. Specificity of anti-HMG sera demonstrated by "western" blots. Chromosomal proteins (left panel) were separated by electrophoresis in 18% polyacrylamide gels containing 0.1% SDS. The proteins were transferred from the gel to DBM paper and the paper treated with the various antisera (indicated in the panels) followed by $^{125}$I-protein A and autoradiography. The various panels show a photograph of the autoradiograms. The anti-HMG was elicited against the globular region of chicken erythrocyte H5 (see Reference 17).

HMG-1 and -2 is particularly relevant when antibodies are used to study the cellular localization or function of the two proteins. HMG-1 derived from different sources are more similar to each other than HMG-1 and -2 obtained from the same source.[9,15,20] Thus, antibodies elicited against HMG-1 purified from calf thymus can be used to study HMG purified from various sources.

HMG-17 and -14 are immunologically distinct from HMG-1 and -2. Antisera to HMG-1 does not react with HMG-17 and anti-17 does not react with HMG-1 or -2.[9,21]

## III. ANTIGENIC DETERMINANTS OF HMG-1 AND HMG-2

Insights into the nature of antigenic determinants of HMG proteins will help us to evaluate the reliability of experiments in which antibodies are used. This is particularly true when the antibodies are used to localize and study the function of a protein inside the nucleus. Steric and conformational effects may determine whether the antibody can recognize and bind to an antigenic determinant. It was of interest to determine whether the antigenic determinants are mostly sequential or steric determinants. By definition, sequential determinants are those due to an amino acid sequence in a random coil form, while steric determinants are those due to the conformation of the antigen.[22] Antibodies toward sequential determinants are expected to react with any peptide that has an identical or very similar sequence.

Upon controlled trypsin or pepsin digestion of either HMG-1 or -2 a gradual loss of antigenicity is observed (Figure 2). The rate at which the antigens lost their ability to bind antibodies was the same regardless of whether anti-HMG-1 or anti -2 was used. The activity present after various times of digestion was due to immunoreactive peptides rather than to residual, intact protein. This conclusion was reached from experiments in which digests of the protein were subjected to electrophoresis on polyacrylamide gels, the separated peptides transferred to DBM paper, and the immunoreactive species on the paper visualized by autoradiography after treatment with antisera and $^{125}$I-protein A. An example of such a study is presented in Figure 3. Already at very early stages of digestion most of the intact HMG is lost and a series of peptides is generated. Most of the peptides visualized by the Coomassie blue stain also bind antibody. After more prolonged digestion smaller peptides are generated

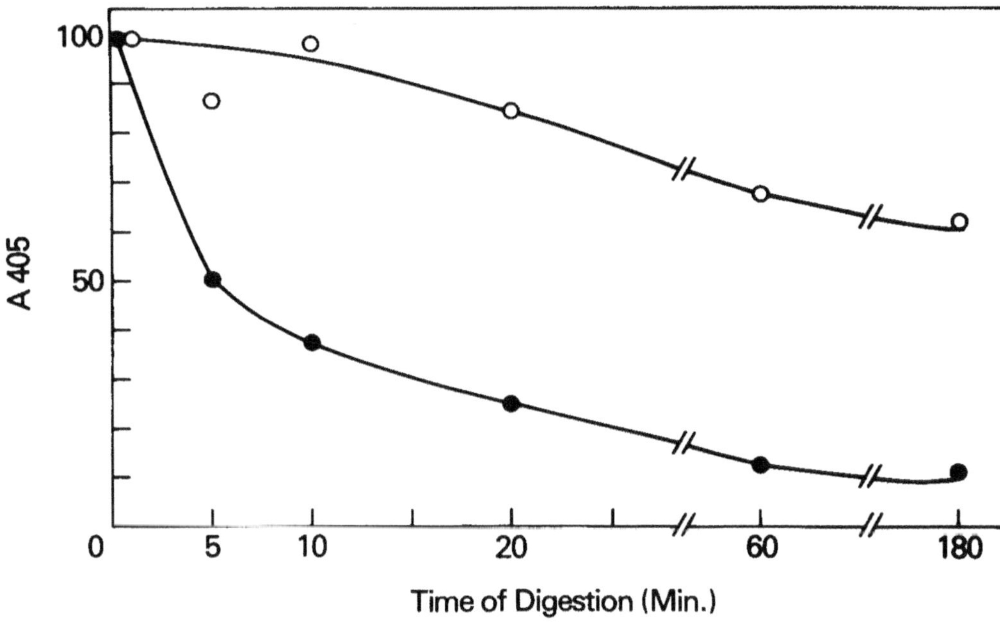


FIGURE 2. Gradual loss of antigenic activity of HMG-1 upon pepsin and trypsin digestion. The protein was digested with the various enzymes and aliquots were removed after various digestion times. The proteolytic activity was destroyed and the remaining antigenic activity measured by ELISA. (○) Activity remaining after pepsin digestion; (●) activity remaining after trypsin digestion. For experimental detail, see Reference 13.

and some of these also bind antibodies. Obviously, a large portion of the antibodies present in anti-HMG-1 and -2 sera are directed against sequential determinants. This situation suggests that the polyclonal antibodies will not be useful in detecting possible conformational changes occurring in these proteins. On the other hand, the presence of antibodies against sequential determinants increases the possibility that among the various types of antibodies present in polyclonal sera some will be able to bind to their target regardless of the exact state in which the HMG is found in the cell.

Recently, it has been reported that HMG proteins are glycoproteins and contain a variety of carbohydrates.[23] It is therefore possible that a portion of the antibodies present in the sera may have specificity toward the carbohydrate moiety. If this is the case, then the sera may react with molecules which are completely unrelated to HMG proteins.

## IV. CELLULAR LOCALIZATION OF HMG PROTEINS

The finding that antibodies elicited by HMG proteins purified from calf thymus cross-react with HMG derived from various sources[15,20] suggests that the antisera can be used as a cytological reagent in a variety of experimental systems.

We have examined the cellular distribution of HMG in Chinese hamster lung cells (line V-79), Fisher rat liver cells (line TR-12), embryonic bovine trachea cells (EBTr-NB1-4), and rat hepatoma (HTC cells).[12,24] In all these cells indirect immunofluorescence studies revealed a distinct cytoplasmic fluorescence pattern. There is a striking contrast between this immunofluorescence pattern and that obtained with antihistone antibodies which give a distinct nuclear fluorescence.[12,24] In metaphase chromosomes the histone antisera intensely stain the entire length of all the chromosomes while the anti-HMG sera preferentially stain the cellular material around the chromosomes. The chromosomes are relatively depleted of fluorescence.[24] These studies have been done with affinity purified antibodies, thereby minimizing the possibility that the cytoplasmic immunofluorescence is due to the presence

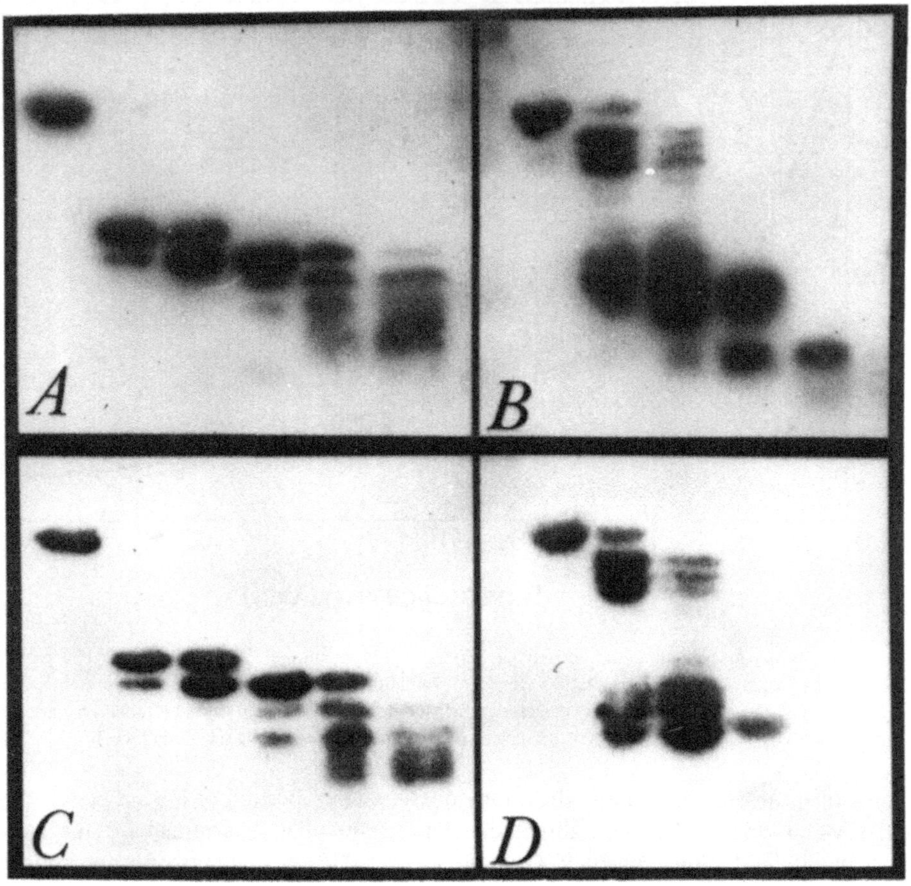


FIGURE 3. Immunoreactive peptides in HMG-2. Proteolytic digests of HMG-2 were electrophoresed in 18% SDS polyacrylamide gels and the peptides transferred to DBM paper and treated with either anti-HMG-1 (A, B) or anti-HMG-2 (C, D). The peptides to which antibodies were bound are visualized by $^{125}$I-protein A and autoradiography (A, C) Pepsin digests for 0, 2, 5, 30, 60, and 180 min; (B, D) tryptic digests for 0, 2, 5, 30, and 60 min. For experimental details, see Reference 13.

of antibodies which react with cytoskeletal elements. The results indicate that the cellular localization of the HMG proteins is different from that of the histones.

Since the anti-HMG sera were elicited against proteins isolated from purified chromatin, the question arises whether the antibodies can recognize and bind to HMG complexed in the chromosomal conformation. Evidence that antibodies to HMG-1 and -2 can bind to chromatin purified from calf thymus appears in Figure 4, which shows the results of an experiment in which the antisera were added to chromatin, the chromatin and bound antibodies precipitated, and the supernatant after centrifugation examined for residual anti-HMG activity. It can be seen that the anti-HMG activity was reduced after the sera were incubated with chromatin. Furthermore, nucleosomes prepared from calf thymus and HeLa cells bind anti-HMG sera and inhibit the binding of anti-HMG to HMG protein.[15] In addition, anti-HMG binds to fractions obtained by sucrose gradient centrifugation of a micrococcal nuclease digest of HeLa chromatin (Figure 5). It can be seen that anti-HMG-1 binds to material remaining at the top of the gradient (pool 1, 2). The HMG-1 and -2 are released from the chromatin by micrococcal nuclease digestion probably because they are associated with the linker region between the nucleosomes.[25] Some HMG-1 or -2 remains associated with core particles (pool 3) and mononucleosomes (pool 4). These may correspond to the fraction of HMG proteins which are more tightly bound to chromatin.[26] Anti-HMG-17, however, does

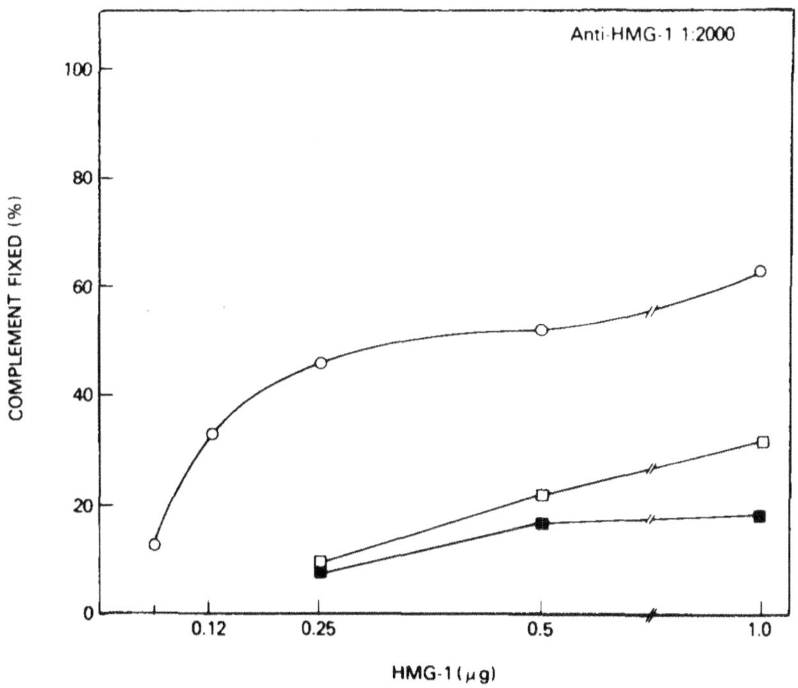


FIGURE 4. Anti-HMG-1 binds to calf thymus chromatin. Antisera diluted 1:2000 was incubated with either 0, (○), 100, (□), or 200 (■) mg chromatin. The chromatin and the adsorbed antibodies were removed from the incubation mixture by centrifugation at 0.15 *M* NaCl and the relative amounts of unadsorbed antibodies left in the supernatant determined by microcomplement fixation. For details, see Reference 9.

bind very little to the fractions which are not associated with chromatin or mononucleosomes. Thus, we conclude that the antibody, which by immunofluorescence detects HMG in the cytoplasm, can recognize HMG antigenic determinants present in chromatin and in nucleosomes.

The question arises whether the cytoplasmic fluorescence represents the true cellular distribution of HMG or is due to some artifactual situation. Several types of experiments suggest that HMG is found exclusively in the nucleus: (1) immunofluorescence and immunoperoxidase staining done in other laboratories detected HMG-1 and -2 mainly in the nucleus;[27,28] (2) immunofluorescence using extremely short washings and formaldehyde fixation using the sera described here also detected more HMG in the nucleus than in the cytoplasm;[27] (3) enucleation studies using cytochalasin B indicated that HMG could be detected exclusively in the nucleus;[30] and (4) radioactively labeled HMG proteins were encapsulated into red blood cell ghosts and these ghosts were fused with somatic cells. Autoradiography of the fused cells detected most of the grains over the nucleus. However, fusion of these cells with cells not previously loaded with HMG revealed that the HMG can redistribute among the nuclei in the fused cells suggesting that the nuclear pool is at least transiently in equilibrium with a cytoplasmic HMG pool.[31]

Faced with such diverse experimental data suggesting that most of the HMG is found in the nucleus, it was important to further examine the reason for the cytoplasmic immunofluorescence. Cell fractionation studies revealed that HMG proteins are found in the cytoplasm.[12,24] This however, could be due to leakage from the nuclei during the fractionation procedure. Leakage of protein from the nucleus during fixation for immunofluorescence could also account for the cytoplasmic fluorescence. Other possibilities include (1) the presence of nonspecific cross-reacting proteins in the cell, and (2) the presence of an HMG

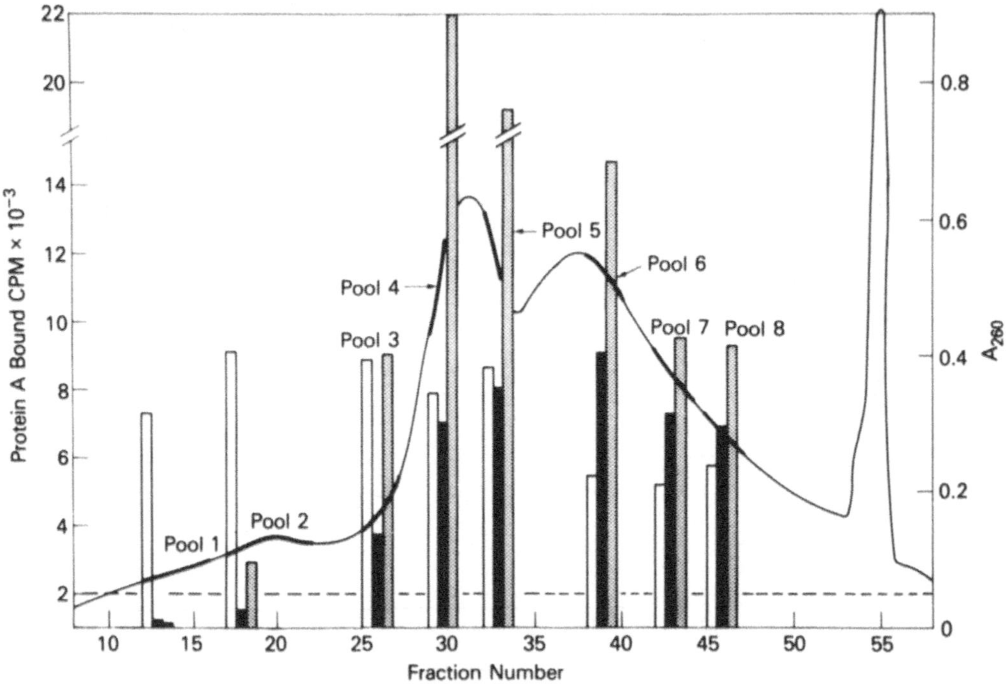


FIGURE 5. Relative distribution of nonhistone proteins and histone H4 in a sucrose gradient used to fractionate a micrococcal nuclease digest of HeLa chromatin. The gradient was pooled into several fractions. Pool 4 contains mostly the 145-bp core particle. Samples from each pool were assayed by ELISA for their content of HMG-1 (open bar) HMG-17 (solid bar) and H4 (dotted bar). The dotted line shows nonspecific binding obtained by using control normal sera. Left side is the top of the gradient. For experimental details, see Reference 21.

precursor in the cytoplasm which upon activation migrates into the nucleus. The existence of cross-reactive cellular proteins or precursors was investigated in three types of experiments (Bustin, unpublished). In the first type, total proteins from HeLa cells were separated by electrophoresis in polyacrylamide gels and the separated proteins were transferred from gels to DBM paper. The paper was then treated with antibodies and $^{125}$I-protein A. In the second type of experiment, the proteins in HeLa cells were labeled with radioactive amino acids and then immunoprecipitated with anti-HMG-1, and in the third, RNA from rat thymus was translated using radioactive amino acids and the translation products immunoprecipitated. In these experiments no cross-reacting proteins or putative precursors were detected. These experiments suggest that the cytoplasmic fluorescence is not due to cross-reaction with a particular cytoplasmic component. However, if the antisera were enriched in antibodies reacting with a protein modification (i.e., poly ADP-ribose, carbohydrate, etc.), then the presence of these on many different proteins in the cytoplasm could give an apparent cytoplasmic fluorescence.

The most plausible explanation for an artifactual cytoplasmic fluorescence is leakage of the protein from the nucleus during cell fixation. We have tried to avoid this problem by microinjecting proteins, antibodies, and functional antibody fragments into living cells. Direct and immediate observation of the injected cells allows determination of the in vivo cellular distribution of the microinjected molecules. To facilitate such an experiment the microinjected molecules are labeled with fluorescein or rhodamine and the microinjected cells studied with a microscope that is equipped with a SIT-video camera and a video monitor which displays the field observed in the microscope. In initial experiments fluoresceinated HMG was injected into the cytoplasm. Three hours after injection most of the fluorescence was localized in the nucleolus. However, control experiments using fluorescent-labeled cytochrome *c* indicated that in some cases the microinjected protein will not migrate to its true cellular location. Injection of fluoresceinated antibodies has the advantage that appropriate controls, using nonimmune IgG can be performed. We have found that intact IgG molecules do not cross the nuclear membrane but that $F(ab)_2$ molecules do.[32] Therefore, experiments designed to test the cellular localization of a protein require injection of $F(ab)_2$. Injection of control $F(ab)_2$ into either the nucleus of cytoplasm of human KD fibroblasts revealed that the $F(ab)_2$ rapidly equilibrates between the nucleus and cytoplasm. When $F(ab)_2$ derived from antihistone H2A was injected into the nucleus it stayed at the sight of injection, while when injected into the cytoplasm it rapidly moved into the nucleus. These experiments demonstrate that in the living cell the microinjected $F(ab)_2$ can home onto and bind to its antigen. $F(ab)_2$ derived from anti-HMG-1 stayed at the site of injection regardless of whether this was the nucleus or the cytoplasm.[33]

Taken together, the results obtained in various laboratories suggest that there is HMG both in the nucleus and in the cytoplasm of most cells. The relative amount of these proteins in the two cellular compartments may depend on cell type, cell cycle, or other factors. The cellular role of proteins, such as HMG, which are relatively conserved during evolution and which may shuttle between the nucleus and cytoplasm is of considerable interest.

## V. FUNCTIONAL STUDIES ON HMG PROTEINS USING ANTIBODIES: CHARACTERIZATION OF HMG mRNA

The histones that are the major class of chromosomal proteins are unique in that the mRNA for all histone classes is not polyadenylated, though the mRNA for most other cellular proteins is polyadenylated. It was of interest to determine whether the mRNA for HMGs, which by definition are chromosomal proteins, is like that of the chromosomal histones (i.e., not polyadenylated) or like that of most cellular proteins (i.e., polyadenylated). Total RNA was isolated from rat thymus and separated into poly $A^+$ and poly $A^-$ by chromatography on polydeoxythymidine-cellulose columns. Both RNA fractions were translated using a reticulocyte lysate translation system. Affinity purified antibodies were used to precipitate the synthesized polypeptides. The results indicated that the mRNA for HMG-1 and -2 is polyadenylated.[34] Thus, the primary transcripts of these proteins are processed differently than the transcripts of the histones. During these studies we noted that the newly translated proteins were very difficult to precipitate, in that unusually high amounts of affinity purified antibodies were necessary. In contrast purified HMG-1 and -2 which were labeled in vitro by reductive methylation were immunoprecipitated very efficiently. Possibly the antibodies that are elicited against the proteins purified from chromatin recognize some posttranslational modifications which are not present on the newly synthesized proteins.

## VI. HMG PROTEINS IN POLYTENE CHROMOSOMES

Polytene chromosomes, by virtue of their structure and size, offer a convenient experimental system to study the organization of a chromosomal component. These are essentially large chromosomes in which 1000 or more chromatids are aligned precisely.[35] Tightly coiled chromatin regions appear as bands while more loosely packed regions appear as interbands. Furthermore, in these chromosomes, transcriptional activity can be induced in defined genetic loci. Antibodies against a variety of proteins have been used to visualize the location of the proteins by immunofluorescence. Among these is the antibody to RNA polymerase which prominently stains interbands and transcriptionally active puffs.[36] Antibodies to histone stain mostly bands, and puffed regions are selectively depleted of immunofluorescence.[37,38] This is probably due to steric hindrance of histone antigenic determinants by components found in the transcriptionally active regions.[39] *Chironomus thummi* contains a protein which by enzyme-linked immunoassay and by "western" blotting cross-reacts with calf thymus HMG.

The distribution of accessible antigenic sites of this protein in *C. thummi* polytene chromosomes was visualized by immunofluorescence. The results indicate that (1) HMG-1 is distributed in a distinct banding pattern along the entire length of the chromosomes; (2) the banding pattern obtained with fluorescent antibody does not strictly correspond to that observed by phase-contrast microscopy; and (3) the amount of HMG-1 increases and the fluorescent banding pattern changes during the development of the organism. These findings suggest that the protein may be involved in the modulation of the structure of selected loci in the chromosome.[39] In transcriptionally active puffs the antigenic determinants of HMG-1 and -2 are more accessible than the determinants of histones. However, the distribution of HMG-1 (i.e., the immunofluorescence pattern) does not correlate with transcriptionally active chromosomal regions nor does it change upon treatment of the chromosomes with NaCl, a treatment which induces transcription as measured by $^3$H-uridine incorporation.

## VII. IN VIVO FUNCTION OF HMG STUDIED BY ANTIBODY MICROINJECTION

Immunochemical and immunocytological techniques allow detection and study of the antigens while present in their "native" state in the cell. Cell fixation techniques are designed to preserve the specimen in a state which, ideally, is identical to that present in the living cell. Immunocytological studies with fixed cells allow localization and identification of specific components or specific structures in the cell. However, they do not allow direct studies of the in vivo function of a particular component. Direct microinjection of antibody molecules into living cells allows functional studies on defined antigens under in vivo conditions. Several studies indicate that antibodies microinjected into cells seem to retain their antigenic binding activity.[1-4] We have reported that microinjection of antibodies to histones into the nuclei of oocytes of *Pleurodeles waltlii* brings about a retraction of the transcription loops of the lampbrush chromosomes present in those nuclei.[5] Similar effects were observed with antibodies to HMG-1. Anti-HMG-1 IgG was injected into the nuclei, the nuclei were opened, and the appearance of the lampbrush chromosomes examined. As early as 30 min after injection most of the transcription loops disappeared. Four hours after injection the chromosomes shrunk almost tenfold.[29] These results suggest that the large size of the lampbrush chromosomes is a consequence of transcription and not vice versa. In addition, the results indicate that free HMG-1, and/or -2 (i.e., not complexed with antibodies) is necessary for the maintenance of the transcriptionally active state of the lampbrush chromosome. Interestingly, transcription of the extrachromosomal, amplified, ribosomal genes is not affected.

Similar experiments have been done in somatic cells. In these experiments intact IgG, $F(ab)_2$ or Fab molecules are microinjected into the cells. The IgG and the IgG fragments are fluoresceinated so as to permit positive identification of each microinjected cell. After microinjection, the uridine pool in the cell is depleted by incubation with glucosamine for 1 hr. At this time $^3$H-uridine is added and the nascent RNA labeled for 1 hr. The slides are then processed for autoradiography and examined under fluorescence optics (to determine which cells have been microinjected) and under bright field optics to visualize the autoradiographic grains. Numerous control experiments with nonimmune IgG and $F(ab)_2$ indicated that the microinjection of control IgG does not affect transcription (see Figure 6). Antibodies to histones, when injected into the nucleus do, and when injected into the cytoplasm do not, inhibit transcription. This is because the IgG cannot traverse the nuclear membrane.[32] $F(ab)_2$ and Fab derived from antihistone when injected into either the nucleus or cytoplasm stops uridine incorporation. The fact that the Fab fragment inhibited the reaction is of specific importance since it indicates that the inhibition of $^3$H-uridine incorporation is not due to cross-linking by the bivalent antibody of the histones in adjacent nucleosomes. These results indicate that antibodies specific to nuclear proteins which are intimately associated with

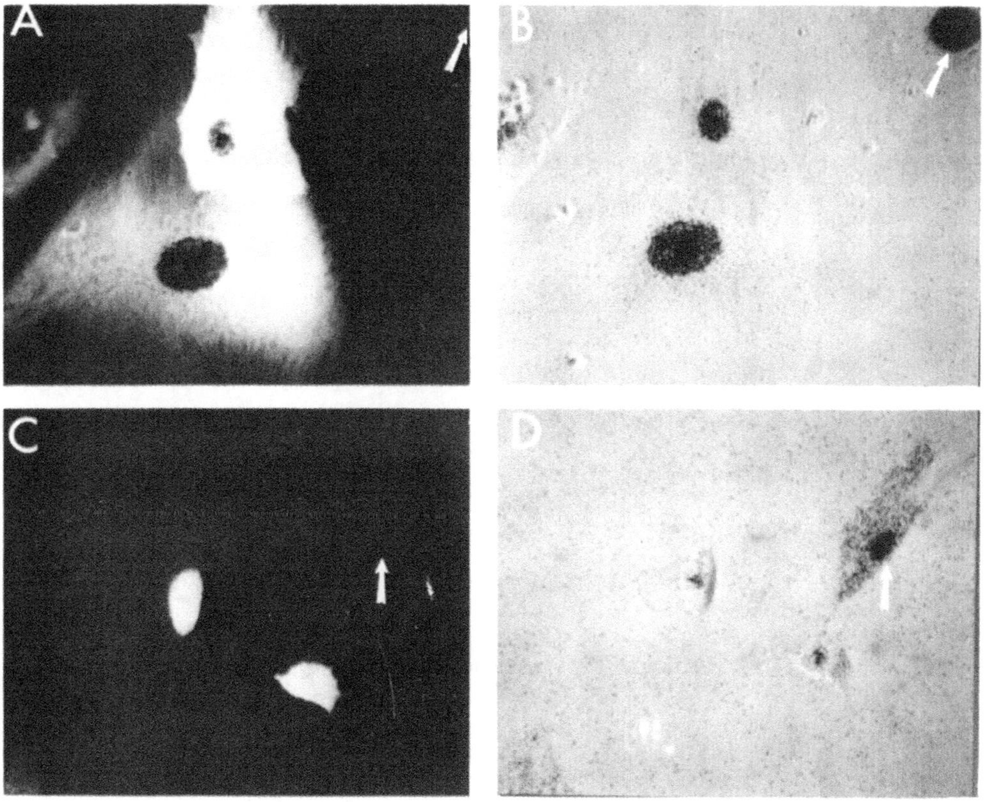


FIGURE 6. Microinjection of functional antibodies into living cells affect transcription. KD human fibroblasts were microinjected with fluoresceinated antibodies. After microinjection the uridine pool was depleted, the cells incubated with $^3$H-uridine, and processed for autoradiography. (A, B) Microinjected with control, nonimmune IgG; (C, D) microinjected with anti-HMG-17 IgG. (A, C) Fluorescence micrographs (the microinjected cells are fluorescent); (B, D) corresponding phase. The dark nuclei are those which incorporated the $^3$H-uridine.

DNA may interfere with transcription. Microinjection of antibodies to HMG-1 into the nucleus of the cell did not inhibit transcription while antibodies to HMG-17 did. Table 2 lists some of the results obtained when antibodies to HMG or other chromosomal proteins have been injected into both somatic cells and nuclei of oocytes.

Taken at face value these results suggest that neither HMG-1 nor -2 are intimately associated with the chromatin fiber or if they are, their presence does not affect transcription. Obviously, it is also possible that the antibody elicited by the HMG extracted from calf thymus does not recognize human HMG-1 or -2. However, the fact that the antibody recognizes HMG-1 present in the polytene chromosomes of *Chironomus* and the lampbrush chromosomes of *Pleurodeles waltii* as well as HMG from chicken and mouse indicates that the antibody can recognize HMG-1 from various species. Another consideration in these studies is the relative amounts of HMG-1, -2, and -17 in the cell. In most cells there is more HMG-1 and -2 than -17. The solutions injected contained approximately 2.5 mg/mℓ IgG. Since the average volume injected into a cell is $5 \times 10^{-11}$ mℓ, it can be calculated that about $5 \times 10^6$ molecules IgG have been introduced into the cell. This may not be sufficient to bind to all the HMG-1 and -2 present in a cell. On the other hand, sufficient anti-HMG-17 antibodies are introduced to neutralize most of the cellular HMG-17. It has been suggested that HMG-17 is associated with transcribable regions in the genome.[11] The antibody microinjection experiments described here support this notion.

**Table 2**
**EFFECT OF MICROINJECTION OF ANTIBODY ON TRANSCRIPTION[a]**

| Antibody specificity | Fragment injected | Locus injected[b] | Effect |
|---|---|---|---|
| **KD cells ($^3$H-U incorp.)** | | | |
| Nonimmune | IgG | N | 0 |
| | | C | 0 |
| Nonimmune | $F(ab)_2$ | N | 0 |
| | | C | 0 |
| Antihistone | IgG | N | Inhibition |
| | | C | 0 |
| Antihistone | $F(ab)_2$ | N | Inhibition |
| | | C | Inhibition |
| Anti-HMG-1 | IgG | N | 0 |
| | | C | 0 |
| Anti-HMG-1 | $F(ab)_2$ | N | 0 |
| | | C | 0 |
| Anti-HMG-17 | IgG | N | Inhibition |
| | | C | 0 |
| **Lampbrush chromosome (loop retraction)** | | | |
| Nonimmune | IgG | N | 0 |
| Antihistone | IgG | N | Loops retract<br>Ribosomal genes not affected |
| Antihistone | IgG | C | No effect |
| Anti-HMG-1 | IgG | N | Loops retract<br>Ribosomal genes not affected |

[a] Effect measured as $^3$H-uridine incorporation as detected by autoradiography; 0: no effect; in lampbrush chromosomes the transcriptional loops retract. The extrachromosomal, amplified ribosomal genes are not affected. The experiments involving microinjection into KD cells were done by Dr. L. Einck, N. I. H.; the microinjections into oocytes were done by Dr. U. Scheer, Heidelberg.

[b] Locus injected: N = nucleus; C = cytoplasm.

## VIII. PITFALLS AND LIMITATIONS

The usefulness of serological techniques in studying the specificity, intracellular organization, and biological function of HMG (or other cellular components) is dependent on the quality and specificity of the antisera used. When polyclonal antisera are elicited it is imperative to use highly purified and well-characterized immunogens. This is particularly true for chromosomal proteins which are poor immunogens. Most immunological assays do not allow absolute proof that the antisera contain antibodies of the desired type. Radioimmune and enzyme-linked immunoassays can detect antigens in nanogram quantities. Thus, there could be a situation where a poor immunogen appears pure by all accepted criteria (polyacrylamide gel, sequence, etc.), yet is contaminated by very small amounts of highly immunogenic components. The antisera will contain antibodies mainly against the contaminants, a situation which by all assays used until recently, will be overlooked. Therefore, it is extremely important to perform immunotransfer experiments in which the protein is transferred from the gels to DBM or nitrocellulose paper and the paper probed with antibodies and $^{125}I$-protein A. The main antigenic bands will then be visible. However, when denaturing gels are used, conformational determinants (for definition, see Section III) may be lost.

Monoclonal antibodies circumvent the problem of initial antigen purification. The most severe complication in using these antibodies is that they are specific for a particular segment of a protein. If, in the intact cell that segment is sterically hindered or if its conformation in the isolated protein is different from that present in the protein complexed in its "native" state, then the monoclonal antibody will not be useful for many studies. Furthermore, it has been reported that in some cases monoclonal antibodies display cross-reactivity with unrelated proteins. Thus, the use of monoclonal antibodies may cause complications due to the presence of antibodies specific for a sequential determinant which may react with similar stretches of amino acids in other proteins.[8] Antibodies against conformational determinants may be more specific unless they are directed against common structures (α helix, β structures, etc.) The reagent of choice woud seem to be affinity purified polyclonal antibodies. These are a mixture of antibodies against various portions of the antigens. Even if part of the antibodies present in this mixture have unexpected cross-reactivities, the great heterogeneity of the antibodies present in the preparation ensures specificity for the antigen. The major drawback of affinity purified antibodies is that often the antibodies with the highest affinity for the antigen are lost because they cannot be recovered from the affinity column. In addition, the treatment used to remove the antibody molecules from the affinity column leads to denaturation and reduced yields.

The anti-HMG antibodies can be used for immunocytochemical studies to visualize the location of the protein in chromatin and chromosomes. The various precautions which have to be taken in experiments involving immunofluorescence have already been mentioned.[7] While the occurrence of specific antibody binding indicates that the antigen is present at the loci where the antibody is detected, absence of antibody binding may simply mean that the antigenic determinants are not available for antibody binding. This may be due to various conformational or steric factors. The problem may be of particular relevance in studies of chromosomal proteins because these often are cross-complexed with DNA, RNA, and other proteins. Immunological reactions occur under mild conditions, require relatively low amounts of materials, and are specific. As such they are particularly suitable for the study of the organization of HMG proteins and other chromosomal components[7] in their native state, thereby providing information which is not easily available by other techniques. Awareness of various pitfalls and limitations in the application of serological techniques to chromatin structure and function will ensure the reliability of the results.

## IX. SUMMARY

Antibodies against chromosomal proteins HMG-1 and -2 purified from calf thymus or other sources can be elicited and used for various studies on these proteins. Provided the proteins are pure, the antibodies elicited are specific and do not cross-react with other chromosomal proteins. However, HMG-1 and -2 cross-react with each other reflecting their known similarity in primary sequence. Antisera elicited by HMG-2 seem to display remarkable specificity for the immunogen; it cross-reacts very little with HMG-1. The sequence and structure similarity between HMG-1 and -2 suggest that most of the monoclonal antibodies produced against either of the two proteins will be specific for shared antigenic determinants. Indeed, two clones obtained so far did not distinguish between the proteins. The anti-HMG-1 and -2 sera contain antibodies directed toward sequential determinants in these proteins. Binding of antibodies to small molecular weight peptides derived from HMG by proteolytic digestion could be visualized by "western" blotting.

The antisera were used to assess sequence homologies between HMGs purified from various sources. The information is not as reliable as that obtained from sequence analysis.

However, it provides more information than that obtained from peptide mapping or determination of terminal amino acids. HMG proteins derived from various sources cross-react immunologically, suggesting that this chromosomal protein is relatively conserved during evolution. By analogy with histones and other highly conserved proteins, it is concluded that in the cell these proteins perform a crucial function for which the entire primary sequence is necessary. The immunological cross-reactivity of the HMG from various sources suggests that the antisera elicited against proteins purified from calf thymus can be used as a "universal" reagent to study HMG from various sources.

Immunofluorescence studies reveal that in a variety of cells there are substantial amounts of HMG in the cytoplasm. This conclusion is supported by microinjection studies and by biochemical isolation of HMG from cytoplasmic preparation. These results, however, are controversial since several investigators using various experimental approaches concluded that in the living cell HMG is found exclusively in the nucleus. It is possible that HMGs readily leak out of the nucleus. In this respect they differ from the histones which seem to be more tightly bound to the chromatin. However, several studies suggest that there may be two states of HMG—some tightly bound to chromatin and some which are loosely bound shuttling between the nucleus and the cytoplasm.

Antibodies to HMG have been used for various studies on the cellular function and organization of these proteins. HMG location in polytene chromosomes has been visualized by immunofluorescence. Their involvement in transcription has been probed by microinjection of antibodies and functional antibody fragments into somatic cells and into nuclei of oocytes. Preliminary characterization of the mRNA coding for these proteins has been achieved by immunoprecipitation with affinity purified antibodies.

The anti-HMG antibodies can be used to obtain insights into the structure and cellular function of these proteins. Serological approaches complement other physical and biochemical techniques used in the study on HMG proteins and, in some cases, can provide information which presently cannot be obtained by other methods. Rigorous characterization of the antisera and awareness of various pitfalls and limitations of the immunochemical techniques used will ensure the reliability of the data obtained.

## ACKNOWLEDGMENTS

I wish to thank Drs. B. Dunn and L. Einck for proofreading and Mrs. H. Jordan for typing the manuscript.

## REFERENCES

1. **Bona, M., Scheer, U., and Bautz, E. K. F.,** Antibodies to DNA polymerase II inhibit transcription in lampbrush chromosomes after microinjection into living oocytes, *J. Mol. Biol.*, 151, 81, 1981.
2. **Antman, K. H. and Livingston, D. M.,** Intracellular neutralization of $SV_{40}$ tumor antigens following microinjection of specific antibody, *Cell*, 19, 627, 1980.
3. **Lin, J. J. C. and Feramisco, J. R.,** Disruption of the in-vivo distribution of the intermediate filaments of fibroblasts through the microinjection of specific monoclonal antibodies, *Cell*, 24, 185, 1981.
4. **Rungger, D., Rungger-Brandle, E., Chaponnier, C., and Gabbiani, G.,** Intranuclear injection of anti-actin antibodies into xenopus oocytes block chromosome condensation, *Nature (London)*, 282, 320, 1979.
5. **Scheer, U., Sommerville, J., and Bustin, M.,** Injected histone antibodies interfere with transcription of lampbrush chromosome loops in oocytes of pleurodeles, *J. Cell Sci.*, 40, 1, 1979.
6. **Einck, L. and Bustin, M.,** Antihistone antibodies microinjected into fibroblast inhibit DNA synthesis, *Fed. Proc. Fed. Am. Soc. Exp. Biol.*, 41, 514, 1982.
7. **Bustin, M.,** Immunological approaches to chromatin and chromosome structure and function, *Curr. Top. Microbiol. Immunol.*, 88, 105, 1979.

8. **Nigg, E. A., Walter, G., and Singer, S. J.,** On the nature of crossreactions observed with antibodies directed to defined epitopes, *Proc. Natl. Acad. Sci. U.S.A.*, 79, 5939, 1982.
9. **Bustin, M., Hopkins, R. B., and Isenberg, I.,** Immunological relatedness of high mobility group proteins from calf thymus, *J. Biol. Chem.*, 253, 1694, 1978.
10. **Spiker, S., Mardian, J. K. W., and Isenberg, I.,** Chromosomal HMG proteins occur in three eukaryotic kingdoms, *Biochem. Biophys. Res. Commun.*, 82, 129, 1978.
11. **Weisbrod, S. and Weintraub, H.,** Isolation of a subclass of nuclear proteins responsible for conferring a DNAase I-sensitive structure on globin chromatin, *Proc. Natl. Acad. Sci. U.S.A.*, 76, 630, 1979.
12. **Bustin, M. and Neihart, N. K.,** Antibodies against chromosomal HMG proteins stain the cytoplasm of mammalian cells, *Cell*, 16, 181, 1979.
13. **Bustin, M., Dunn, B., Gillette, R., Mendelsohn, E., and Soares, N.,** Antigenic determinants of chromosomal proteins HMG-1 and HMG-2, *Biochemistry*, 21, 6773, 1982.
14. **Bustin, M., Reisch, J., Einck, L., and Klippel, T. H.,** Auto antibodies to nucleosomal proteins: antibodies to HMG-17 in autoimmune diseases, *Science*, 215, 1245, 1982.
15. **Romani, M., Vidali, G., Tahourdin, C. S. M., and Bustin, M.,** Solid phase radioimmunoassay for chromosomal components, *J. Biol. Chem.*, 255, 468, 1980.
16. **Renart, J., Reiser, J., and Stark, G. R.,** Transfer of proteins from gels tó diazobenzyloxymethyl paper and detection with antisera, *Proc. Natl. Acad. Sci. U.S.A.*, 76, 3116, 1979.
17. **Allan, J., Smith, B. J., Dunn, B., and Bustin, M.,** Antibodies against the folding domain of histone H5 cross react with $H_1^0$ but not with $H_1$, *J. Biol. Chem.*, 257, 10533, 1982.
18. **Champion, A. B., Prager, E. M., Wacher, D., and Wilson, A. C.,** in *Microcomplement Fixation in Biochemical and Immunological Taxonomy of Animals*, Wright, D. A., Ed., Academic Press, New York, 1974, 387.
19. **Walker, T. M., Gooderhan, K., Hastings, J. R. B., Mays, E., and Johns, E. W.,** The primary structures of non-histone chromosomal proteins HMG-1 and HMG-2, *FEBS Lett.*, 122, 264, 1980.
20. **Romani, M., Rodman, F. C., Vidali, G., and Bustin, M.,** Serological analysis of the species specificity in the high mobility group chromosomal proteins, *J. Biol. Chem.*, 254, 2918, 1979.
21. **Tahourdin, C. S. M., Neihart, N. K., Isenberg, I., and Bustin, M.,** Immunochemical detection of chromosomal protein HMG-17 in chromatin subunits, *Biochemistry*, 20, 910, 1981.
22. **Crumpton, M. J.,** Protein antigens: the molecular bases of antigenicity and immunogenicity, in *The Antigens*, Vol. 2, Sela, M., Eds., Academic Press, New York, 1974, 1.
23. **Reeves, R., Chang, D., and Chung, S. C.,** Carbohydrate modifications of the high mobility group proteins, *Proc. Natl. Acad. Sci., U.S.A.*, 78, 6704, 1981.
24. **Isackson, P. J., Bidney, D. L., Reeck, G. R., Neihart, N. K., and Bustin, M.,** High mobility group chromosomal proteins isolated from nuclei and cytoplasm of cultured hepatoma cells are similar, *Biochemistry*, 13, 4466, 1980.
25. **Levy, V. B., Wong, N. C. W., and Dixon, A.,** Selective association of trout specific H6 protein with chromatin regions susceptible to DNase I and DNase II, *Proc. Natl. Acad. Sci., U.S.A.*, 74, 2810, 1977.
26. **Mathew, C. G. P., Goodwin, G. H., and Johns, E. W.,** Studies on the association of the high group non-histone chromatin proteins with isolated nucleosomes, *Nucleic Acid Res.*, 6, 167, 1979.
27. **Smith, B. J., Robertson, D., Birbeck, M. S. C., Goodwin, G. H., and Johns, E. W.,** Immunochemical studies of high mobility group non-histone chromatin proteins HMG-1 and HMG-2, *Exp. Cell Res.*, 115, 420, 1978.
28. **Bhullar, B. S., Hewitt, T., and Candido, P. M.,** The large high mobility group proteins of a rainbow trout are localized predominantly in the nucleus and nucleolo of a cultured trout cell line, *J. Biol. Chem.*, 256, 8801, 1981.
29. **Sheer, U., Kleinschmidt, T., and Bustin, M.,** Unpublished.
30. **Gordon, J. S., Bruno, T., and Lucas, J. J.,** Heterogeneous binding of high mobility group chromosomal proteins to nuclei, *J. Cell Biol.*, 88, 373, 1981.
31. **Rechsteiner, M. and Kuehl, L.,** Microinjection of the non-histone chromosomal protein HMG-1 into bovine fibroblasts and Hela cells, *Cell*, 16, 901, 1979.
32. **Einck, L. and Bustin, M.,** Functional antibody fragments microinjected into tissue culture cell cytoplasm transverse the nuclear envelope, *Fed. Proc. Fed. Am. Soc. Exp. Biol.*, 40, 1571, 1981.
33. **Einck, L. and Bustin, M.,** Unpublished.
34. **Bustin, M., Neihart, N. K., and Fagan, J. B.,** m-RNA of chromosomal proteins HMG-l and HMG-2 are polyadenyated, *Biochem. Biophys. Res. Commun.*, 101, 893, 1981.
35. **Bostock, C. J. and Sumner, A. T.,** *The Eukaryotic Chromosome*, Elsevier, Amsterdam, 1978.
36. **Jamrich, M., Greenleaf, A. L., and Bautz, E. K. F.,** Localization of RNA polymerase in polytene chromosomes of Drosophila melanogaster, *Proc. Natl. Acad. Sci. U.S.A.*, 74, 2079, 1977.
37. **Kurth, P. D., Mourdrianakis, E. N., and Bustin, M.,** Histone localization in polytene chromosomes by immunofluorescence, *J. Cell Biol.*, 78, 910, 1978.

38. **Bustin, M., Kurth, P. D., Moudrianakis, E. N., Goldblatt, D., Sperling, R., and Rizzo, W. B.,** Immunological probes for chromatin structure, *Cold Spring Harbor Symp., Quant. Biol.*, 32, 379, 1978.
39. **Kurth, P. D., Reisch, J. C., and Bustin, M.,** Selective exposure of antigenic determinants in chromosomal protein upon gene activation in polytene chromosomes, *Exp. Cell Res.*, 143, 257, 1983.

*Note*: This review covers the literature published up to November 1982.

Chapter 6

# ORGANIZATION OF THE SITES FOR DNA ATTACHMENT TO THE NONHISTONE PROTEINACEOUS NUCLEAR SKELETON

**Sergei V. Razin, Victor V. Chernokhvostov, Olga V. Yarovaya, and Georgii P. Georgiev**

## TABLE OF CONTENTS

## I. INTRODUCTION AND HISTORY

As was shown in earlier experiments in which isolated cell nuclei were fractionated, part of the nuclear proteins remained in the insoluble fraction after deoxyribonucleoprotein has been removed by high salt extraction.[52,53,77] A combination of biochemical and cytochemical analyses has revealed that the residual material consists of nuclear membranes, nucleoli, and some extranucleolar material scattered throughout nuclei.[78]

By electron microscopy[27] it was shown that nuclei sequentially treated with 0.14 *M* NaCl, DNase I and 2 *M* NaCl, lost all of their DNA and histones but still retained their envelope, a nucleolus and thin (~100 Å) threads distributed within the internal volume and sometimes attached to the nucleoli and membranes. The filaments were associated with ~200-Å granules, the latter being lost after RNAase treatment (Figure 1b). When the 2 *M* NaCl extraction was omitted, the threads were thicker, reaching 600 to 900 Å in diameter. In this case, a significant part of DNA remained within the nuclei (Figure 1a).

We suggested that the filaments were skeletal structures of the nuclei to which ribonucleoprotein granules, as well as deoxyribonucleoprotein fibrils of chromosomes, were attached. We called these filaments "nucleonemas". A hypothetical model is presented in Figure 1c.[27] Thereafter similar results were obtained by several authors.[6,7,15,56,71,79]

The internal nuclear material surviving the nuclease and 2 *M* NaCl treatment was designated as a "nuclear network"[6,23,56,71] or a "nuclear matrix".[9] The latter term is now commonly used in the literature.

Berezney and Coffey[6-9] studied in detail the protein composition of residual nuclei referred to as a "nuclear matrix" and demonstrated that it was rather simple. The most prominent bands had molecular weights of 69,000, 66,000, and 62,000. However, these data were later revised, and now the protein composition of the nuclear matrix is generally considered to be complex.[10,17,24,39,47,68] These proteins with molecular weights of 69,000, 66,000, and 62,000 were found to form the basic structure of the nuclear lamina.[29,30] The nuclear lamina is a protein layer located at the inner side of the nuclear membrane. It is resistant to high salt extraction of nuclei and is considered as a most stable part of the nuclear matrix.[39] Preparations of the nuclear lamina free of contaminations with other parts of the nuclear matrix can be obtained by some variations in the isolation procedure.[39] In fact, the nuclear lamina had been isolated and shown to form a residual nuclear structure unrelated to the nuclear matrix.[1,2] Investigations into the protein composition and ultrastructure of residual nuclei demonstrated the existence of large DNA domains in dehistonized nuclei and metaphase chromosomes.[3,5,18,19,61] The ends of the domains seemed to be fixed in some way. After rather extensive deproteinization, nuclear DNA still remained in the supercoiled state.[5,18,19,36]

Electron microscopy of dehistonized metaphase chromosomes and nuclei directly revealed large DNA loops attached to a protein-containing scaffold.[3,50,61] The size of the loops was estimated to vary from 30 to 90 kb. Finally, experiments in which nuclei were subjected to mild digestion with nucleases demonstrated that a fraction of total DNA remained bound to nuclear residues after a 2 *M* NaCl treatment. The amount of this DNA was in good correlation with the size of DNA fragments, suggesting the DNA thread was attached to a nuclear skeleton at average intervals of 50 to 100 kb.[12,63,75] The average distances between DNA attachment points were found to be identical in interphase nuclei and metaphase chromosomes of the same cell line.[63] It was suggested that DNA fragments between two attachment sites might correspond to chromosomal functional units, transcriptons, or replicons.[28] However, evidence for these suggestions is only hypothetical. On the other hand, it appears that the nuclear matrix is involved in DNA replication, transcription, hnRNA processing, and transport (for review see References 11, 37, 48, 51, and 70). These data suggest that the nuclear matrix is a complex structure involved not only in fixation of DNA loops, but also in DNA and hnRNA functioning. This may be in good agreement with the complex protein com-

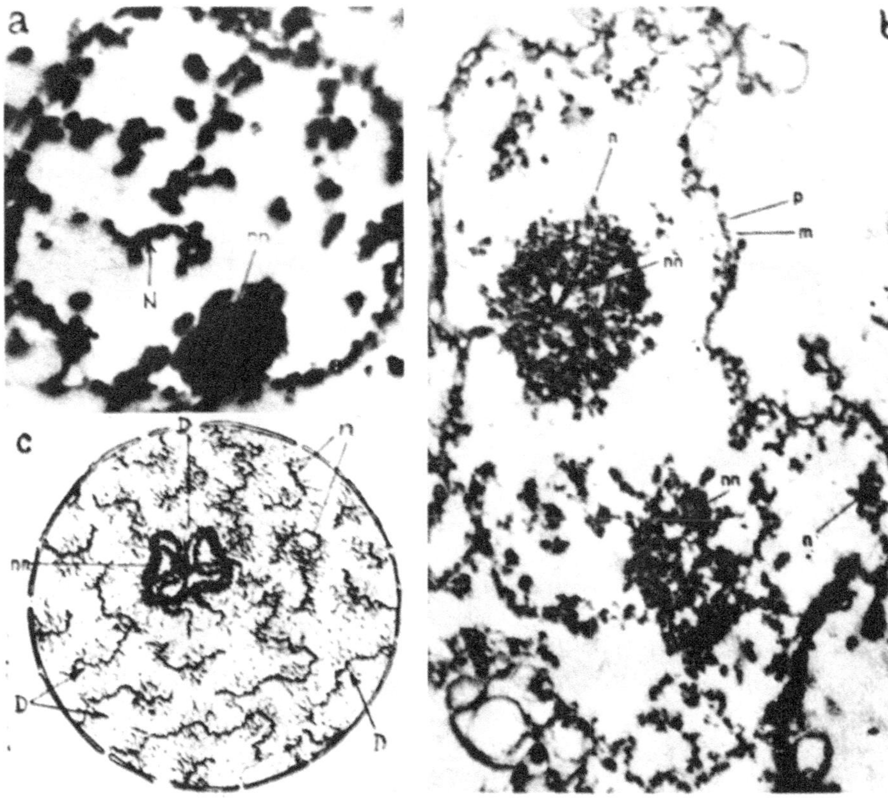


FIGURE 1. Electron microscopy of the isolated rat liver nuclei. (a) Treatment with DNAase I and then with 0.14 *M* NaCl; (b) treatment with DNAase I and then with 0.14 *M* and 2 *M* NaCl; (c) the hypothetical model of cell nucleus based on the results obtained. n = Nucleonemas, D = DNP fibrils, nn = nucleolus, m = nuclear membrane with pores (p). N = thick fibrils formed after DNAase I and 0.14 *M* NaCl treatment.

position of the nuclear matrix. The scaffold of metaphase chromosomes freed of cytoskeleton contaminations was found recently to have a much simpler protein composition.[46] One may assume that in mitosis the nuclear skeleton performs only one major function, namely, DNA binding and folding.

In this chapter, we will mainly consider the aspect of DNA attachment to the nuclear skeletal structures and discuss the nature of proteins and DNA sequences located at the sites of their interaction. For convenience, we will refer to the nuclei or metaphase chromosomes treated with nucleases and dehistonized as a nuclear or chromosomal matrix, and to the protein part of the matrix as a nuclear or chromosomal skeleton. The nuclear skeleton is subdivided into the nuclear lamina, residual nucleoli, and internal filaments, i.e., nucleonemas. The content of DNA in nuclear or chromosomal matrix preparations varies depending on the intensity of nuclease treatment. This DNA fraction is referred to as "matrix" DNA.

## II. STRUCTURAL HETEROGENEITY OF THE SITES OF DNA ATTACHMENT TO THE NUCLEAR SKELETON

Since the first experiments,[15,27,56,71,77-79] the procedure used for isolation of the nuclear matrix includes treatment of the isolated nuclei with nuclease(s) and solutions with low and high ionic strength. In most cases, modifications of Berezney and Coffey[6-9] are used. In the original procedure developed by these authors, the isolated nuclei were extracted with a solution of low ionic strength, digested with endogenous nucleases, extracted with 2 *M* NaCl, and digested with DNAase I and pancreatic RNAase.

In our recent experiments,[63] we exploited the following procedure for nuclear matrix isolation:

1. The cells were lysed with nonionic detergents (0.5% NP 40 and 0.1% Triton® X-100) in a buffer containing 100 m*M* NaCl, 50 m*M* KCl, 10 m*M* Tris-HCl, pH 7.0, 0.1 m*M* EDTA, and 10% glycerol. The nuclei were purified by differential centrifugation.
2. The purified nuclei were treated with increasing amounts of micrococcal nuclease, or with the same amount of nuclease but for increasing time intervals, in the buffer for cell lysis (without detergents) after addition of $CaCl_2$ up to 5 m*M*.
3. The digested nuclei were extracted with 2 *M* NaCl in 10 m*M* Tris-HCl, pH 9.0, 10 m*M* EDTA, 0.1% NP 40, and the nuclear matrix was separated from extracted proteins and separated from DNA by centrifugation through a layer of 15% glycerol in the latter solution.
4. In some cases, the nuclease and 2 *M* NaCl treatments were repeated with dehistonized preparations of the nuclear matrix.

All operations were performed at 0 to 4°C and all solutions contained 0.1 m*M* phenylmethylsulfonyl fluoride (PMSF).

The protein composition of the nuclear matrix preparation obtained in such a way from mouse L cells is presented in Figure 2. The preparations also contained DNA, the relative amount of which decreased with an increase in the duration of nuclease treatment followed by a decrease in the size of DNA fragments remaining attached to the nuclear skeleton. The average distance between two attachment sites of a DNA fibril was calculated on the basis of measuring the average size of bound DNA fragments and the percentage of total DNA remaining in the nuclear matrix. This was found to be ~60 kb. Extensive nuclease treatment yielded short fragments (150 bp predominantly) of matrix DNA.[63]

Nuclear matrix preparations were also obtained from mouse Ehrlich ascites carcinoma cells. The protein composition of these cells was very similar to that of L cells (Figure 2). However, the size distribution of DNA fragments was quite different. Whatever the intensity of nuclease treatment, the DNA bound to the nuclear skeleton survived as very long fragments, ~10 kb or more (Figure 3).[66] These fragments did not contain single-strand breaks since their molecular weight remained high in alkaline agarose gels (Figure 3B). After partial digestion with nuclease, about 0.02% of total DNA survived the treatment and remained bound to the skeleton. The average size of these fragments was equal to ~10 kb. Fragments of such a size can be detected in matrix preparations from L cells, but there they constituted only a minor portion of the total skeleton-associated DNA.

The number of nuclease-protected long DNA fragments in the nuclear matrix of Ehrlich ascites carcinoma cells remained unchanged when nuclei were preincubated and extracted in the presence of 2-mercaptoethanol.

Simple calculations show that each diploid nucleus should contain about 120 sites, the structure of which protects the long stretch of DNA against nuclease digestion. If they are distributed randomly among chromosomes, each chromosome should contain about three such sites only. It is clear that they cannot be responsible for the attachment of 60 to 100-kb DNA loops to the nuclear skeleton.

The absence of short DNA fragments associated with the skeleton from the nuclei of Ehrlich ascites carcinoma cells isolated in a buffer containing 0.1 m*M* EDTA can be attributed to partial degradation of certain elements of the matrix, for example, of inner filaments (nucleonemas) which are known to be very fragile structures.[39] To check such a possibility, we included several additional components, in particular bivalent cations, that are commonly used for stabilization of nuclei, into the medium for cell lysis. Before nuclease treatment, the nuclei were washed with the standard isolation buffer. The most striking effect was

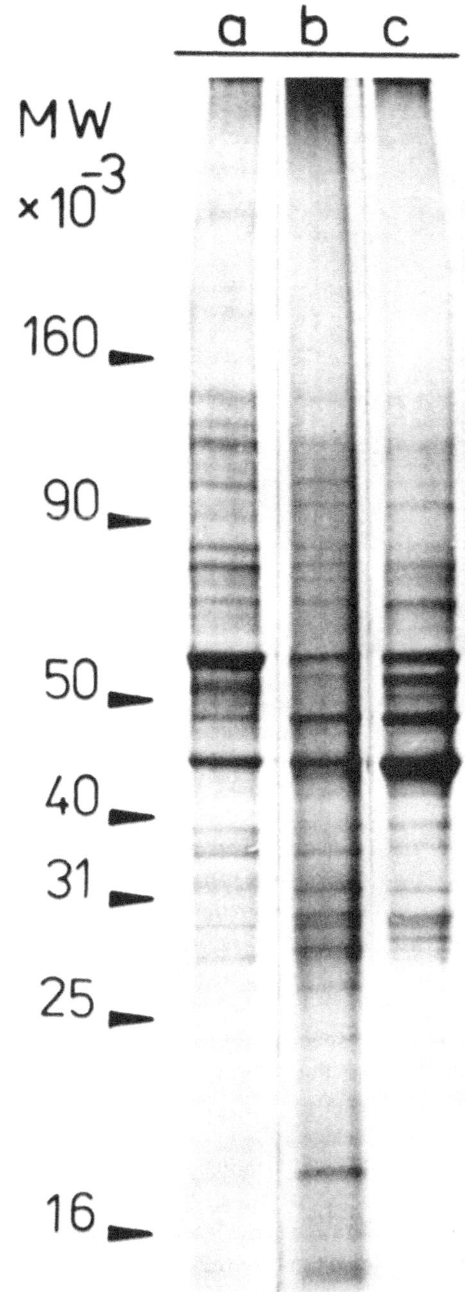


FIGURE 2. Electrophoretic patterns of proteins of different nuclear matrix preparations. Lane (a) Nuclear matrix from L cells; lane (b) nuclear matrix from Ehrlich ascites carcinoma cells lysed in the medium lacking $Cu^{2+}$ ions; lane (c) nuclear matrix from Ehrlich ascites carcinoma cells lysed in the medium containing $Cu^{2+}$ ions.

obtained upon the addition of $Cu^{2+}$ ions, which were shown previously[39] to stabilize the diffuse matrix. Figure 4 shows the electrophoretic profile of skeleton-associated DNA fragments obtained after treatment of nuclei with increasing doses of micrococcal nuclease. One can detect again the fragments of high molecular weight DNA, but now they constitute only a minor fraction. Most of the matrix DNA is represented by relatively low molecular weight

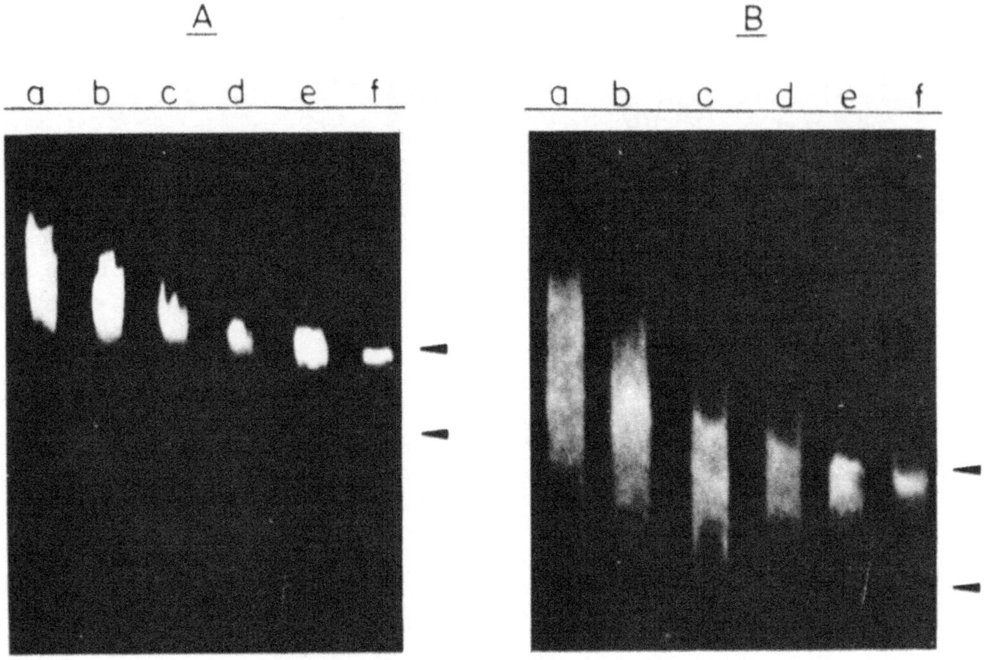


FIGURE 3. Electrophoretic distribution in 0.7% neutral (A) or alkaline (B) agarose gels of matrix DNA from Ehrlich ascites carcinoma cells lysed in the medium lacking $Cu^{2+}$ ions. The comparative amount of matrix DNA was as follows: lane (a) 1.57%; lane (b) 0.73%; lane (c) 0.29%; lane (d) 0.11%; lane (e) 0.04%; lane (f) 0.02%. The arrows indicate the positions of markers (10,800 and 5400 bp).

fragments distributed among discrete bands corresponding in size to the oligonucleosome series. The average size of DNA diminishes proportionally to the DNA digestion and the decrease of the DNA content in the matrix. After partial digestion, most of the fragments are recovered in the 140-bp band (Figure 5). The total number of such fragments per diploid nucleus is ~$10^5$ (Table 1) which corresponds to one fragment per 50-kb segment of DNA. The figure is close to that obtained for L cells.[63,64]

The effect of $Cu^{2+}$ ions is rather specific. For example, $Zn^{2+}$ ions possess only a slight protective effect. After cell lysis in the presence of Zn,$^{2+}$ short DNA fragments can be detected in the nuclear matrix preparations (Figure 4), but their content is about 10 times as low as after treatment with $Cu^{2+}$ (Table 1). $Mg^{2+}$ ions do not protect the structures involving short DNA fragments. Only high molecular weight (~10 kb) DNA fragments are detected in the nuclear matrix prepared from cell lysed in the presence of $Mg^{2+}$ (Figure 4) which includes, after partial digestion, the same fraction of total DNA (0.02%) as the nuclear matrix from cell lysed either in the original buffer containing 0.1 m*M* EDTA or in a $Cu^{2+}$-containing buffer.

What is the role of histones in the protection of skeleton-associated DNA against nuclease treatment? In these experiments nuclei were dehistonized by extraction with 2 *M* NaCl and then digested with micrococcal nuclease. Before DNA analysis, the residual nuclei were again passed through a glycerol gradient containing 2 *M* NaCl. The results were essentially the same. Again, two types of fragments with the size of ~10 kb and ~140 bp could be detected in the preparation of nuclei isolated in the presence of $Cu^{2+}$. The only difference was that no discrete bands of DNA from the oligonucleosome series could be detected; some heterogeneous material was accumulated and then after partial digestion, converted into more homogeneous ~140-bp fragments (Figure 5). In the nuclear matrix isolated from cells lysed in a medium lacking $Cu^{2+}$, only long DNA fragments were found.

The question arises of how the observed phenomena are related to the in vivo situation.

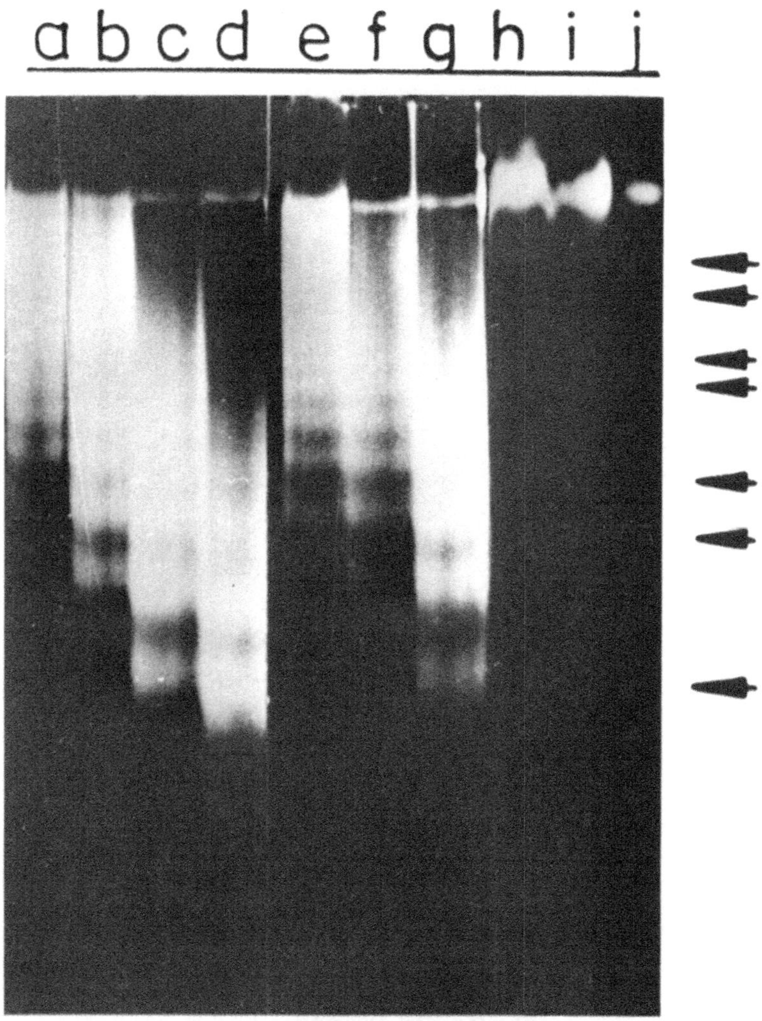


FIGURE 4. Electrophoretic distribution in a neutral 1.2% agarose gel of matrix DNA fragments obtained from the Ehrlich ascites carcinoma cell nuclei isolated in the medium containing lanes (a-d) $Cu^{2+}$ ions; lanes (e-g) $Zn^{2+}$ ions; lanes (h-j) $Mg^{2+}$ ions. The comparative amount of matrix DNA was lane (a) 3.5%; lane (b) 2.3%; lane (c) 1.5%; lane (d) 1%; lane (e) 0.45%; lane (f) 0.27%; lane (g) 0.16%; lane (h) 3.8%; lane (i) 0.7%; lane (j) 0.03%. The arrows indicate the positions of markers (1960, 1910, 1620, 1320, 930, 880, and 590 bp).

One might suggest that both short and long "matrix" DNA fragments are protected by specific association with nuclear skeleton proteins against nuclease digestion. However, the survival of a small DNA fraction may also be explained (1) by the resistance of DNA itself to nuclease, (2) by contamination of nuclear matrix preparations with nonlysed cells or aggregated nuclei whose DNA is less accessible to nuclease, and (3) by its protection by nonhistone protein aggregation. Several control experiments were conducted to check all these possibilities.

The first explanation was ruled out by the observations that all protected DNA was readily digested after deproteinization to acid-soluble products by either micrococcal nuclease or DNAase I.

To exclude the second possibility, we purified nuclear matrix preparations by density

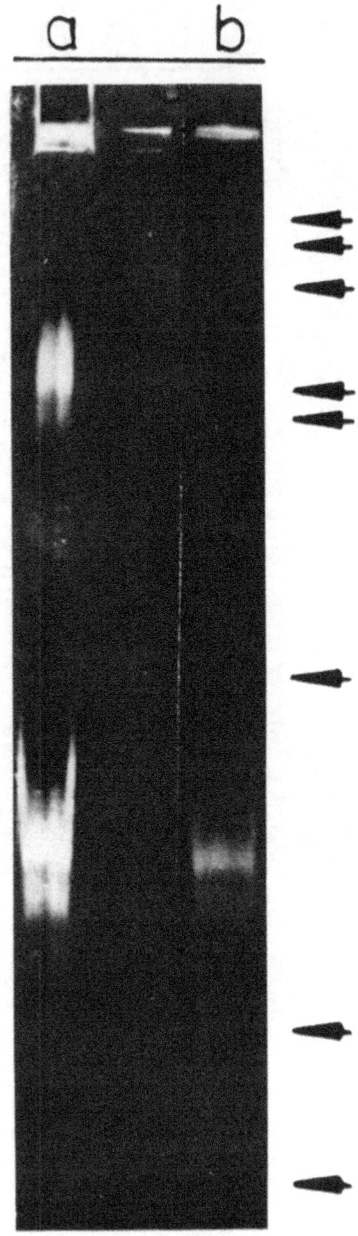


FIGURE 5. Electrophoretic distribution in an 8% acrylamide gel of matrix DNA fragments obtained after limit digestion with micrococcal nuclease of Ehrlich ascites carcinoma cell nuclei isolated in the medium containing $Cu^{2+}$ ions. Lane (a) nuclei were treated with micrococcal nuclease before high salt extraction; lane (b) nuclei were treated with micrococcal nuclease both before and after high salt extraction. The arrows indicate the positions of markers (1744, 1084, 766, 543, 525, 237, 109, and 83 bp).

centrifugation in a metrizamide gradient. It has been demonstrated in model experiments that the nuclear matrix can easily be separated from both unlysed cells and unextracted nuclei by density centrifugation in a 0.4 to 0.6 *M* metrizamide gradient (Figure 6a). When our preparations of the nuclear matrix were subjected to such centrifugation, no contami-

**Table 1**
**AVERAGE DISTANCE BETWEEN THE APPARENT ATTACHMENT SITES OF DNA TO THE NUCLEAR SKELETON IN NUCLEI ISOLATED BY DIFFERENT METHODS**

| Sample | Comparative amt. of matrix DNA (%) | Ave. distance between 2 attachment sites (kb) |
|---|---|---|
| Nuclei isolated in the $Cu^{2+}$-containing medium | 5.5 | 46 |
| | 3.2 | 50 |
| | 2.1 | 57 |
| Nuclei isolated in the $Cu^{2+}$-containing medium and extracted (before nuclease treatment) with buffer solution containing 1% of 2-mercaptoethanol | 0.4 | 250 |
| | 0.2 | 275 |
| | 0.1 | 242 |
| Nuclei isolated in the $Zn^{2+}$-containing medium | 0.4 | 575 |
| | 0.18 | 555 |
| | 0.11 | 545 |

nations with unlysed cells or nuclei could be detected (Figure 6b). Additional purification of the nuclear matrix in the metrizamide density gradient did not change in any way the size distribution of DNA fragments remaining attached to the nuclear skeleton after partial digestion of the nuclei with micrococcal nuclease (not shown).

The third possibility is not easily ruled out. However, it should be mentioned that (1) both short and long DNA fragments have discrete sizes after partial digestion of the nuclei (~140 bp and ~10 kb, respectively); (2) under the standard conditions when cell lysis is performed in the $Cu^{2+}$-containing medium, the apparent number of short and long DNA fragments in nuclear matrix preparations was found to be quite reproducible and almost identical for different cell types of the same species; (3) in some cells (for example, L cells), the addition of $Cu^{2+}$ is not necessary for obtaining short fragments;[63] (4) $Cu^{2+}$ ions added to the medium after cell lysis cannot induce the attachment of short fragments;[66] and (5) $Cu^{2+}$ ions are present in nuclei in vivo, probably being bound to chromatin proteins.[14] Besides these considerations, one can find that the average distance between two attachment sites coincides quite well with the average size of DNA loops determined by independent methods.[5,50,61] Furthermore, the protected DNA seems to be nonrandom.[38,62-64] (See also below.) Therefore, on the whole, the results obtained suggest that, in vivo, two types of structures provide attachment to the nuclear skeleton and protection against nuclease digestion for long and short "matrix" DNA fragments. Attachment sites for long DNA segments are very rare in nuclei but appear to be very stable. They remain unchanged under various preparative conditions for nuclei and nuclear matrices. In contrast, attachment sites involving short DNA fragments are numerous and, possibly, regularly spaced throughout the whole DNA. These attachments are very fragile and easily dissociated during the course of isolation. At least in the case of Ehrlich carcinoma cells, the use of $Cu^{2+}$ ions during the cell lysis is an obligatory requirement for preserving them.

The nature of the stabilizing effect produced by $Cu^{2+}$ ions is not clear. It is possible that $Cu^{2+}$ ions act as an oxidizing agent causing, in particular, the formation of disulfide bonds between proteins and thus stabilizing the whole matrix structure.[39] However, in our case, neither sodium tetrathionate nor $H_2O_2$ at a concentration inducing S-S bond formation in proteins protected the attachment sites including short DNA fragments in the matrix of Ehrlich carcinoma cells.[66]

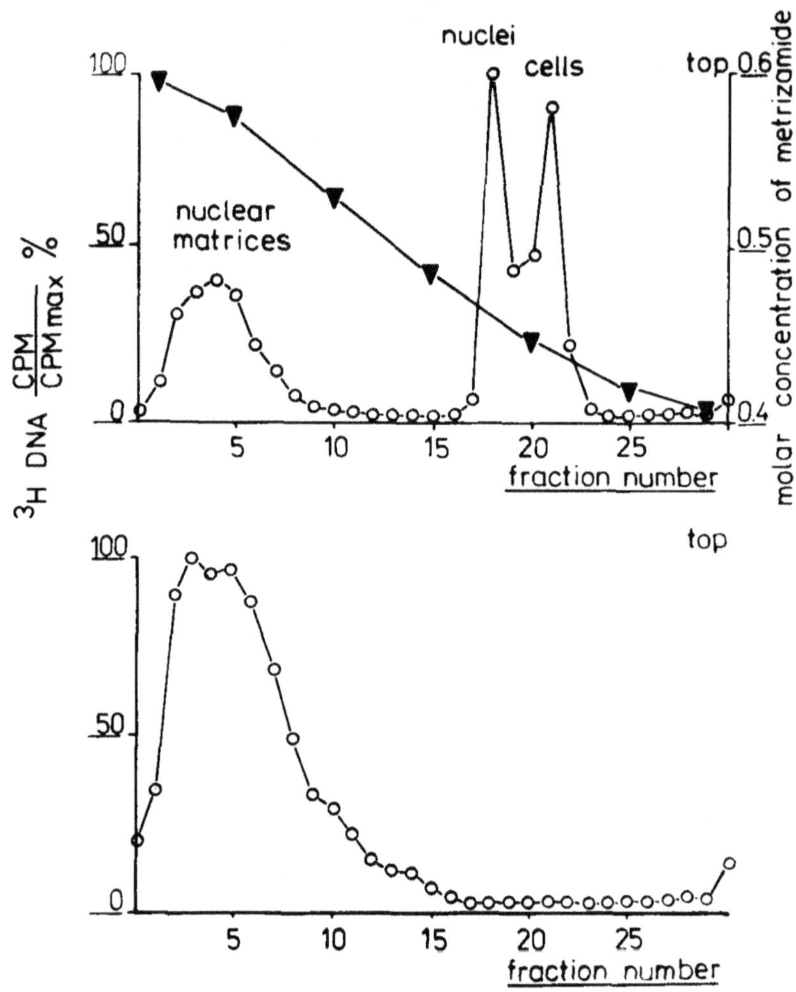


FIGURE 6. Purification of nuclear matrices in a metrizamide density gradient. Lane (a) Centrifugation in a metrizamide linear gradient (from 0.4 to 0.6 *M*) of a mixture containing nuclear matrices, unextracted nuclei, and unlysed cells; lane (b) centrifugation of nuclear matrices isolated by the standard procedure in a metrizamide linear gradient (from 0.4 to 0.6 *M*). All the centrifugations were carried out in a Beckman SW50.1 rotor for 2 hr at 35,000 rpm and 4°C.

On the other hand, the reducing agent 2-mercaptoethanol (1%) added to the nuclei isolated in the presence of $Cu^{2+}$ diminished fivefold the number of short DNA fragments recovered in the nuclear matrix (Table 1). However, the effect of 2-mercaptoethanol can be explained not only by the reduction but also by the binding of $Cu^{2+}$ ions.[74] Such an explanation of several $Cu^{2+}$ effects was discussed by Laemmli et al.[44,46] and seems to be most attractive now. However, we cannot exclude the possibility that $Cu^{2+}$ ions act by inhibiting some specific proteinases which destroy certain skeletal elements. In that case, the variations among tissues can be interpreted in terms of different concentrations of such proteinases.

## III. PROTEINS INTERACTING WITH DNA AT THE SITES OF DNA ATTACHMENT TO THE NUCLEAR SKELETON

As was mentioned already, the polypeptide composition of the nuclear matrix is very complex. However, it is clear that only a fraction of proteins is involved in the formation of DNA attachment sites. This fraction can be a minor one since the overall protein patterns are very similar in matrix preparations containing and lacking short DNA fragments (Figure 2).

To analyze the problem, one should remove the bulk of matrix proteins from those directly interacting with DNA. We studied proteins which remained bound to matrix DNA after solubilization of the latter.[65] Unfortunately, the interactions responsible for DNA attachment to the nuclear skeleton were rather stable and DNA could be solubilized only after the action of strong dissociating agents such as (1) 2% SDS or 2% sarcosyl, (2) 6 *M* guanidine hydrochloride (Gu·HCl), or (3) a mixture containing 2 *M* NaCl and 2 *M* urea. The first two reagents completely destroyed the nuclear matrix while the third one did not, as was observed by light and electron microscopy. Treatment with 6 *M* urea alone (without NaCl) as well as with a mixture containing 2 to 5 *M* urea and 0.5 to 1.5 *M* NaCl did not solubilize the skeleton-attached DNA. The presence of NaCl at a 2-*M* concentration was critical. Possibly, both ionic and hydrophobic interactions are involved in DNA binding.

To isolate proteins remaining bound to DNA and to purify them from other solubilized proteins of the nuclear matrix, the material was ultracentrifuged in a $Cs_2SO_4$ density gradient containing the same dissociating agent that had been used for solubilization.[65] In these experiments, the matrix was prepared from L cells lysed in the presence of $Cu^{2+}$ ions (to standardize the isolation procedure). The nuclease treatment was rather mild to obtain relatively long skeleton-associated DNA fragments. This was important for resolution of the DNA-protein band from the band containing free proteins. As shown in Figure 7, protein-label remained associated with DNA fragments solubilized from nuclear matrices by 2% sarcosyl treatment and banded in $Cs_2SO_4$ density gradient. This protein-label was found to be resistant to DNAase and RNAase treatments and at least partly sensitive to proteinase treatment followed by phenol-chloroform extraction. Much more protein remained bound to DNA after solubilization with 6 *M* Gu·HCl or with a mixture containing 2 *M* urea and 2 *M* NaCl. As a result, the density of the complexes was shifted toward the lower one comparing to free DNA.

For further analysis, the bound proteins were additionally labeled with $^{125}I$ in vitro[32] and the DNA was digested with DNAase I. Thereafter, the material was fractionated by electrophoresis in the presence of SDS[43] and autoradiographed.

The results presented in Figure 8 also show, for comparison, the electrophoretic patterns of total matrix proteins of L cells. It appears that from seven to eight different polypeptides are tightly bound to the skeleton-associated DNA and cannot be dissociated by sarcosyl. Their molecular weights are 60, 52, 43, 37, 31, 27, 18, and 14 kD. The latter polypeptide is detected only in some experiments and can be a degradation product of heavier components. None of the polypeptides coincide with any of the major bands of total matrix preparations. A number of additional polypeptides are recovered after dissociation with guanidine hydrochloride. The most prominent among them are those with molecular weights of 55, 36, and 20 kD, some of which (55 and 36 kD) can be seen in preparations of total matrix proteins and therefore may be considered as major proteins of the nuclear skeleton. A similar set of proteins was found to be associated with DNA fragments solubilized from the nuclear matrix with a mixture containing 2 *M* NaCl and 2 *M* urea (Figure 8).

Proteins which were not dissociated from DNA after ionic detergent treatment were found in many eukaryotic cells.[16,36,42,55,76] Nakane et al.[36,55] showed that the 50- and 60-kD proteins associated with DNA isolated in 2% SDS were important for the supercoiled state of this DNA. It was suggested that these proteins were located at the basement of DNA loops. The proteins with molecular weights of 60 and 52 kD (p60 and p52) in our preparations seem to be identical with the corresponding proteins described by Ide and Nakane. These authors also found that the treatment of DNA with mercaptoethanol in SDS eliminated DNA supercoiling as well as removed both p60 and p50.[55]

We performed similar experiments treating the samples of skeleton-associated DNA with

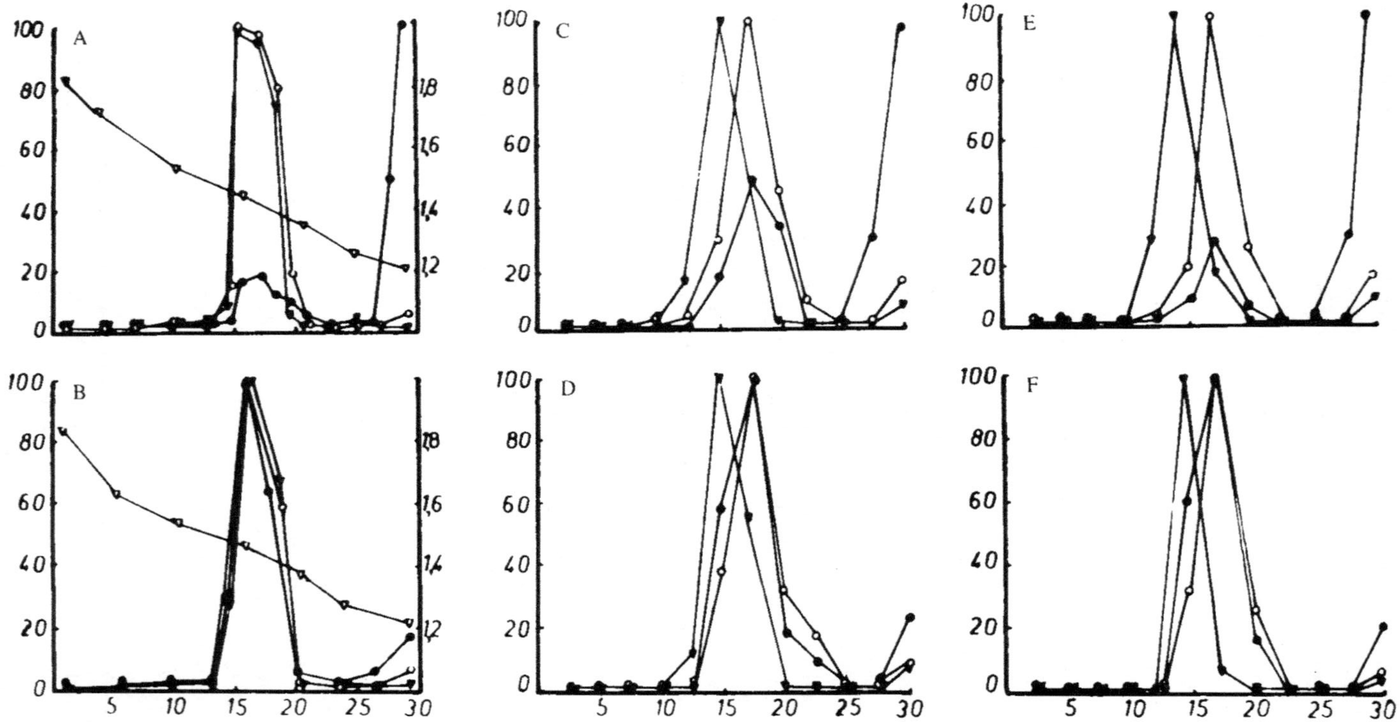


FIGURE 7. Purification of matrix DNA-protein complexes in a $Cs_2SO_4$ density gradient (A, C, E) and re-centrifugation of isolated matrix DNA-protein complexes in the same conditions (B, D, F). The DNA was solubilized from the nuclear matrix preparations with the aid of (A, B) 2% sarcosyl; (C, D) 6 *M* Gu·HCl; (E, F) a mixture containing 2 *M* urea and 2 *M* NaCl. Left ordinate = radioactive label cpm/cpm max %; right ordinate = density of CsCl g/cm$^3$; abscissa = fraction number. (○) Indicates matrix DNA ($^3H$ label); (●) indicates matrix proteins ($^{14}C$ label); (△) indicates $Cs_2SO_4$ density curve; (▲) indicates exogenous deproteinized DNA ($^{32}P$ label) which was added to the samples as an internal marker. The top fractions containing the highest protein label are omitted from A, C, and E to obtain a scale suitable for comparisons of the protein label remaining associated with the DNA band.

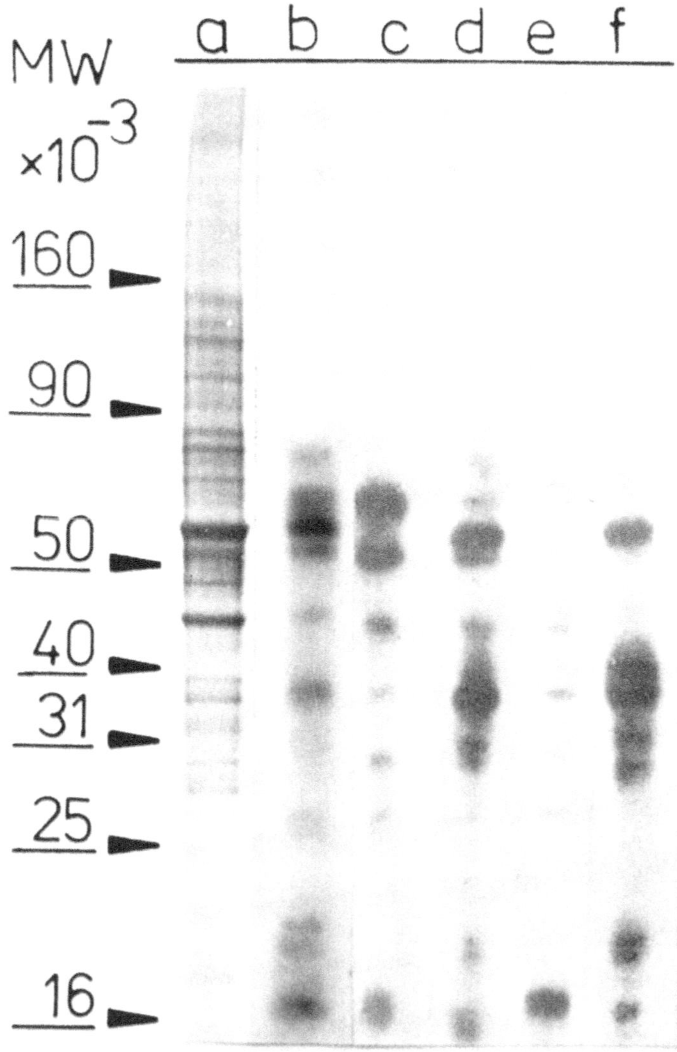


FIGURE 8. Electrophoretic patterns of proteins remaining bound to the DNA fragments solubilized from nuclear matrices by different treatments. Lane (a) Total proteins of nuclear matrices from L cells, Coomassie staining; lanes (b-f) proteins remaining bound to matrix DNA after treatment with (b) 2 *M* urea and 2 *M* NaCl, (c) sarcosyl, (d) 6 *M* Gu·HCl, (e) 2% sarcosyl and 2% 2-mercaptoethanol, (f) 6 *M* Gu·HCl and 2% 2-mercaptoethanol, autoradiographs.

1% mercaptoethanol in 1% sarcosyl. The major part of p52 and p60 was dissociated from DNA (Figures 8 and 9). About 10% of total p52 and p60, however, remained bound to the DNA even after repeated treatment with 1% mercaptoethanol in the presence of 1% sarcosyl followed by ultracentrifugation in a $Cs_2SO_4$ density gradient. One may suppose that the heterogeneity of p52 and p60 depends, for example, on their modification.

All other polypeptides (p43, p37, p31, p27, and p18) are resistant to mercaptoethanol treatment. The apparent amount of p18 seems to be higher in the preparation treated with sarcosyl and 2-mercaptoethanol as compared to the preparation treated with sarcosyl without mercaptoethanol (Figures 8 and 9). Taking into consideration the molecular weight, one may suggest that p18 is a histone. As was reported by some authors, part of the histones cannot be dissociated from DNA even after the treatment with high salt solutions containing

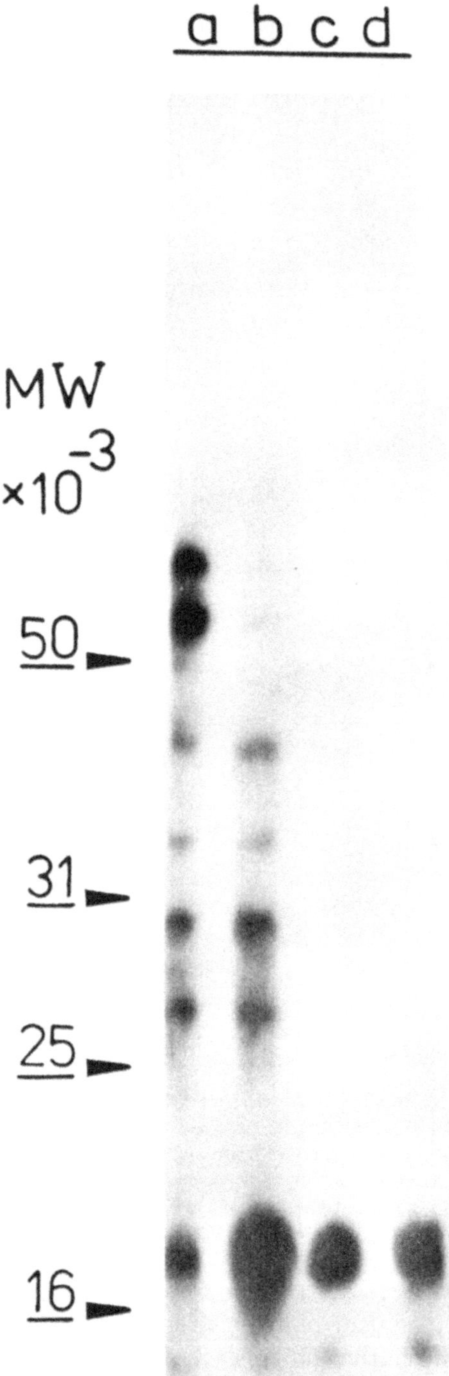


FIGURE 9. Autoradiographs of electrophoresed proteins tightly bound to matrix DNA (lanes a and b) and to soluble DNA (lanes c and d). (a) Proteins remaining bound to matrix DNA fragments after 2% sarcosyl or (b) 2% sarcosyl, 2% 2-mercaptoethanol treatment; (c) proteins remaining bound to soluble DNA fragments after 2% sarcosyl or (d) 2% sarcosyl, 2% 2-mercaptoethanol treatment.

urea.[73] To check this possibility, we performed acid-urea gel electrophoresis,[59] and found that all of the proteins did not migrate, in contrast to histones. According to this test, p18

**Table 2**
**CONTENT OF PROTEIN-BOUND DNA IN THE DNA SAMPLES FROM NUCLEAR MATRIX AND REMOVED FRACTION HAVING DIFFERENT SIZES AND OBTAINED BY DIFFERENT DISSOCIATION PROCEDURES[a]**

Binding to Nitrocellulose Filter, % of Input

| | Nuclear matrix DNA | | | | Removed DNA | | | | |
|---|---|---|---|---|---|---|---|---|---|
| | Sarcosyl prep. | | Gu·HCl prep. | | Sarcosyl prep. | | Gu·HCl prep. | | |
| Size of DNA (kb) | m-eth − | m-eth + | m-eth − | m-eth + | m-eth − | m-eth + | m-eth − | m-eth + | Sheared phage λ DNA |
| 0.5 | 46 | | 97 | | 4 | | 2 | | |
| 1.0 | 49 | 47 | 89 | 93 | 7 | 6 | 5 | 6 | 2 |
| 5.0 | 61 | 53 | 100 | 100 | 18 | 15 | 20 | 19 | 3 |
| 10.0 | 85 | | 100 | | 42 | | 36 | | |

[a] All the figures in the table are averaged from the results of three independent experiments.

is not a histone, but the final conclusion can be made only after a peptide analysis or immunological studies are done.

Our data suggest that proteins tightly associated with matrix DNA can be involved in the attachment of DNA to the nuclear skeleton. To further analyze this possibility, two questions are posed. Are these proteins associated with matrix DNA only?, and what fraction of matrix DNA is tightly associated with the proteins?

First, we studied proteins tightly bound to DNA free of nuclear residues after mild nuclease treatment and dehistonization.[65] The procedure used was exactly the same as in the case of matrix DNA. One can see (Figure 9) that the only protein component remaining attached to the DNA solubilized after sarcosyl treatment was p18 (sometimes together with p14). None of six other proteins (p60, p52, p43, p37, and p27) could be detected when equal amounts of matrix DNA and soluble DNA were compared. Hence, all of these proteins are exclusively bound to skeleton-associated DNA segments.

It is noteworthy that again only p18 is detected in the DNA after treatment with guanidine hydrochloride (not shown). Thus, the fraction of proteins loosely bound to DNA, which could be removed with sarcosyl but not with guanidine hydrochloride is complexed only with skeleton-attached DNA segments.

Second, we studied the binding of DNA obtained in a $Cs_2SO_4$ density gradient on nitrocellulose filters.[65] It is known that only those fragments of double-stranded DNA which are associated with proteins are bound to the filters in certain conditions.[67,72] The results are presented in Table 2. One can see that at least every other fragment of matrix DNA contains tightly bound proteins that cannot be detached by sarcosyl treatment. However, the method is semiquantitative and if only one polypeptide chain is associated with each DNA fragment, the filter assay would be less than 100% efficient even in a model experiment. The actual figures usually do not exceed 50%.[67] Thus, it is likely that each fragment of matrix DNA carries at least one tightly bound polypeptide resistant to sarcosyl treatment. The percentage of DNA bound to a filter does not depend on the size of DNA fragment, thus indicating that the polypeptide(s) is/are located at, or close to, the point of DNA attachment to the nuclear skeleton.

The removal of p52 and p60 by pretreatment with 2-mercaptoethanol has no effect on the results of filter binding tests. Thus, the fragments from which these proteins have been removed still contain some other protein(s). It can be p18 which is most abundant among the proteins remaining bound to matrix DNA after 2% sarcosyl-2% 2-mercaptoethanol treatment. When DNA is solubilized by guanidine hydrochloride treatment, virtually all DNA fragments are detained on a nitrocellulose filter. This result is not surprising since matrix DNA fragments isolated from guanidinium chloride contain more proteins than those solubilized with sarcosyl, as described above.

In preparations of DNA free of skeletal structures, the percentage of the material retained on a nitrocellulose filter depends on the size of fragments. It increases proportionally to an increase in the DNA size. A simple calculation shows that each 30-kb segment of DNA (or 15-kb segment if a 50% efficiency of the method is assumed) is associated with tightly bound p18 (the only abundant protein present in such preparations). Again, in accordance with the above electrophoretic data, the DNA treated with guanidine hydrochloride does not differ from that treated with sarcosyl.

It is interesting that the quantity, electrophoretic patterns, and distribution of tightly bound proteins between matrix DNA and DNA are the same in metaphase chromosomes and interphase nuclei.[65] The proteins remaining bound to the DNA solubilized by sarcosyl treatment from the chromosomal matrix are shown in Figure 10. Hence, the size of DNA attachment to the skeletal elements, which are realized through proteins tightly bound to DNA, are organized in a very similar manner in metaphase chromosomes and in interphase nuclei, thus suggesting the structural role of these attachment sites.

The final question is with which type of DNA fragments, long or short, tightly bound proteins are associated. Matrix preparations obtained after a rather intensive digestion of nuclei with micrococcal nuclease were dissociated by sarcosyl treatment, and the DNA was fractionated into long and short fragments by ultracentrifugation in a 5 to 20% sucrose gradient containing 1% sarcosyl, 20 m*M* Tris-HCl, pH 7.5, 1 *M* NaCl, 0.1 m*M* EDTA, and 0.1 m*M* PMSF. Then the proteins were labeled and analyzed as described above. All the proteins resistant to sarcosyl treatment were associated with short DNA fragments. No proteins tightly bound to long matrix DNA fragments were detected. However, since the amount of long "matrix" DNA fragments available for analysis was much lower than the amount of short "matrix" DNA fragments, we cannot now draw any final conclusion.

## IV. THE NATURE OF DNA SEGMENTS INVOLVED IN THE ATTACHMENT TO THE NUCLEAR SKELETON

The question about the properties of DNA segments involved in the attachment to the nuclear skeleton has been much debated during the last years. However, the results obtained by different authors are controversial.[4,13,20,38,49,47,60,62-64,69] Recently, we reexamined the question by separately analyzing long and short fragments of matrix DNA isolated from Ehrlich ascites carcinoma cells nuclei after partial digestion with micrococcal nuclease. First, we investigated the distribution of both matrix DNA fractions in a CsCl density gradient and the pattern of their cleavage with different restriction endonucleases. Long DNA fragments were banded in a CsCl density gradient as a sharp peak with a buoyant density of 1.69 g/cm$^3$ typical of AT-rich mouse satellite DNA,[41] being virtually the only satellite DNA of the mouse genome. Mild digestion of long "matrix" DNA fragments with EcoRII restriction endonuclease yielded a series of fragments whose sizes were multiples of 245 bp (Figure 11). Such a pattern is also typical of mouse satellite DNA.[34] Consequently, one may conclude that the large segments of skeleton-associated DNA are represented by AT-rich satellite DNA. It should be pointed out that only about 0.2% of the entire satellite DNA is protected against nuclease by the skeletal structures. Possibly this is not a random part of satellite DNA. The specific interaction of satellite DNA with the skeleton can depend on an adjacent nonsatellite sequence which is quite a minute fraction to be detected by overall analysis of our DNA preparations.

The buoyant density of short DNA fragments is characteristic of the main-band DNA (1.70 g/cm$^3$) or slightly higher.[63] No specific cuts can be detected upon treatment of short

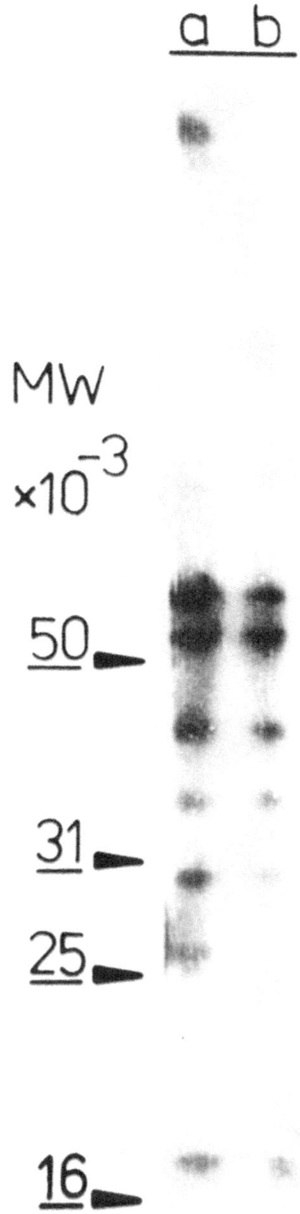


FIGURE 10. Comparison of proteins tightly bound to nuclear matrix DNA and chromosomal matrix DNA. Lane (a) Proteins associated with the nuclear matrix DNA, autoradiograph; lane (b) proteins associated with chromosomal matrix DNA, autoradiograph.

matrix DNA fragments with different restriction endonucleases (including EcoRII). Therefore, this major fraction of matrix DNA is rather heterogeneous and is not represented by satellite DNA. Similar results have been obtained earlier with short nuclease-protected DNA fragments from metaphase chromosomes.[38]

The further strategy for study of short matrix DNA fragments was to clone these fragments and then to analyze properties of many representatives from this fraction. The oligo G, oligo C tailing with an insertion into the PstI site of pBR322 plasmid was used for preparing the library of clones containing short segments of nuclease-protected, skeleton-associated DNA

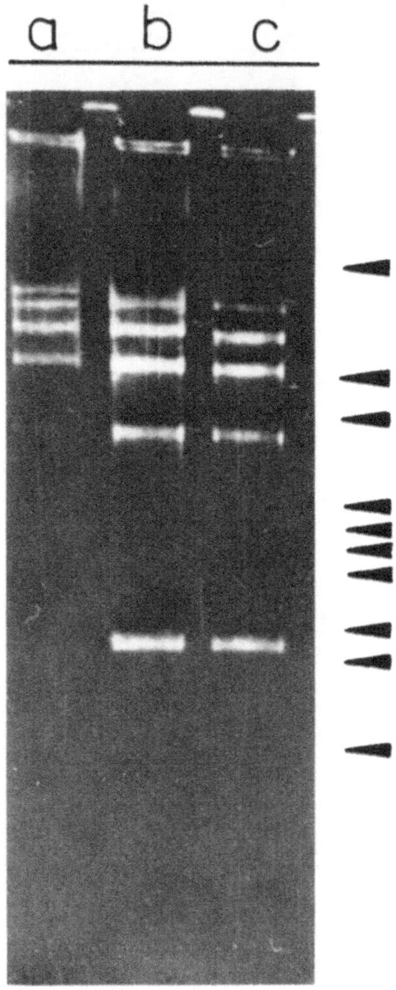


FIGURE 11. Electrophoretic distribution in an 8% acrylamide gel of EcoRII digestion products of matrix DNA from Ehrlich ascites carcinoma cell nuclei isolated in the medium lacking $Cu^{2+}$ ions. Lanes (a-c) Digestion of matrix DNA with increasing amounts of EcoRII. The arrows indicate the positions of markers (1762, 720, 540, 372, 348, 329, 300, 259, 227, and 179 bp).

from mouse Ehrlich ascites carcinoma cells. The cloned fragments were characterized in different ways. First, they were hybridized with highly labeled total mouse DNA. Some fragments did not bind, possibly due to their content of low repeated or single-copy DNA sequences. Other fragments bound measurable amounts of total DNA (0.025%) and these should correspond to repetitive DNA.

The major line of our study is to determine the DNA sequence in a number of insertions in order to find out whether some specific sequences or motives can be detected in matrix DNA. The work is in progress now, but the sequences of the first six fragments analyzed possess some features in common though they are quite different in sequences (Table 3). The major similarity is that all of them contain a number of oligopyrimidine/oligopurine stretches distributed nonrandomly between two DNA strands. This can create a specific DNA configuration recognizable by proteins involved in DNA binding to the skeleton. However, more cloned fragments should be analyzed before final conclusions are drawn.

**Table 3**

| |
|---|
| 5′-CGTCA·TTTTT·CAAGT·CGTCA·AGTGA·ATGTT·TTTTC·CTTTT·TTCCA·TGATT·TTTCA·GTTTT·TCTTG·CCATA·TTTTA |
| 5′-CCTCC·CTGGG·TACTG·CTCCA·AACCT·CCTAG·TTCTA·ACAGT·GGTGA·AGGGT·CAGGT·AGTGA·GAGTG·TGAGG·CCTAT·GAGCC·CCCCT·GTGGG·TGCCC·AGGGA·GCTAG·GGTTA·TCTCT·G |
| 5′-TTGGG·CCAGT·TTTGT·GTTGT·ATACC·TTACA·GAATT·TTCAA·AAATT·ACATT·AGAAA·TAATT·GTTTT·GTTTT·ACCCT·TTTTG·AACTT·ATTTT·TCGAG·AAAGG·GTCCT·CCTTC·CTTCC·TTT |
| 5′-CCCCC·TTTTT·GGCCA·GGCTG·CCTTA·CATGT·CTTTT·TCTTT·GTTTT·TCAGC·CCCCT·TTCTC·TTCTA·ACACA·ATGAG·CTACC·TACTT·TACAC·ACC |
| 5′-CCCCT·GGGGT·TTGTG·CTTAA·AAGCT·CACCC·TGAGG·AAGGC·AAGGG·CGCGC·ATAGG·GATCT·GGGGC·TCATT·TTTTT·GTATG·TCGCC·A |
| 5′-CCAGC·TAGAA·CAGCA·AGCTC·TGAAG·CAAAC·CCCAA·ATGTC·CCCAG·ACAGA·ATGAA·GTCAG·GTCCA·TAAAA·ACAAA·CCAAC·CGTAC·A |

## V. DISCUSSION: THE ORGANIZATION OF TWO TYPES OF SITES FOR DNA ATTACHMENT TO THE NUCLEAR SKELETON

The results obtained indicate that there are two types of DNA-nuclear skeleton associations. Attachment sites of the first type are very rare ($\sim 10^2$ per diploid nucleus) and highly resistant to the isolation procedure. Long (10 kb) stretches of satellite DNA are involved in their formation. Attachment sites of the second type are numerous ($\sim 10^5$ per diploid nucleus) but quite fragile. The DNA present within these structures is heterogeneous in sequences though seems to possess some features in common.

Recently, Lebkowski and Laemmli[45] also suggested the existence of two types of DNA attachment to the nuclear skeleton on the basis of quite different observations. They have found that, in nuclei, DNA is associated with the skeleton through two types of interactions: resistant to $Cu^{2+}$-binding substances and sensitive to them. Metaphase chromosomes have only bonds of the second type.[46] It is very likely that $Cu^{2+}$-sensitive attachments suggested by Laemmli correspond to those protecting short DNA fragments in our experiments.

One might speculate that stable attachments which do not depend on the presence of $Cu^{2+}$ ions involve the nuclear lamina. Both attachments of the first type[38] and lamina[31] seem to be absent from metaphase chromosomes. Jeppesen and Bankier[38] did not detect long DNA fragments protected against micrococcal nuclease in metaphase chromosomes extracted with a 2-*M* NaCl solution.

The nuclear lamina is a very stable component of the nuclear matrix which survives any modification in the matrix isolation procedure. On the other hand, the filaments of the inner matrix, or nucleonemas, are very sensitive to them. For example, these filaments cannot be detected in nuclear matrices isolated in a medium containing 2-mercaptoethanol.[39] They also disappear after treatment with $Cu^{2+}$-binding reagents (Laemmli, personal communication). Under these conditions, the lamina is the only visible component of the nuclear matrix. Furthermore, several authors[25,26,35,40] have demonstrated that highly repetitive DNA is associated with nuclear envelope preparations containing mainly nuclear lamina. Electron-microscopic data obtained with serial slices of interphase nuclei support the idea of chromosome attachment to the nuclear lamina.[54] The rare but stable interactions of the first type can be involved in a specific fixed arrangement of unfolded chromosomes within the interphase nucleus.

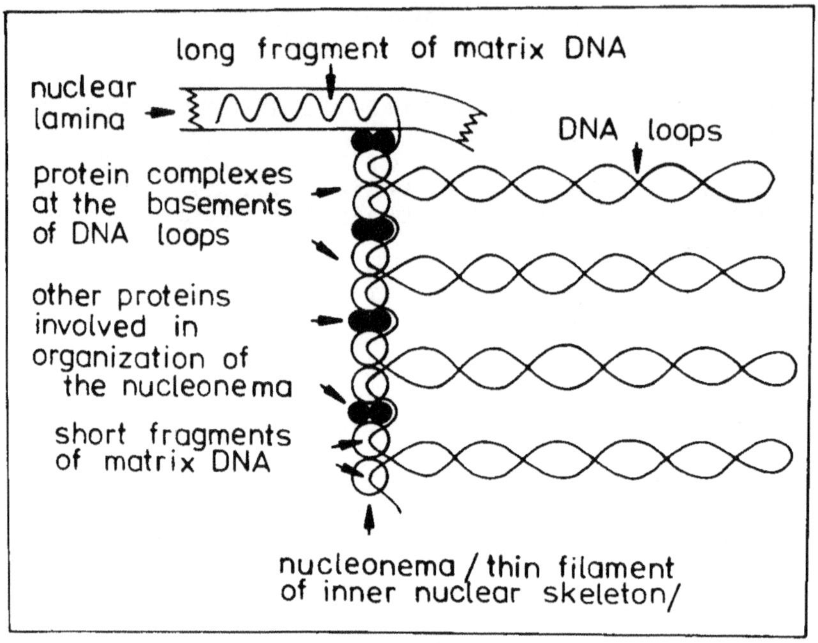


FIGURE 12. The proposed general scheme for DNA attachment to nuclear skeleton through type-I and -II attachment sites.

In our opinion, attachments of the second type must involve nucleonemas since these attachment sites and nucleonemas[39] respond in a similar manner to the treatment with 2-mercaptoethanol and other $Cu^{2+}$-binding agents. In contrast to the lamina, nucleonemas are found both in interphase nuclei and in metaphase chromosomes. It is possible to visualize the scaffold of a metaphase chromosome by means of silver staining. The material stained is located in the center of the chromosome. In an interphase nucleus, it is dispersed in the inner part of the nucleus (Laemmli, personal communication).

The interaction of DNA loops with nucleonemas is also supported indirectly by early electron microscopic observations (see Section I and Figure 1). The average distance between two attachment points calculated from our data is about 50 to 60 kb (see Table 1). This value corresponds quite well to the average size of superhelical DNA loops in histone-depleted nuclei.[5] It is likely that attachment sites of the second type are responsible for the fixation of DNA loop ends (Figure 12).

One may wonder why attachment sites of the second type are so fragile whereas proteins involved in the interactions are tightly bound to DNA. A possible explanation is that the integrity of nucleonemas depends on the interaction of some proteins other than those tightly bound to DNA. Possibly, nucleonemas should be considered as threads composed of protein complexes bound to the ends of topological loops in DNA. Being very stable, these complexes can be bound to one another and to the nuclear lamina by rather weak bonds (see Figure 12). Some $Cu^{2+}$-containing proteins are possibly involved in maintaining the integrity of nucleonemas. Recently, Lewis and Laemmli[46] have found two such proteins with molecular weights of 130 and 170 kD in the highly purified scaffolds of metaphase chromosomes in HeLa S3 cells. The interaction of these proteins with $Cu^{2+}$ ions drastically affects both the scaffold appearance and the sedimentation properties of histone-depleted metaphase chromosomes. Although the authors considered the proteins described as the only abundant components of the chromosomal scaffold, actually all the proteins tightly bound to DNA can be omitted from the analysis since the samples were not treated with DNAase before electrophoresis.

According to our hypothesis, the nucleonema must consist of repeating units. Recently, Engelhardt et al.[22] obtained some electron microscopic evidence for this suggestion. The formation of nucleonemas from rather independent stable subunits joined to one another by bonds sensitive to $Cu^{2+}$-containing reagents appears, at the moment, as a most attractive explanation for the lability of nucleonemas.

As we found in early electron microscopic studies, nucleonemas are connected to the nuclear envelope in many places.[27] It is possible that the contacts of chromosomes with the nuclear lamina are realized not only through attachment sites of the first type but also through independent contacts of nucleonemas with the lamina.

The fine organization of nucleonemas (involving DNA-protein interactions) is not yet clear. Possibly, proteins most tightly bound to DNA (i.e., p18, p27, p31, p37, and p43) play a key role. Among them, p18 is dominant. The latter is associated not only with attached DNA fragments but also with DNA fragments removed from the matrix (one polypeptide per 15- to 30-kb segment of DNA, on the average).

It is possible that the sites of p18 attachment to DNA are potential sites of DNA-nuclear skeleton interactions. Normally, each fourth of them is converted to the real one by an addition of several other nonhistone proteins, in particular, of p52 and p60. The formation of disulfide bonds can also be involved. It is interesting that p18 iodination increases after cleavage of p52 and p60 by 2-mercaptoethanol-sarcosyl treatment (Figure 9). If the above suggestion is true, then a number of potential attachment sites should exist in chromatin, which suggests the possibilities for total and local variations in chromatin condensation.

In experimental conditions, we could induce such variations by preextracting the nuclei with 0.35 *M* NaCl. After this treatment, the average distance between two apparent attachment sites was diminished twofold.[66] Recently, Nordelli and coauthors[58] described variations in the size of DNA loops in vivo looking at cells at different development stages. The size of the loops increased at later stages.

DNA attachments to the nuclear skeleton can be heterogeneous. We have found that about 10% of p52 and p60 are not removed by 2-mercaptoethanol-sarcosyl treatment.[65] At the same time, about one fifth of the standard number of short DNA fragments can be recovered in the matrix obtained from nuclei preincubated in a buffer solution containing 1% mercaptoethanol before nuclease digestion and high salt extraction. Recently, Hartwig[33] has shown that 2-mercaptoethanol treatment drastically increases the size of DNA topological loops, which is in agreement with our results.

Many questions concerning the organization of sites for DNA attachment to the nuclear skeleton and their functional significance remain open. It is not clear whether these sites include certain functional elements of the genome or they are randomly distributed. It was suggested by Dingman[21] that the attachment sites could correspond to replication origins. However, no experimental proof has yet been obtained for this suggestion besides the above-mentioned correlation between the sizes of DNA loops and replicons at different stages of development.[58]

Along with constant attachment sites, a number of temporary associations between DNA and the skeleton can appear in the course of chromosome functioning. For example, replicative forks and transcriptional complexes seem to be bound to the nuclear skeleton. The organization of such "functional" attachment sites has not yet been studied at all.

Finally, it is possible that some components other than proteins and DNA can be involved in the organization of DNA attachment sites. These can be sphingomyelins (some preliminary evidence has already been obtained in our laboratory) and probably polysaccharides whose content in the nuclear matrix is rather high.

## REFERENCES


1. **Aaronson, R. P. and Blobel, G.,** On the attachment of the nuclear pore complex, *J. Cell Biol.*, 62, 746, 1974.
2. **Aaronson, R. P. and Blobel, G.,** Isolation of nuclear pore complexes in association with a lamina, *Proc. Natl. Acad. Sci. U.S.A.*, 72, 1007, 1975.
3. **Adolph, K. W., Cheng, S. M., and Laemmli, U. K.,** Role of nonhistone proteins in metaphase chromosome structure, *Cell*, 12, 805, 1977.
4. **Basler, J., Hastie, N. D., Pietras, D., Matsui, S., Sandberg, A. A., and Berezney, R.,** Hybridization of nuclear matrix attached deoxyribonucleic acid fragments, *Biochemistry*, 20, 6921, 1981.
5. **Benyajati, C. and Worcel, A.,** Isolation, characterization and structure of the folded interphase genome of Drosophila melanogaster, *Cell*, 9, 393, 1976.
6. **Berezney, R. and Coffey, D. S.,** Identification of a nuclear structural protein network in rat liver nuclei, *Fed. Proc. Fed. Am. Soc. Exp. Biol.*, 33, 1395, 1974.
7. **Berezney, R. and Coffey, D. S.,** Identification of a nuclear protein matrix, *Biochem. Biophys. Res. Commun.*, 60, 1410, 1974.
8. **Berezney, R. and Coffey, D. S.,** Nuclear protein matrix: association with newly synthesized DNA, *Science*, 189, 291, 1975.
9. **Berezney, R. and Coffey, D. S.,** Nuclear matrix. Isolation and characterization of a framework structure from rat liver nuclei, *J. Cell Biol.*, 73, 616, 1977.
10. **Berezney, R.,** Effect of protease inhibitors on matrix proteins and the association of replicating DNA, *Exp. Cell Res.*, 123, 411, 1979.
11. **Berezney, R.,** The nuclear matrix: a structural milieu for the intranuclear attachment and replication of eucaryotic DNA, in *International Cell Biology 1980-1981*, Schweiger, L., Ed., Springer-Verlag, Basel, 1981, 214.
12. **Berezney, R. and Buchholtz, L. A.,** Dynamic association of replicating DNA fragments with the nuclear matrix of regenerating liver, *Exp. Cell Res.*, 132, 1, 1981.
13. **Bowen, B. C.,** DNA fragments associated with chromosome scaffolds, *Nucleic Acids Res.*, 9, 5093, 1981.
14. **Bryan, S. E., Vizard, D. L., Beary, D. A., Labiche, R. A., and Hardy, K. L.,** Partitioning of zinc and copper within subnuclear nucleoprotein particles, *Nucleic Acids Res.*, 9, 5811, 1981.
15. **Busch, H. and Smetana, K.,** The nuclear ribonucleoprotein network and the nuclear residue, in *The Nucleolus*, Busch H. and Smetana, K., Eds., Academic Press, New York, 1970, 361.
16. **Capesius, I., Krauth, W., and Werner, D.,** Proteinase K-resistant and alkaly-stable bound proteins in higher plant DNA, *FEBS Lett.*, 110, 184, 1980.
17. **Comings, D. E. and Okada, T. A.,** Nuclear proteins. III. The fibrillar nature of the nuclear matrix, *Exp. Cell Res.*, 103, 341, 1976.
18. **Cook, P. R. and Brazell, I. A.,** Supercoils in human DNA, *J. Cell Sci.*, 19, 261, 1975.
19. **Cook, P. R., Brazell, I. A., and Jost, E.,** Characterization of nuclear structures containing superhelical DNA, *J. Cell Sci.*, 22, 203, 1976.
20. **Cook, P. R. and Brazell, I. A.,** Mapping sequences in loops of nuclear DNA by their progressive detachment from the nuclear cage, *Nucleic Acids Res.*, 8, 2895, 1980.
21. **Dingman, C. W.,** Bidirectional chromosome replication: some topological considerations, *J. Theor. Biol.*, 43, 187, 1974.
22. **Engelhardt, P., Plagens, U., Zbarsky, I. B., and Filatova, L. S.,** Granules 25—30 nm in diameter — a basic constituent of the nuclear matrix, nuclear envelope and chromosome scaffold, *Proc. Natl. Acad. Sci. U.S.A.*, in press.
23. **Faiferman, I. and Pogo, A. O.,** Isolation of a nuclear ribonucleoprotein network that contains heterogeneous RNA and is bound to the nuclear envelope, *Biochemistry*, 14, 3808, 1975.
24. **Fisher, P. A., Berrios, M., and Blobel, G.,** Isolation and characterization of a proteinaceous subnuclear fraction composed of nuclear matrix, peripheral lamina and nuclear pore complexes from embryos of Drosophila melanogaster, *J. Cell Biol.*, 92, 674, 1982.
25. **Franke, W. W. and Krien, S.,** Nuclear membrane attached DNA is enriched in repetitive sequences, *Naturwissenschaften*, 59, 37, 1972.
26. **Franke, W. W., Deumbling, B., Zentgraf, H., Falk, H., and Rae, P. M. M.,** Nuclear membranes from mammalian liver. Characterization of membrane attached DNA, *Exp. Cell Res.*, 81, 365, 1973.
27. **Georgiev, G. P. and Chentsov, Yu. S.,** On the structure of cell nuclei: experimental electron microscopic investigation of isolated nuclei, *Proc. Natl. Acad. Sci. U.S.S.R.*, 132, 199, 1960.
28. **Georgiev, G. P., Nedospasov, S. A., and Bakayev, V. V.,** Supranucleosomal levels of chromatin organization, in *The Cell Nucleus*, Vol. 6 (Part C), Busch, H., Ed., Academic Press, New York, 1978, 4.
29. **Gerace, L., Blum, A., and Blobel, G.,** Immunocytochemical localization of the major polypeptides of the nuclear pore complex-lamina fraction. Interphase and mitotic distribution, *J. Cell Biol.*, 79, 546, 1978.

30. **Gerace, L. and Blobel, G.,** The major lamina polypeptides comprise a peripheral nuclear shell, *Eur. J. Cell Biol.*, 22(1), 120, 1980.
31. **Gerace, L. and Blobel, G.,** The nuclear envelope lamina is reversibly depolymerized during mitosis, *Cell*, 19, 277, 1980.
32. **Glover, J. S., Salter, D. N., and Shepperd, B. P.,** A study of some factors that influence the iodination of ox insulin, *Biochem. J.*, 103, 120, 1967.
33. **Hartwig, M.,** Organization of mammalian chromosomal DNA: supercoiled and folded circular DNA subunits from interphase cell nuclei, *Acta Biol. Med. Ger.*, 37, 421, 1978.
34. **Hörz, W. and Zachau, H. G.,** Characterization of distinct segments in mouse satellite DNA by restriction nucleases, *Eur. J. Biochem.*, 73, 383, 1977.
35. **Humbert, J.,** Qualitative and quantitative studies of the nuclear membrane complex isolated from eukaryotic cells, *Experimentia*, 30, 704, 1974.
36. **Ide, I., Nakane, M., Anzai, K., and Andoh, T.,** Supercoiled DNA folded by nonhistone proteins in cultured mammalian cells, *Nature (London)*, 258, 445, 1975.
37. **Jackson, D. A., McCready, S. J., and Cook, P. R.,** RNA is synthesized at the nuclear cage, *Nature (London)*, 292, 552, 1981.
38. **Jeppesen, P. G. N. and Bankier, A. T.,** A partial characterization of DNA fragments protected from nuclease degradation in histone-depleted metaphase chromosomes of the Chinese hamster, *Nucleic Acids Res.*, 7, 49, 1979.
39. **Kaufmann, S. H., Coffey, D. S., and Shaper, J. H.,** Considerations in the isolation of rat liver nuclear matrix, nuclear envelope and pore complex lamina, *Exp. Cell Res.*, 132, 105, 1981.
40. **Kay, R. R. and Johnston, I. R.,** Repetitive DNA associated with rodent liver nuclear envelopes, *Experimentia*, 30, 472, 1974.
41. **Kit, S.,** Equilibrium sedimentation in density gradients of DNA preparations from animal tissues, *J. Mol. Biol.*, 3, 711, 1961.
42. **Krauth, W. and Werner, D.,** Analysis of the most tightly bound proteins in eukaryotic DNA, *Biochim. Biophys. Acta*, 564, 390, 1979.
43. **Laemmli, U. K.,** Cleavage of structural proteins during the assembly of the head of bacteriophage T4, *Nature (London)*, 227, 680, 1970.
44. **Lebkowski, J. S. and Laemmli, U. K.,** Evidence for two levels of DNA folding in histone-depleted HeLa interphase nuclei, *J. Mol. Biol.*, 156, 309, 1982.
45. **Lebkowski, J. S. and Laemmli, U. K.,** Non-histone proteins and long-range organization of HeLa interphase DNA, *J. Mol. Biol.*, 156, 325, 1982.
46. **Lewis, C. D. and Laemmli, U. K.,** Higher order metaphase chromosome structure: evidence for metalloprotein interactions, *Cell*, 29, 171, 1982.
47. **Long, B. H., Huang, C.-Y., and Pogo, A. O.,** Isolation and characterization of the nuclear matrix in Friend erythroleukemia cells: chromatin and hnRNA interactions with the nuclear matrix, *Cell*, 18, 1079, 1979.
48. **Mariman, E. C. M., van Eckelen, C. A. G., Reinders, R. J., Berns, A. J. M., and van Venrooij, W. J.,** Adenoviral heterogeneous nuclear RNA is associated with the host nuclear matrix during splicing, *J. Mol. Biol.*, 154, 103, 1982.
49. **Matsumoto, L. H.,** Enrichment of satellite DNA on the nuclear matrix of bovine cells, *Nature (London)*, 294, 481, 1980.
50. **McCready, S. J., Akrigg, A., and Cook, P. R.,** Electronmicroscopy of intact nuclear DNA from human cells, *J. Cell Sci.*, 39, 53, 1979.
51. **McCready, J. J., Godwin, J., Mason, D. W., Brazell, I. A., and Cook, P. R.,** DNA is replicated at the nuclear cage, *J. Cell Sci.*, 46, 365, 1980.
52. **Mirsky, A. E. and Pollister, A. W.,** Chromosin, a deoxyribonucleoprotein complex of the cell nucleus, *J. Gen. Physiol.*, 30, 117, 1946.
53. **Mirsky, A. E. and Ris, H.,** The chemical composition of isolated chromosomes, *J. Gen. Physiol.*, 31, 7, 1947.
54. **Murray, A. B. and Davies, H. G.,** Three-dimensional reconstruction of the chromatin bodies in the nuclei of mature erythrocytes from the newt triturus cristatus: the number of nuclear envelope-attachment sites, *J. Cell Sci.*, 35, 59, 1979.
55. **Nakane, M., Ide, T., Anzai, K., Ohara, S., and Andoh, T.,** Supercoiled DNA folded by non-histone proteins in cultured mouse carcinoma cells, *J. Biochem.*, 84, 145, 1978.
56. **Narayan, K. S., Steele, W. J., Smetana, K., and Busch, H.,** Ultrastructural aspects of the ribonucleoprotein network in nuclei of Walker tumor and rat livers, *Exp. Cell Res.*, 46, 65, 1967.
57. **Nelkin, B. D., Pardoll, D. M., and Vogelstein, B.,** Localization of SV40 genes within supercoiled loop domains, *Nucleic Acids Res.*, 8, 5623, 1980.
58. **Nordelli, M., Michelli, G., Carri, M. T., and Marilley, M.,** Supercoiled loop organization of the Xenopus laevis genome, *Biology of the Cell*, Abstr. 1st Eur. Congr. Cell Biology, 45, 108, 1982.

59. **Panyim, S. and Chalkley, R.,** The heterogeneity of histones. I. A quantitative analysis of calf histones in very long polyacrylamide gels, *Biochemistry*, 8, 3972, 1969.
60. **Pardoll, D. and Vogelstein, B.,** Sequence analysis of nuclear matrix associated DNA from rat liver, *Exp. Cell Res.*, 128, 466, 1980.
61. **Paulson, J. R. and Laemmli, U. K.,** The structure of histone-depleted metaphase chromosomes, *Cell*, 12, 817, 1977.
62. **Razin, S. V., Mantieva, V. L., and Georgiev, G. P.,** DNA adjacent to attachment points of deoxyribonucleoprotein fibril to chromosomal axial structure is enriched in reiterated base sequences, *Nucleic Acids Res.*, 5, 4737, 1978.
63. **Razin, S. V., Mantieva, V. L., and Georgiev, G. P.,** The similarity of DNA sequences remaining bound to scaffold upon nuclease treatment of interphase nuclei and metaphase chromosomes, *Nucleic Acids Res.*, 7, 1713, 1979.
64. **Razin, S. V., Mantieva, V. L., Nedospasov, S. A., and Bakayev, V. V.,** Higher orders of DNA packing in chromosomes, *Eur. J. Cell Biol.*, 22, 85, 1980.
65. **Razin, S. V., Chernokhvostov, V. V., Roodyn, A. V., Zbarsky, I. B., and Georgiev, G. P.,** Proteins tightly bound to DNA in the regions of DNA attachment to the skeletal structures of interphase nuclei and metaphase chromosomes, *Cell*, 27, 65, 1981.
66. **Razin, S. V., Yarovaya, O. V., and Georgiev, G. P.,** Two types of DNA attachments to nuclear skeleton in Ehrlich ascites carcinoma cells, *J. Cell Biochem.*, submitted.
67. **Riggs, A. D., Suzuki, H., and Bourgeois, S.,** Lac repressor-operator interaction. I. Equilibrium studies, *J. Mol. Biol.*, 48, 67, 1970.
68. **Riley, D. E. and Keller, J. M.,** The polypeptide composition and ultrastructure of nuclear ghosts isolated from mammalian cells, *Biochim. Biophys. Acta*, 444, 899, 1976.
69. **Robinson, S. I., Nelkin, B. D., and Vogelstein, B.,** The ovalbumin gene is associated with the nuclear matrix of chicken oviduct cells, *Cell*, 28, 99, 1982.
70. **Shaper, J. H., Pardoll, D. M., Kaufmann, S. H., Barrack, E. R., Vogelstein, B., and Coffey, D. S.,** The relationship of nuclear matrix to cellular structure and function, *Enzyme Regulation*, 17, 213, 1979.
71. **Smetana, K., Steele, W. J., and Busch, H.,** A nuclear ribonucleoprotein network, *Exp. Cell Res.*, 31, 198, 1963.
72. **Thomas, T. L. and Patel, G. L.,** Optimal conditions and specificity of interaction of a distinct class of nonhistone chromosomal proteins with DNA, *Biochemistry*, 15, 1481, 1976.
73. **Uschewa, A., Avramova, Z., and Tsanev, R.,** Tightly bound somatic histones in mature ram sperm nuclei, *FEBS Lett.*, 138, 50, 1982
74. **Vortisch, V., Kroneck, P., and Hemmerich, P.,** Model studies on the coordination of copper in enzymes. IV. Structure and stability of cuprous complexes with sulfur-containing ligands, *J. Am. Chem. Soc.*, 98, 2821, 1976.
75. **Wanka, F., Mullenders, L. H. F., Bekers, A. G. M., Pennings, L. J., Aelen, J. M. A., and Eygensteyn, J.,** Association of nuclear DNA with a rapidly sedimenting structure, *Biochem. Biophys. Res. Commun.*, 74, 739, 1977.
76. **Werner, D. and Petzelt, C.,** Alkali-stably bound proteins in eukaryotic and prokaryotic DNAs show common characteristics, *J. Mol. Biol.*, 150, 297, 1981.
77. **Zbarsky, I. B. and Debov, S. S.,** Proteins of cell nuclei, *Dokl. Akad. Nauk SSSR*, 63, 795, 1948.
78. **Zbarsky, I. B. and Georgiev, G. P.,** Cytological characteristics of protein and nucleoprotein fractions of cell nuclei, *Biochim. Biophys. Acta*, 32, 301, 1959.
79. **Zbarsky, I. B., Dmitrieva, N. P., and Yermolaeva, L. P.,** On the structure of tumour cell nuclei, *Exp. Cell Res.*, 27, 573, 1962.

Chapter 7

# THE NUCLEAR MATRIX AND THE ORGANIZATION OF NUCLEAR DNA

**Bert Vogelstein, Donald Small, Sabina Robinson, and Barry Nelkin**

## TABLE OF CONTENTS

## I. INTRODUCTION

Understanding the way in which the immense amount of DNA contained in a eukaryotic nucleus is organized is one of the greatest challenges in biology. Structurally, this organization is responsible for packaging a huge amount of a highly charged polymer into a relatively small nuclear volume. Functionally, this organization must ensure a mechanism for the orderly progress of DNA replication in each cell cycle and for the highly orchestrated expression of specific genes throughout development.

Three hierarchical levels of nuclear DNA organization have so far been discovered. The first two, those of nucleosomes[1,2] and supernucleosomal clusters[3-6] are maintained largely by histones. The third level of organization, topologically constrained domains of DNA (loops), is maintained by nonhistone proteins, and therefore is a suitable subject for this volume.

## II. THE CONCEPT OF SUPERCOILED LOOPS

The organization of DNA into topologically constrained domains seems to be a ubiquitous feature of the organization of the eukaryotic genome. The first demonstrations of such loops were made by Callan and Gall, who showed that DNA is the continuous component of the loops in lampbrush chromosomes.[7,8] Later, it was shown by sedimentation studies and electron microscopy that if the histones were removed from the nucleus or from chromosomes, the DNA did not simply unfold and disperse. Instead, the DNA seemed to be organized into discrete domains anchored to a residual nuclear skeleton (also called matrix, scaffold, or cage).[9-17] This domain concept was strongly supported by studies of intact nuclei, in which the rate of digestion and extraction of chromatin upon nuclease digestion was carefully measured.[18]

Our studies began with the development of a method for conveniently visualizing these topologically constrained loops of DNA in interphase nuclei. The method was of critical importance for furthering our studies because it allowed us to morphologically follow the structures during various experimental manipulations. The method was as follows. Cells grown on coverslips were permeabilized with a detergent-containing solution; nuclei remained attached to the coverslips through their connections to cytoskeletal elements.[19-21] The nuclei were then extracted while still on the coverslip, with a buffer containing 2 *M* NaCl.[22] This treatment removes greater than 98% of the histones and most other nuclear proteins (see References 23 to 25 and our unpublished data). Greater than 99% of the nuclear DNA remains with these structures, and can be visualized by staining with a DNA-binding fluorochrome. After relaxation of supercoiling, the loops could be seen to form a halo surrounding the nuclear matrix (see below). The halo contained more than 80% of the total nuclear DNA; the remainder was within the confines of the matrix. The diameter of the DNA halo surrounding the nuclear matrix was a measure of the relative superhelical density of the DNA loops that comprised the halo. This has been shown in the following ways.

### A. Titration with Ethidium Bromide in Isotonic Buffer

As shown in Figure 1, the diameter of the DNA halo is strikingly dependent on the ethidium bromide concentration. At low concentrations of ethidium bromide, there is a small halo formed by the negatively supercoiled DNA loops. Ethidium bromide unwinds negatively supercoiled loops and as the ethidium bromide concentration is raised, the loops unwind, forming a larger halo. At an ethidium bromide concentration of 0.4 to 0.6 μg/mℓ in isotonic buffer, the loops are fully relaxed and the halo assumes its maximum size. At higher ethidium bromide concentrations, the ethidium bromide continues to unwind the DNA, resulting in supercoiling in the opposite (positive) direction. The diameter of the DNA halo then progressively diminishes in size.

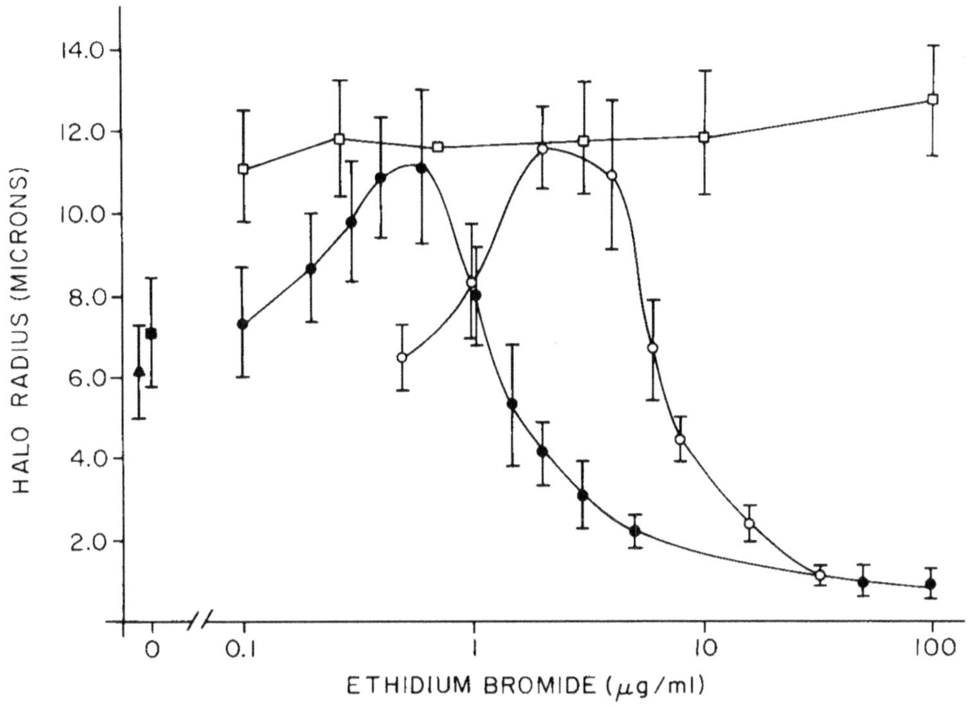


FIGURE 1. Dependence of the halo diameter on ionic conditions and ethidium bromide concentration. Trypsin dissociated 3T3 cells were placed in 35-mm Lux® plastic dishes to a density of 1 to 2 × $10^4$ cells per dish. Dishes were preferred over coverslips because of the slight drying that sometimes occurred with coverslips during the eventual transfer of coverslips to slides for microscopy. 16 to 24 hr after plating, the medium was removed from the dish and the cells washed twice at 2°C with isotonic (IS) buffer (100 m*M* NaCl, 50 m*M* KCl, 2 m*M* EDTA, 20 m*M* Tris, pH 7.05). This and all subsequent treatment of cells was performed with 3 mℓ of the appropriate buffer. Cells were then detergent treated by incubation in isotonic buffer containing 0.5% NP-40 (Bethesda Research Laboratories, Bethesda, Md.) at 2° for 1 min followed by a second incubation for 4 min in the same solution. The treated nuclei which remained attached to the dishes were washed in isotonic buffer and then extracted at 2°C in high salt (HS) buffer (2 *M* NaCl, 2 m*M* EDTA, 20 m*M* Tris, pH 7.05) twice for 2.5 min each. The nuclei were washed with IS buffer at 2°C, then incubated at room temperature in IS buffer containing 10 m*M* $MgCl_2$ plus 10 μg/mℓ DAPI or IS buffer containing various concentrations of ethidium bromide. After 1 min, this solution was removed and 0.05 mℓ of the same solution containing DAPI or ethidium bromide added to the dish. A coverslip was placed in the dish and the nuclei immediately observed. At no stage were the nuclear preparations allowed to dry. The halo diameters of 25 nuclei from each dish were measured. To do this, a nucleus was positioned in the middle of the field under phase optics; fluorescence illumination was then initiated and the measurement made within the first 2 sec of observation with the aid of an ocular micrometer using a Nikon® Labophot microscope with a G filter (for ethidium bromide) or a U filter (for DAPI). The measurements were performed in a blind fashion in that the observer making the measurements was not aware of the way in which the nuclei had been treated; 25 structures could be measured in 5 to 6 min. (○) HS buffer with indicated concentrations of ethidium bromide; (■) IS buffer plus 10 μg/mℓ DAPI; (▲) IS buffer plus 10 m*M* $MgCl_2$ plus 10 μg/mℓ DAPI; (●) IS buffer with indicated concentration of ethidium bromide; and (□) IS buffer with indicated concentrations of ethidium bromide after exposure of the structures to the ultraviolet light source of the microscope for 30 sec. Halo radius was measured from the edge of the matrix to the edge of the halo. Means and standard deviations (25 structures per point) are shown. Except for □, all of the observations were made within 2 sec after the initiation of ultraviolet illumination.

The halo can also be visualized by staining with 4,6-diamidino-2-phenylindole (DAPI), an agent that binds specifically to DNA, but does not intercalate into it (or unwind it) as does ethidium bromide.[26] As expected, the halo diameter is insensitive to the DAPI concentration. The diameter of the DAPI stained halo is very similar to the diameter of the ethidium bromide treated halo at low ethidium bromide concentrations (≤0.1 μg/mℓ). The size of the halo measured by DAPI reflects the negative supercoiling of nuclear DNA once the histones have been removed. Note in Figure 1 that the size of this halo is considerably larger (7.0 ± 1.3 μm) than the size of the halo produced by the positively supercoiled DNA loops induced by high concentrations of ethidium bromide (1.0 ± 0.2 μm). This is consistent with the fact that high concentrations of ethidium bromide can induce a higher superhelical density (in the positive sense) than exists in nuclear DNA after removal of the histones.

### B. Response to Ionic Environment

The data discussed above were generated by measuring the response of structures to ethidium bromide in a buffer containing approximately 150 m*M* monovalent salts. Also included in Figure 1 is an ethidium bromide titration performed in the high salt buffer (2 *M* NaCl) in which the structures were originally prepared. This curve is similar to the one reported previously for matrix-halo structures prepared in a similar fashion.[22] Comparison of the data at the two salt concentrations shows that the curve is shifted to the left at lower salt concentrations; the concentration of ethidium bromide required to fully relax the DNA loops is 0.4 to 0.6 μg/mℓ in the isotonic buffer, whereas it is 2 to 4 μg/mℓ in the presence of 2 *M* NaCl. There are two reasons for this shift. First, the affinity constant of ethidium bromide for DNA is approximately fivefold higher in the isotonic buffer as compared to the 2 *M* NaCl buffer.[27] Quantitatively, this effect is the more important. Second, the superhelical density of DNA varies with ionic strength (DNA is more negatively supercoiled at higher ionic strengths).[28] Hence, the higher ethidium bromide concentration required to maximally extend the halo of DNA loops in the higher ionic strength buffer is consistent with the known physicochemical properties of ethidium bromide and topologically constrained DNA.

Magnesium has also been shown to affect the superhelical density of topologically constrained DNA.[28] In an experiment in which 100 matrix-halo structures stained with DAPI were examined with or without $MgCl_2$, a statistically significant decrease in halo diameter was found in the presence of magnesium (Figure 1).

### C. Response to Nicking

Supercoiling cannot exist in nicked DNA. Consequently, if the DNA in the matrix-halo structures is nicked by exposure to low concentrations of nucleases or exposure to ultraviolet light, the nicked DNA loops form a halo which is insensitive to the concentration of ethidium bromide. In Figure 1, the data are shown for nicking with ultraviolet light; a similar curve was generated after treatment with DNAase I. The size of this halo was similar to that formed by maximal relaxation of the loops induced by titration with ethidium bromide (Figure 1). If nuclease treatment is more extensive, the halo is gradually cleaved, leaving the residual nuclear matrix.[22]

### D. Response to Topoisomerase I

When the matrix-halo structures are treated with a highly purified preparation of topoisomerase I from HeLa cells, (generously provided by Dr. L. F. Liu, Johns Hopkins University School of Medicine) the loops become relaxed and the diameter of the DAPI-stained halo increases from 7.4 to 11.6 μm (Figure 2, Table 1). The size of the halo formed is similar to that of the halo produced by nicking with nucleases (Table 1). However, relaxation induced by the topoisomerase can easily be distinguished from relaxation induced by nicking with nucleases. In the former case, a high concentration of ethidium bromide is observed to positively supercoil the relaxed DNA loops, resulting in a very small halo (Table 1). After treatment with DNAase I, however, the halo diameter is still large when stained with a high concentration of ethidium bromide (Table 1), because nicked loops cannot be positively supercoiled.

The results noted above show that by several criteria (response to ethidium bromide, ionic conditions, ultraviolet light, nucleases, and topoisomerases) the halo surrounding the residual nuclear matrix behaves as expected for a structure composed of topologically constrained

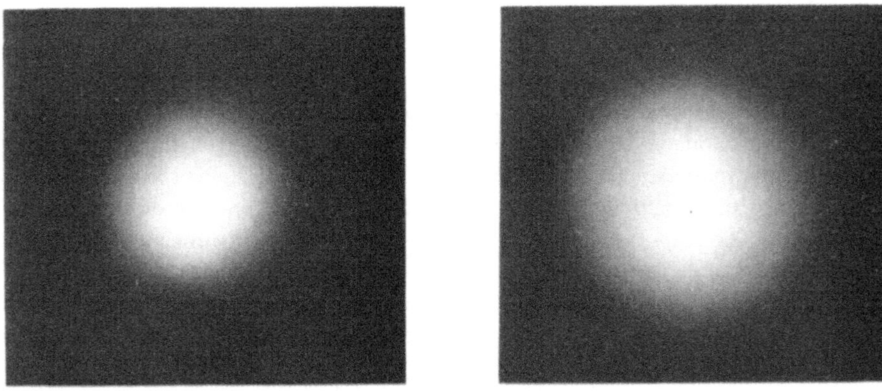

FIGURE 2. Matrix-halo structures before and after relaxation by topoisomerase I. Left: matrix-halo structures incubated in isotonic buffer for 5 min at 37°C, then stained with DAPI. Right: matrix-halo structures incubated in isotonic buffer for 5 min at 37°C with topoisomerase I, then stained with DAPI.

**Table 1**
**THE EFFECT OF EXOGENOUSLY ADDED TOPOISOMERASE I AND DNAase I ON THE HALO DIAMETER**

| | Halo radius (μm)[a] | |
|---|---|---|
| **Enzyme incubated with matrix-halo structures** | **Structures stained with DAPI after incubation with enzyme** | **Structures treated with EBr after incubation with enzyme** |
| None[b] | 7.4 ± 1.6 | 1.0 ± 0.2 |
| Topoisomerase I[c] | 11.6 ± 2.4 | 0.9 ± 0.2 |
| DNAase I[d] | 11.9 ± 1.8 | 12.2 ± 1.4 |

[b] Matrix-halo structures were prepared as described in the legend to Figure 1. They were then incubated in isotonic buffer at 37°C for 5 min, and then stained with DAPI (10 μg/mℓ) in isotonic buffer containing 10 m*M* $MgCl_2$, or with ethidium bromide (100 μg/mℓ) in isotonic buffer, at room temperature, and immediately measured.

[c] Matrix-halo structures were incubated in isotonic buffer containing HeLa cell topoisomerase I (100 U/mℓ) for 5 min at 37°C and then stained with DAPI or treated with ethidium bromide.

[d] Matrix-halo structures were incubated in isotonic buffer containing 10 m*M* $MgCl_2$ and DNAase I (0.1 μg/mℓ) for 5 min at 37°C and then stained with DAPI or treated with ethidium bromide as above.

DNA loops. Several observations indicated that the loops actually are anchored to protein components of the nuclear matrix in a specific way, rather than simply representing the aggregation of long chains of DNA: (1) protein disruptive agents (proteases, ionic detergents, urea, guanidine) immediately disrupt the structures so that the DNA is completely dispersed. (2) In contrast to the above, DNA endonucleases do not disrupt the matrix-halo structures and disperse the DNA, as would be expected if the DNA were simply entangled. Low levels of endonuclease treatment relax the DNA loops, resulting in a stable halo insensitive to ethidium bromide. Further nuclease treatment results in a somewhat larger halo because of linearization of loops that are still anchored to the matrix through one end. With still further nuclease treatment, the halo gradually decreases in size and intensity until only a small halo remains. (3) The halo diameter is remarkably constant from nucleus to nucleus in a given cell type and the diameter is invariant through the cell cycle (even though the amount of DNA per nucleus increases by twofold from $G_1$ to $G_2$). (4) The matrix-halo structures are invariant features of all active cells (i.e., capable of entering S phase or expressing genes) so far examined. The cells examined include those from insects,[29] slime molds,[30] rodents,[22,31] birds,[32] humans,[33] and plants.[34] (5) There is no correlation of halo diameter with C-value in these various nuclei, as one might expect if the halo were the result of aggregation induced by high DNA concentrations. In fact, there is a very strong correlation between halo diameter and replicon size,[33,34] which is difficult to easily explain if the halo represents nonspecific DNA aggregation.

Equipped with a way to microscopically analyze the status of DNA loops, we set out to determine the organization of DNA with respect to these loops. We were intrigued with the idea that the loops represented functional units of DNA organization, and our initial studies have been concerned with determining how the DNA organization in loops is affected by the processes of replication and gene expression.

## III. REPLICATION

### A. Experimental Results

Autoradiographic experiments revealed a surprising pattern of nascent DNA with respect to DNA loops.[22] Long labeling periods with [$^3$H]-thymidine, which uniformly labeled the DNA, showed that most (80%) of the DNA in a nucleus was located in the halo region of matrix-halo structures. Short pulses of [$^3$H]-thymidine, which label only nascent DNA, showed a strikingly different pattern: almost all of the labeled DNA was localized in the central matrix region. The autoradiographic grains gradually moved out to the halo region in pulse-chase experiments. A stable distribution was obtained after a 45-min chase period — approximately the time required to complete a replicon (Figure 3). Since the cells used were in logarithmic growth (unsynchronized), these results suggested that DNA replication complexes are located on the nuclear matrix and that loops of DNA are reeled through the complexes as they are replicated. Electron microscopic autoradiography of thin sections showed that the nascent DNA was distributed throughout the matrix in a pattern similar to that of the unextracted nucleus.[35] These autoradiographic results were supported by biochemical experiments wherein the loops of DNA were gradually and progressively cleaved from their matrix anchorage points with DNAase I.[35] The DNA at the base of the loops was markedly enriched in newly replicated DNA (Figure 4). This nascent DNA could be removed from its matrix attachment points with a brief chase protocol, as in the autoradiographic experiments. The pattern of replication with respect to DNA loops was found to be a general one among eukaryotes, because studies with *Physarum polycephalum* have given results equivalent to those with mammalian cells.[30]

Numerous control experiments have been done to diminish the possibility that nascent DNA is artifactually binding to the nuclear matrix during the extraction procedures by virtue of features unique to replicating DNA. Reconstruction experiments have demonstrated no preferential affinity of matrix components for single-stranded DNA, or nascent DNA, or replicating chromatin, or nascent DNA complexed with matrix proteins, when added at any (or all) of the extractions steps.[30,35,36] Autoradiographic experiments have shown that the same percentages of nuclei and matrices are labeled, eliminating the possibility of a random adherence of extracted nascent DNA to all matrix structures.[36] We have been able to generate labeled single-stranded DNA in nuclei *in situ* enzymatically. This single-stranded DNA was not enriched in nuclear matrix preparations which were clearly enriched for nascent DNA.[36]

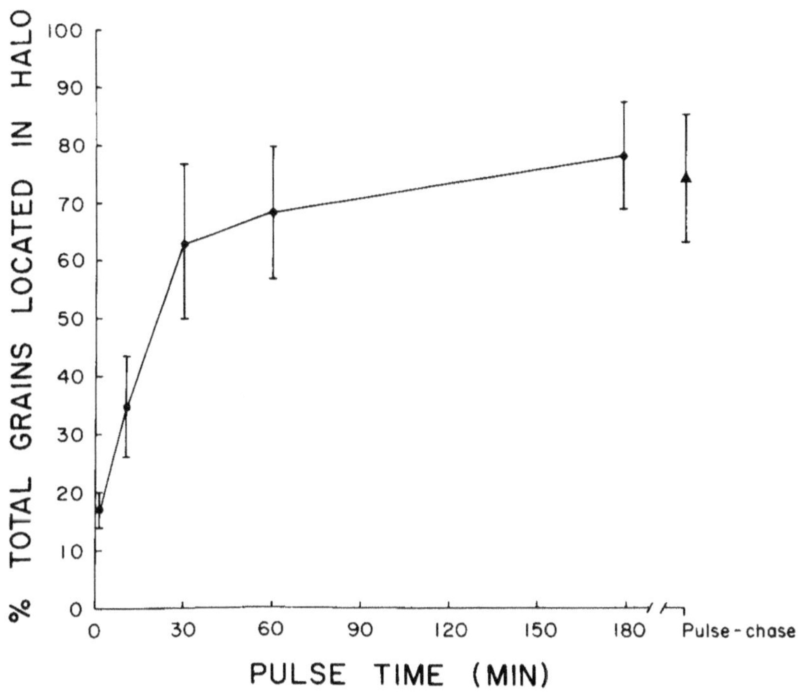


FIGURE 3. Movement of newly replicated DNA from the nuclear matrix to the halo regions of matrix-halo structures. 3T3 cells were pulse labeled for the indicated time (abscissa) with [$^3$H]-thymidine. Matrix-halo structures were prepared from these cells and autoradiographed. The number of grains within the halo region (outside the matrix border) and the number of grains within the matrix were determined for each of 25 nuclei for each point. The data are expressed as (number of grains within the halo regions/total number of grains) × 100. Mean and standard deviations are indicated. The data point from a 1-min pulse, 60-min chase is also shown. (From Vogelstein, B. et al., *Cell*, 22, 79, 1980. With permission.)

Size analysis showed that nascent DNA is not simply protected by matrix components; the replicating DNA sequences are as sensitive to nuclease digestion as is the nonreplicating DNA after removal of the histones (References 30 and 35 and unpublished data).

**B. Model**

These studies, and similar ones performed by other investigators, suggested specific models of DNA replication.[35,37-40] In these models, each loop of DNA functions as a replicon, the loops being reeled through fixed replication complexes on the nuclear matrix during DNA replication. The idea that DNA moves through replication complexes is certainly not new (e.g., see References 41 to 43). However, such models have several conceptually satisfying features when compared to the more conventional view of mobile enzymes moving along the DNA helix. First, it has been demonstrated that many of the enzymes required for DNA replication may exist as a complex.[44-47] Energetically, it would appear less favorable for a large and unwieldy complex to move along a thin DNA helix than the reverse. Second, with fixed site models, new loops of DNA are automatically generated through the DNA synthesis itself, thus intrinsically preserving the ordered arrangement of DNA loops during this period of the cell cycle. Third, an intriguing feature of eukaryotic DNA replication is the fact that once replicated in a given S phase, a replicon origin will not become reactivated until the subsequent S phase.[48-50] One explanation for this finding suggested by Blumenthal et al. is that " . . . such a restriction could result from changes during the cycle . . . in the chromatin structure."[48] Perhaps this postulated change in the chromatin structure is simply that replicon origins are oriented within loops in such a way that they are only in contact

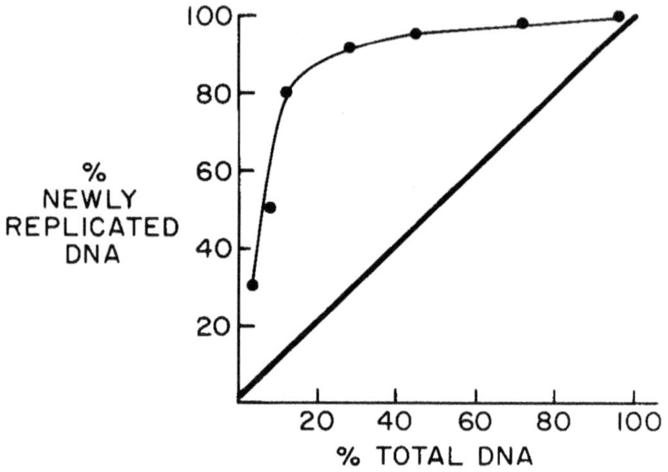


FIGURE 4. Nonrandom association of newly replicated DNA with the nuclear matrix of mammalian cells. 3T3 cells were labeled for 48 hr with [$^{14}$C]-thymidine and then for 30 sec with [$^{3}$H]-thymidine. Nuclear matrices containing progressively less of the total nuclear DNA were made from these cells by varying the extent of DNAase I digestion.[31] The percentage of total nuclear DNA ([$^{14}$C]-thymidine radioactivity) and newly replicated DNA ([$^{3}$H]-thymidine radioactivity) relative to that contained in the matrices in which no DNAase was used was assessed by TCA precipitation of the pelleted nuclear matrices. The diagonal line represents the expected result if the distribution of newly replicated DNA with respect to the nuclear matrix were random.

with replication complexes on the nuclear matrix at specific times during the cell cycle. This idea is experimentally testable.

## IV. TRANSCRIPTIONALLY ACTIVE GENES

There have been several lines of evidence suggesting that the nuclear matrix plays a role in mediating gene expression (see below). In 1979, we began a series of experiments to determine whether specific genes were organized nonrandomly with respect to supercoiled loops and the nuclear matrix. The basic strategy of these experiments was to treat matrix-halo structures with DNA endonucleases so that only DNA sequences contiguous with potential anchoring sequences would remain bound to the matrix. Sequences that were far removed from putative anchoring sequences would be severed from the DNA that remained anchored to the matrix. The anchored DNA could then be purified and analyzed for its content of specific sequences by hybridization techniques.

### A. Considerations in the Isolation Procedure

As with any experiment involving fractionation of cellular components, the fractionation procedure itself is of critical importance both technically and conceptually. We have used several variations of procedures designed to obtain "matrix-halo structures", i.e., a salt-resistant residual nuclear skeleton to which dehistonized supercoiled loops are anchored. Several points are important about the procedure we currently use.

First, nuclei are isolated in a solution containing relatively physiological concentrations of cations, i.e., potassium and sodium with a combined concentration of 150 m*M* and magnesium at a concentration of approximately 5 m*M*. Although we have in the past obtained similar results with EDTA containing buffers or with more hypotonic ones, some nuclei swell considerably, hence distorting skeletal components, in these latter solutions. In particular, the morphology of nuclei and matrix-halo structures is dramatically altered when incubated in solutions of very low ionic strength and in the absence of divalent cations. This may explain the apparent absence of matrix-like components in procedures designed to maximally spread nuclear contents (e.g., "Miller spreads"[51]).

Second, the proteins of the matrix seem to be organized in a very specific spatial conformation. Reactive sulfhydryl groups lie close enough together that they can be cross-linked over very short intermolecular distances (e.g., via air oxidation or sodium tetrathionate).[52] Hence, precautions must be taken to minimize the exposure of the structures to potential oxidants. Many recent studies of the nuclear matrix[32,53-58] have taken such precautions by using rapid procedures for nuclear isolation, freshly made detergent solutions, and by including, where appropriate, sulfhydryl blocking reagents. However, reducing agents (mercaptoethanol, dithiothreitol) must be avoided because of the possiblity that they may release a copper-containing protein required for stabilization of major structural elements of the matrix.[57,58]

Third, calcium concentrations in excess of 0.5 m*M*, which are often used in nuclear isolation procedures (especially those employing micrococcal nuclease), irreversibly alter the substructure of the nucleus.[57,58] Hence, $Ca^{++}$ must be avoided.

Finally, nuclease digestions must be performed after removal of the histones. Numerous studies have shown that the nucleosomal organization of DNA exerts profound effects on the accessibility of nuclear DNA to nuclease digestion (e.g., References 59 to 62). Since transcribed genes and newly replicated DNA are preferentially sensitive to nuclease digestion while in their native chromosomal conformation, the histones must be removed prior to any examination of the distribution of these sequences with respect to supercoiled loops and the nuclear matrix.

### B. Association of Active Genes with the Nuclear Matrix

We began our examination of active genes with the analyses of SV40 genes in SV40 transformed 3T3 cells.[31] We found that the SV40 genes were significantly enriched in DNA purified from the matrices. Moreover, the enrichment increased as more DNA was cleaved from the halo, showing that the SV40 genes were very close to the matrix anchorage sites. Control experiments with mouse alpha and beta globin genes showed that the enrichment was not due to technical artifacts involved in the hybridization analyses. This association of SV40 genes with the nuclear matrix was found in several different lines of SV40-transformed cells, each with a unique SV40 integration site.[31]

Although the matrix association of SV40 genes in several different cell lines suggested that this might be due to their transcription, we could not test this directly because of the unavailability of cell lines containing SV40 genes which are not transcribed. We needed to test a gene that was transcribed in one cell type but not in another. We chose to analyze the ovalbumin gene, which is actively transcribed in hen oviduct but not in hen liver nuclei. Our initial studies showed that the ovalbumin gene was associated with the nuclear matrix in hen oviduct nuclei but not in hen liver or brain nuclei (Figure 5).[32] Subsequent studies in our laboratory have shown that other egg white protein genes (conalbumin, ovomucoid) are also bound to the nuclear matrices of hen oviduct cells but not to hen liver or brain cells.

These studies and those of others[63,64] showed that transcribed genes are preferentially associated with the nuclear matrix. This raised the question of whether transcriptional complexes anchor these transcribed sequences. Indeed, it had been shown in prokaryotes that transcriptional complexes appear to anchor topologically constrained domains of DNA (at least, those detected in nucleoids),[65-68] so this explanation seemed quite plausible. However, recent experiments have shown that the situation is quite different in eukaryotes. First, in prokaryotes, RNAases destroy the transcriptional complexes and completely disperse the DNA domains anchored to them.[65,67,68] However, in eukaryotes, RNAases have no effect

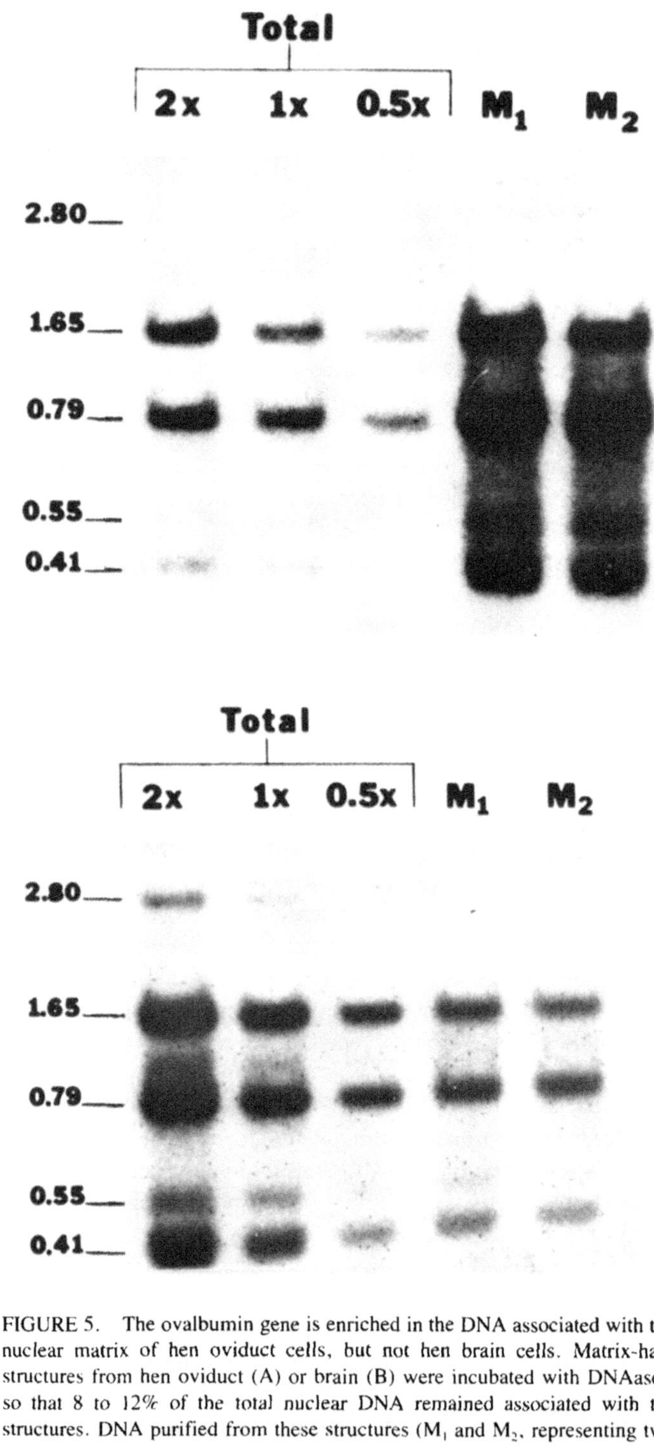


FIGURE 5. The ovalbumin gene is enriched in the DNA associated with the nuclear matrix of hen oviduct cells, but not hen brain cells. Matrix-halo structures from hen oviduct (A) or brain (B) were incubated with DNAase I so that 8 to 12% of the total nuclear DNA remained associated with the structures. DNA purified from these structures ($M_1$ and $M_2$, representing two different preparations), and total nuclear DNA in amounts equal to 0.5-, 1-, and 2-fold used in the $M_1$ and $M_2$ lanes were cleaved with HinfI, electrophoresed through an agarose gel, transferred to nitrocellulose, and hybridized with a probe detecting the ovalbumin gene.

on the morphology of the matrix-halo structures or their biophysical behavior.[69,70] Furthermore, transcribed genes remain anchored to the matrix under conditions where more than

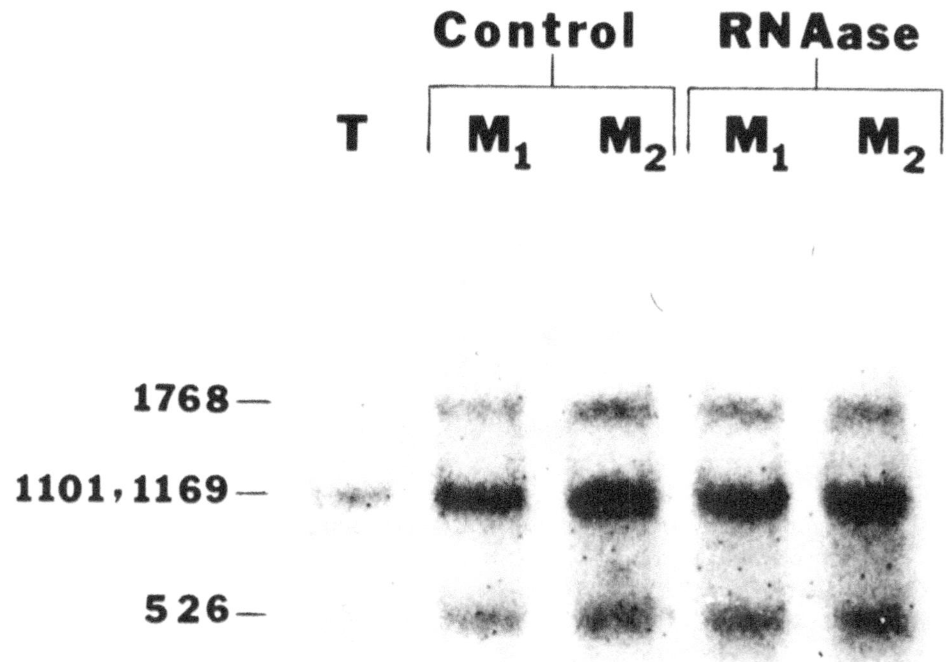


FIGURE 6. RNAase does not affect the distribution of SV40 genes with respect to the nuclear matrices of SV40 transformed cells. Matrix-halo structures from SVB203 cells were prepared. Half of the structures M were incubated with RNAase A (100 μg/mℓ) and RNAase $T_1$ (50 U/mℓ) at 37°C for 10 min and half $M_c$ were incubated at 37°C without RNAases. The structures were then cleaved with DNAase I so that 10 or 12% of the total nuclear DNA remained associated with preparations $M_1$ and $M_2$, respectively, in the control matrices, and 11 or 14% of the total nuclear DNA remained with preparations $M_1$ and $M_2$ of the RNAase-treated matrices. DNA purified from these structures and total nuclear DNA in an amount equal to that used in the $M_1$ and $M_2$ lanes was cleaved with HindIII, electrophoresed through an agarose gel, transferred to nitrocellulose, and hybridized with a probe detecting the SV40 genes. The RNAase treatment removed nearly all of the nascent RNA from the nuclear matrices. This was assessed in the following way: cells were pulse labeled with [$^3$H]-uridine for 2 min immediately prior to harvest and used to make matrix-halo structures. 92% of the labeled RNA remained associated with these structures; this is consistent with experiments performed by other investigators.[75-77] When these structures were treated with RNAases as described above, 99.6% of this RNA was removed from them.

99.7% of nascent RNA (labeled with a 2-min pulse of [$^3$H]-uridine) is removed from the matrix. This exhaustive RNAase treatment has been applied to matrix-halo structures from a variety of species and has been found not to change the distribution of SV40 genes in mouse cells, of ovalbumin genes in oviduct cells, or of a variety of transcribed genes in *Drosophila* Kc cells (e.g., Figure 6). Second, in prokaryotes, active transcription is required for the maintenance of the topologically constrained domains, as can be assessed with RNA synthesis inhibitors.[65,66] In mouse cells, however, treatment with actinomycin D under conditions in which greater than 98% of mRNA synthesis is inhibited does not disrupt the loop organization. Furthermore, this treatment does not change the distribution of SV40 genes and metallothionein genes in mouse cells; they are still anchored to the matrix.[70]

Other recent experiments in our laboratory have further shown that transcription is not required for the association of transcriptionally activated genes with the nuclear matrix. Heat shock is a stimulus often used by investigators to rapidly and quantitatively alter the pattern of gene expression in several types of cells.[71] During heat shock of *Drosophila* cells, it has been shown that RNA polymerase II rapidly dissociates from genes transcribed before the heat shock,[72,73] and that many of these RNA polymerase molecules bind to a discrete subset of genes (the heat shock protein [hsp] genes) which dramatically increase their rate of transcription.[71-73] We have found that hsp genes (hsp 70 and hsp 23) are associated with the nuclear matrix of *Drosophila* cells both before heat shock (when they are transcribed at very low levels) as well as during heat shock (when they are transcribed at high rates).[29] A nontranscribed gene (the chorion gene) is depleted in these same matrix DNA preparations. Furthermore, we have examined the distribution of the actin genes in these heat shocked cells. They are tightly bound to nuclear matrix components before heat shock, when they are actively transcribed. During heat shock, actin gene transcription is eliminated in more than 80% of the cells; despite this, the actin genes remain quantitatively bound to the nuclear matrix.[29]

What is the meaning of the binding of transcriptionally active genes to nuclear matrix components? We believe that this is the first step in a sort of ''treadmill'' system of gene expression. In this treadmill concept, the transcription of genes and subsequent processing of the gene products occurs along a continuous structural supporting framework. This is to be contrasted with a stochastic or ''pinball'' model wherein the various enzymes, genes, gene products, and cofactors involved in gene expression all interact in the nuclear ''sap'' through essentially random interactions dictated by the various affinity constants. The treadmill concept, although obviously speculative, offers a unifying mechanism for the cell to control many aspects of gene expression (from transcription to translation) in a structurally and spatially defined manner. Moreover, such a concept encompasses many of the observations that show an association between genes, gene products, and skeletal elements within the cell.[54,75-86]

## V. CONCLUDING REMARKS

The studies noted in this chapter have demonstrated intriguing associations between transcribing or replicating DNA and the nuclear matrix. These associations appear to be generalized features in eukaryotes, since they have been found in diverse species. We feel that these results are strongly supportive of a role for nuclear structural elements in the control of certain aspects of DNA replication and transcription. However, these studies are obviously incomplete, and only hint at the immense complexities involved. More studies need to be done to rule out the possibility that replicating or transcribing sequences are artifactually binding to matrix components through single-stranded regions. Additional investigations are required to find out when in the history of the cell an activated gene becomes bound to the matrix. Detailed analyses will have to be performed to determine the relationship between nuclear structural elements during interphase and metaphase. Finally, we have no idea as to which proteins (or possibly other components) mediate the anchorage of specific genes to the nuclear matrix. It is only when such proteins, and the specific sequences that they bind, are identified that the full importance and meaning of the associations so far documented will begin to become clear.

## ACKNOWLEDGMENTS

The authors are grateful to S. Smith for expert preparation of the manuscript. This work was supported by training grants GM 07309 and CA-09243 and grant GM 31676 awarded by the NIH, DHHS.

## REFERENCES


1. **Kornberg, R. D.,** Structure of chromatin, *Ann. Rev. Biochem.*, 46, 931, 1977.
2. **Felsenfeld, G.,** Chromatin, *Nature (London)*, 271, 115, 1978.
3. **Finch, J. T. and Klug, A.,** Solenoidal model for superstructure in chromatin, *Proc. Natl. Acad. Sci. U.S.A.*, 73, 1897, 1976.
4. **Renz, M., Nehls, P., and Hozier, J.,** Involvement of histone H1 in the organization of the chromosome fiber, *Proc. Natl. Acad. Sci. U.S.A.*, 74, 1879, 1977.
5. **Worcel, A. and Benyajati, C.,** Higher order coiling of DNA in chromatin, *Cell*, 12, 83, 1977.
6. **Stratling, W. H., Muller, V., and Zentgraf, H.,** The higher order repeat structure of chromatin is built up of globular particles containing eight nucleosomes, *Exp. Cell Res.*, 117, 301, 1978.
7. **Callan, H. G. and Macgregor, H. C.,** Action of deoxyribonuclease on lampbrush chromosomes, *Nature (London)*, 181, 1479, 1958.
8. **Gall, J. G.,** Kinetics of deoxyribonuclease action on chromosomes, *Nature (London)*, 198, 36, 1963.
9. **Cook, P. R. and Brazell, I. A.,** Supercoils in human DNA, *J. Cell Sci.*, 19, 261, 1975.
10. **Ide, T., Nakane, M., Anzai, K., and Andoh, T.,** Supercoiled DNA folded by non-histone proteins in cultured mammalian cells, *Nature (London)*, 258, 445, 1975.
11. **Comings, D. E. and Okada, T. A.,** Nuclear proteins. III. The fibrillar nature of the nuclear matrix, *Exp. Cell Res.*, 103, 341, 1976.
12. **Benyajati, C. and Worcel, A.,** Isolation, characterization, and structure of the folded interphase genome of *Drosophila melanogaster*, *Cell*, 9, 393, 1976.
13. **Pinon, R. and Salts, Y.,** Isolation of folded chromosomes from the yeast *Saccharomyces cerevisiae*, *Proc. Natl. Acad. Sci. U.S.A.*, 74, 2850, 1977.
14. **Hartwig, M.,** Organization of mammalian chromosomal DNA: supercoiled and folded circular DNA subunits from interphase cell nuclei, *Acta Biol. Med. Ger.*, 37, 421, 1978.
15. **McCready, S. J., Akrigg, A., and Cook, P. R.,** Electron microscopy of intact nuclear DNA from human cells, *J. Cell Sci.*, 39, 53, 1979.
16. **Paulson, J. R. and Laemmli, U. K.,** The structure of histone-depleted metaphase chromosomes, *Cell*, 12, 817, 1977.
17. **Okada, T. A. and Comings, D. E.,** Higher order structure of chromosomes, *Chromosoma*, 72, 1, 1979.
18. **Igo-Kemenes, T. and Zachau, H. G.,** Domains in chromatin structure, *Cold Spring Harbor Symp. Quant. Biol.*, 42, 109, 1977.
19. **Lenk, R., Ransom, L., Kaufman, Y., and Penman, S.,** A cytoskeletal structure with associated polyribosomes obtained from HeLa cells, *Cell*, 10, 67, 1977.
20. **Osborn, M. and Weber, K.,** The detergent-resistant cytoskeleton of tissue culture cells includes the nucleus and the microfilament bundles, *Exp. Cell Res.*, 106, 339, 1977.
21. **Lehto, V. P., Virtanen, I., and Kurki, P.,** Intermediate filaments anchor the nuclei in nuclear monolayers of cultured human fibroblasts, *Nature (London)*, 373, 175, 1978.
22. **Vogelstein, B., Pardoll, D., and Coffey, D. S.,** Supercoiled loops and eucaryotic DNA replication, *Cell*, 22, 79, 1980.
23. **Berezney, R. and Coffey, D. S.,** Nuclear matrix: isolation and characterization of a framework structure from rat liver nuclei, *J. Cell Biol.*, 73, 616, 1977.
24. **Peters, J. E. and Comings, D. E.,** Two-dimensional gel electrophoresis of rat liver nuclear washes, nuclear matrix and hnRNA proteins, *J. Cell Biol.*, 86, 135, 1980.
25. **Berezney, R. and Buchholtz, L. A.,** Isolation and characterization of rat liver nuclear matrices containing high molcular weight deoxyribonucleic acid, *Biochemistry*, 20, 4995, 1981.
26. **Muller, W. and Gautier, F.,** Interactions of heteroaromatic compounds with nucleic acids: A-T specific non-intercalating DNA ligands, *Eur. J. Biochem.*, 54, 385, 1975.
27. **LePecq, J. B. and Paoletti, A. J.,** Fluorescent complex between ethidium bromide and nucleic acids physical-chemical characterization, *J. Mol. Biol.*, 27, 87, 1967.
28. **Anderson, P. and Bauer, W.,** Supercoiling in closed circular DNA: dependence upon ion type and concentration, *Biochemistry*, 17, 594, 1978.
29. **Small, D., Nelkin, B. D., Lis, J., Gilmore, D., and Vogelstein, B.,** The association of transcribed sequences with the nuclear matrix of *Drosophila* cells before and during heat shock, submitted.
30. **Hunt, B. F. and Vogelstein, B.,** Association of newly replicated DNA with the nuclear matrix of *Physarum polycephalum*, *Nucleic Acids Res.*, 9, 349, 1981.
31. **Nelkin, B. D., Pardoll, D., and Vogelstein, B.,** Localization of SV40 genes within supercoiled loop domains, *Nucleic Acids Res.*, 8, 5623, 1980.
32. **Robinson, S. I., Nelkin, B. D., and Vogelstein, B.,** The ovalbumin gene is associated with the nuclear matrix of chicken oviduct cells, *Cell*, 28, 99, 1982.

33. **Vogelstein, B., Nelkin, B., Pardoll, D., Robinson, S., and Small, D.,** Supercoiled loops and transcribing genes, in *UCLA Symposium on Gene Regulation*, O'Malley, B. W., Ed., Academic Press, New York, 1983.
34. **Buongiorno-Nardelli, M., Michele, G., Carri, M. T., and Marilley, M.,** A relationship between replicon size and supercoiled loop domains in the eucaryotic genome, *Nature (London)*, 298, 100, 1982.
35. **Pardoll, D., Vogelstein, B., and Coffey, D. S.,** A fixed site of DNA replication in eucaryotic cells, *Cell*, 19, 527, 1980.
36. **Vogelstein, B., Nelkin, B., Pardoll, D., and Hunt, B. F.,** Further evidence for the physiological association of newly replicated DNA with the nuclear matrix, in *The Nuclear Envelope and the Nuclear Matrix*, Mall, G., Ed., Alan R. Liss, New York, 1982, 169.
37. **Berezney, R. and Buchholtz, L. A.,** Dynamic association of replicating DNA fragments with the nuclear matrix of regenerating liver, *Exp. Cell Res.*, 132, 1, 1981.
38. **McCready, S. J., Godwin, J., Mason, D. W., Brazell, I. A., and Cook, P. R.,** DNA is replicated at the nuclear cage, *J. Cell Sci.*, 46, 365, 1980.
39. **Dijkwel, P., Mullenders, L., and Wanka, F.,** Analysis of the attachment of replicating DNA to a nuclear matrix in mammalian interphase nuclei, *Cell*, 19, 527, 1980.
40. **Dvorkin, V. M. and Vanyushin, B. F.,** Replication and kinetics of the reassociation of DNA of the nuclear matrix of the regenerating rat liver, *Biochemistry (USSR)*, 43, 1297, 1978.
41. **Jacob, F., Brenner, S., and Cuzin, F.,** On regulation of DNA replication in bacteria, *Cold Spring Harbor Symp. Quant. Biol.*, 28, 329, 1963.
42. **Comings, D. E. and Kakefuda, T.,** Initiation of deoxyribonucleic acid replication at the nuclear membrane in human cells, *J. Mol. Biol.*, 33, 225, 1968.
43. **Dingman, C. W.,** Bidirectional chromosome replication — some topological considerations, *J. Theor. Biol.*, 43, 187, 1974.
44. **Genta, V. M., Kaufman, D. G., Kaufman, W. K., and Gerwin, B. I.,** Eukaryotic DNA replication complex, *Nature (London)*, 259, 502, 1976.
45. **Barel, E., Baril, B., Elford, H., and Luftig, H. R.,** DNA polymerases and a possible multi-enzyme complex for DNA biosynthesis in eukaryotes, in *Mechanism and Regulation of DNA Replication*, Kober, A. and Kohiyama, M., Eds., Plenum Press, New York, 1974, 275.
46. **Prem veer Reddy, G. and Pardee, A. B.,** Multienzyme complex for metabolic channeling in mammalian DNA replication, *Proc. Natl. Acad. Sci. U.S.A.*, 77, 3312, 1980.
47. **Westergard, O. and Johnson, B.,** Isolation of a replication complex from eukaryotes, *Biochem. Biophys. Res. Commun.*, 55, 341, 1973.
48. **Blumenthal, A. B., Kriegstein, H. J., and Hogness, D. S.,** Units of DNA replication in *Drosophila melanogaster* chromosomes, *Cold Spring Harbor Symp. Quant. Biol.*, 38, 205, 1973.
49. **Kriegstein, H. J. and Hogness, D. S.,** Mechanism of DNA replication in *Drosophila* chromosomes — structure of a replication fork and evidence for bidirectionality, *Proc. Natl. Acad. Sci. U.S.A.*, 71, 135, 1974.
50. **Harland, R. A. and Laskey, E. M.,** Regulated replication of DNA microinjected into eggs of *Xenopus laevis*, *Cell*, 21, 761, 1980.
51. **Miller, O. L., Jr., and Beatty, B. R.,** Visualization of nucleolar genes, *Science*, 164, 955, 1969.
52. **Kaufmann, S. H., Coffey, D. S., and Shaper, J. H.,** Considerations in the isolation of rat liver nuclear matrix, nuclear envelope, and pore complex lamina, *Exp. Cell Res.*, 132, 105, 1981.
53. **Fisher, P. A., Berrios, M., and Blobel, G.,** Isolation and characterization of a proteinacious subnuclear fraction composed of nuclear matrix, peripheral lamina, and nuclear pore complexes from embryos of *Drosophila melanogaster*, *J. Cell Biol.*, 92, 674, 1982.
54. **Vogelstein, B. and Hunt, B. F.,** A subset of small nuclear ribonucleoprotein particle antigens is a component of the nuclear matrix, *Biochem. Biophys. Res. Commun.*, 105, 1224, 1982.
55. **van Eekelen, C. A. G., Salden, M. H. L., Habets, W. J. A., van de Putte, L. B. A., and van Venrooij, W. J.,** On the existence of an internal nucleoprotein structure in HeLa cells, *Exp. Cell Res.*, 141, 181, 1982.
56. **Capco, D. G., Wan, K. M., and Penman, S.,** The nuclear matrix: three dimensional architecture and protein composition, *Cell*, 28, 847, 1982.
57. **Lebkowski, J. S. and Laemmli, U. K.,** Evidence for two levels of DNA folding in histone depleted HeLa interphase nuclei, *J. Mol. Biol.*, 156, 309, 1982.
58. **Lebkowski, J. S. and Laemmli, U. K.,** Non-histone proteins and long range organization of HeLa interphase DNA, *J. Mol. Biol.*, 156, 325, 1982.
59. **Weintraub, H. and Groudine, M.,** Chromosomal subunits in active genes have an altered conformation, *Science*, 193, 848, 1976.
60. **Garel, A. and Axel, R.,** Selective digestion of transcriptionally active ovalbumin genes from oviduct nuclei, *Proc. Natl. Acad. Sci. U.S.A.*, 73, 3966, 1976.

61. **Bloom, K. S. and Anderson, J. N.,** Fractionation of hen oviduct chromatin into transcriptionally active and inactive regions after selective micrococcal nuclease digestion, *Cell,* 15, 141, 1978.
62. **Seale, R. L.,** Assembly of DNA and protein during replication in HeLa cells, *Nature (London),* 255, 247, 1975.
63. **Cook, P. R., Long, J., Hayday, A., Lania, L., Fried, M., Chiswell, D. J., and Wyke, J. A.,** Active viral genes in transformed cells lie close to the nuclear cage, *EMBO J.,* 1, 447, 1982.
64. **Jackson, D. A., McCready, S. J., and Cook, P. R.,** RNA is synthesized at the nuclear cage, *Nature (London),* 292, 552, 1981.
65. **Pettijohn, D. W. and Hecht, R.,** RNA molecules bound to folded bacterial genome stabilize DNA folds and segregate domains of supercoiling, *Cold Spring Harbor Symp. Quant. Biol.,* 38, 31, 1973.
66. **Dworsky, P. and Schachter, M.,** Effect of rifampicin on the structure and membrane attachment of the nucleoid of *Escherichia coli, J. Bacteriol.,* 116, 1364, 1973.
67. **Stonington, O. and Pettijohn, D.,** Folded genome of *Escherichia coli* isolated in a protein-DNA-RNA complex, *Proc. Natl. Acad. Sci. U.S.A.,* 68, 6, 1971.
68. **Worcel, A. and Burgi, E.,** Structure of folded chromosomes of *Escherichia coli, J. Mol. Biol.,* 71, 127, 1972.
69. **Cook, P. R. and Brazell, I. A.,** Spectro fluorometric measurement of the binding of ethidium bromide to superhelical DNA from cell nuclei, *Eur. J. Biochem.* , 84, 464, 1978.
70. **Robinson, S. I., Nelkin, B. D., and Vogelstein, B.,** Factors involved in the association of transcriptionally activated genes with the nuclear matrix, manuscript in preparation, 1984
71. **Ashburner, M. and Bonner, J. J.,** Induction of gene activity in *Drosophila* by heat shock, *Cell,* 17, 241, 1979.
72. **Greenleaf, A. L., Plagens, U., Jamrich, M., and Bautz, K. F.,** RNA polymerase B (or II) in heat inducted puffs of *Drosophila* cells, *Proc. Natl. Acad. Sci. U.S.A.,* 88, 323, 1981.
73. **Sass, H.,** RNA polymerase B in polytene chromosomes: immunofluorescent and autoradiographic analysis during stimulated and repressed RNA synthesis, *Cell,* 28, 269, 1982.
74. **Pederson, T.,** Proteins associated with heterogeneous nuclear RNA in eucaryotic cells, *J. Mol. Biol.,* 83, 163, 1974.
75. **Herman, R., Weymouth, C., and Penman, S.,** Heterogeneous nuclear RNA-protein fibers in chromatin-depleted nuclei, *J. Cell Biol.,* 78, 663, 1978.
76. **Miller, T. E., Huang, C., and Pogo, A. O.,** Rat liver nuclear skeleton and ribonucleoprotein complexes containing hnRNA, *J. Cell Biol.,* 76, 675, 1978.
77. **van Eekelen, C. A. G. and van Venrooij, W. J.,** hnRNA and its attachment to a nuclear-protein matrix, *J. Cell Biol.,* 88, 554, 1981.
78. **Maundrell, K., Maxwell, E. S., Puvion, E., and Scherrer, K.,** The nuclear matrix of duck erythroblast is associated with globin hnRNA coding sequences but not with the major proteins of 40S nuclear RNP, *Exp. Cell Res.,* 136, 435, 1981.
79. **Ciejek, E. M., Nordstrom, J. L., Tsai, M. J., O'Malley, B. W.,** Association of RNA precursors with the nuclear matrix of chick oviduct, *Biochemistry,* 21, 4945, 1982.
80. **Mariman, E. C. M., van Eekelen, C. A. G., Reinders, L. J., Berns, A. J. M., and van Venrooij, W. J.,** Adenovirus heterogeneous nuclear RNA is associated with the host nuclear matrix during splicing, *J. Mol. Biol.,* 154, 103, 1982.
81. **Miller, T. E., Huang, C., and Pogo, A. O.,** Rat liver nuclear skeleton and small molecular weight RNA species, *J. Cell Biol.,* 76, 692, 1978.
82. **Zieve, G. and Penman, S.,** Small RNA species of HeLa cells: metabolism and subcellular localization, *Cell,* 8, 19, 1976.
83. **Wolosewick, J. J. and Porter, K. R.,** Stereo high voltage electron microscopy of whole cells of the human diploid line, W1-38, *Am. J. Anat.,* 147, 303, 1976.
84. **Lenk, R. and Penman, S.,** The cytoskeletal framework and poliovirus metabolism, *Cell,* 16, 289, 1979.
85. **Fulton, A. B., Wan, K. M., and Penman, S.,** The spatial distribution of polyribosomes in 3T3 cells and the associated assembly of proteins into the skeletal framework, *Cell,* 20, 849, 1980.
86. **Cervera, M., Dreyfuss, G., and Penman, S.,** Messenger RNA is translated when associated with the cytoskeletal framework in normal and VSV-infected HeLa cells, *Cell,* 23, 113, 1981.

Chapter 8

# NONHISTONE PROTEINS AND NUCLEAR MATRIX STRUCTURES

**A. Oscar Pogo and Roman Procyk**

## TABLE OF CONTENTS

## I. INTRODUCTION

In recent years, increased attention has been directed toward understanding the nature of a framework of filaments which remains after living cells are extracted with nonionic detergents such as Triton® X-100.[1] The treatment yields a residuum containing the nucleus and a firmly attached mesh of cytofilaments that is designated as the cytoskeleton. This "Triton cytoskeleton" consists largely of intermediate-sized filaments and bundles of microfilaments. If this residuum is digested simultaneously with DNAase I and RNAase A and then treated with high salt, a nuclear network and a residual nucleolus are obtained with cytoskeletal components still solidly attached. The nuclear network must differ from the cytoplasmic network since no microtubules, microfilaments, and intermediate-sized filaments have been detected in the nuclear interior.

Wolosewick and Porter[2], have postulated that another structure, the microtrabeculae, which pervades every nook and cranny of the cytoplasmic space also is a cytoskeletal component. The microtrabeculae is a three-dimensional lattice and differs from microtubules, intermediate filaments, and actin. Penman and co-workers[3] maintain a similar distinction, but Heuser and Kirschner[4] have challenged this concept. They indicated that the microtrabeculae represents agglutinated individual actin and intermediate filaments.

Whether a microtrabeculae-like structure does exist in the nuclear interior remains to be known. As in the cytoplasm, it is possible that nucleoprotein aggregation is produced by the chemicals and treatments to which the nucleus is exposed in order to solubilize DNA, RNA, and histones. At the present time, it is unclear whether the proteinaceous nuclear network results from such manipulations. We intend here to critically evaluate this topic and the current concepts about the nature of nuclear matrix. In addition, since it is usually assumed that nuclear envelope, lamina, and pore complexes are matrix components, we are discussing them in the general context of the nuclear structures that are obtained after whole nuclei are DNA- and histone-depleted. Ultimately, nuclear matrix will be explicable only when biochemistry and physiology are integrated with morphology. At present, what is lacking is sound biochemistry, but despite this shortcoming, recent evidence of nucleoproteins that can be cross-linked through sulfhydryl groups, bind metal ions, and form polymers may further our understanding of the nature of the nuclear matrix.

## II. WHAT IS NUCLEAR MATRIX?*

Nuclear matrix is defined as the residuum that remains after the DNA and RNA are digested, the lipids are dissolved with a nonionic detergent, and the histones and many nonhistone proteins are solubilized by high salt. This is the original definition of nuclear matrix made by Berezney and Coffey.[8] Obviously, the nature and properties of the residuum must depend upon the method used to isolate the nuclei, to digest DNA and RNA, and to solubilize lipids and histones as well as many nonhistone proteins. Since the nature of the residuum varies according to the procedure employed, there are evidently as many kinds of nuclear matrix as methods. At the present time there is no biochemical characterization of nuclear matrix. Only electron microscopic examination shows the morphological features of what is known as nuclear matrix. Thus, the nuclear residuum obtained using the published procedures reveals a thin, electron-dense material at the periphery which contains the lamina and nuclear pore complexes (Figure 1). The interior shows residual nucleoli and an ill-defined tangle of fibrils. Together they are the space-filling meshwork of the nucleus and in every respect they should be designated the nuclear matrix. Unlike the lamina and nuclear pore complexes, the presence of a nuclear network depends upon the procedure used to obtain the nucleus and the means by which DNA is digested.

* Synonyms: nuclear cage,[5] nuclear skeleton,[6] and nuclear scaffold.[7]

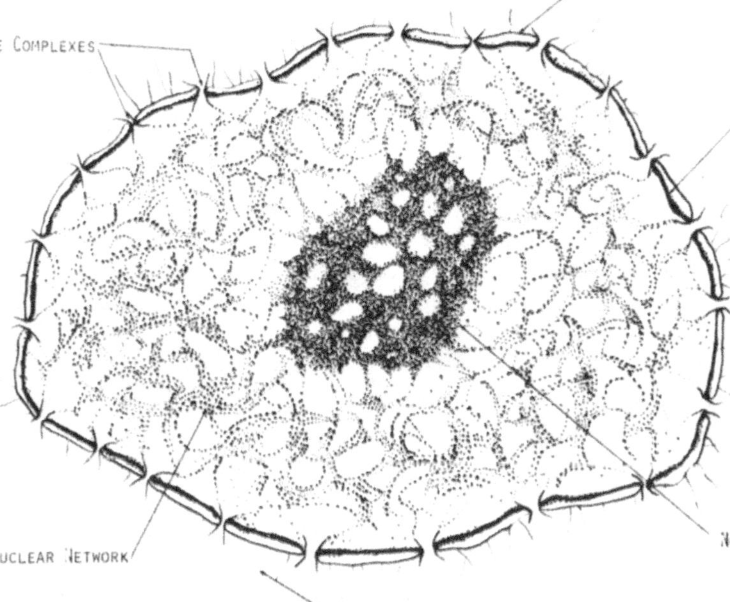


FIGURE 1. Schematic representation of a nuclear residuum.

## III. LAMINA, PORE-COMPLEXES, AND NUCLEAR ENVELOPE AS COMPONENTS OF NUCLEAR MATRIX

The lamina and the pore-complexes are the only well-characterized components of the nuclear residuum. The lamina is defined as a distinct, proteinaceous layer of amorphous material, originally termed the "fibrous lamina" by Fawcett.[9] It is interposed between the inner nuclear membrane and the peripheral chromatin and it occurs in a wide range of eukaryotic cells.[10,11] The term nuclear lamina has been generally adopted for this structure.[12] It varies in thickness in different cell types, and in some cases can only be demonstrated after removal of the nuclear envelope. It has a very simple polypeptide composition.[13-16] Immunofluorescence analyses with specific antisera have allowed these proteins to be localized upon the nucleoplasmic face of the nuclear envelope.[12-14] In most nuclei, lamina is composed of three well-defined proteins, lamins A, B, and C which migrate at approximately 70, 67, and 60 kilodaltons respectively on sodium dodecyl sulfate (SDS) gels.[12] A tryptic mapping of lamins A and B indicates that they are a product of separate genes.[17,18] In contrast, lamins A and C have nearly identical peptide maps which suggests the possibility that lamin C is a cleavage product of lamin A.[18] Cross-linking experiments revealed that lamins are able to form disulfide bonds when either the nuclei or nuclear envelope are treated with $H_2O_2$ or with *o*-phenanthroline-cupric coordination complex.[17,18] It appears that both lamins A and B are present as homotypic oligomers. They therefore contain accessible sulfhydryl groups in such a configuration that upon oxidation, disulfide bonds are readily formed. Lamins A and B yield tetramers and trimers, however lamin C yields monomers.[17] Although constancy in the number of lamins in different cells has been stressed, there is increasing evidence that it can vary.[12,17] For example, *Xenopus laevis* oocyte contains a protein which migrates electrophoretically with rat liver lamin B.[19] The comparable fraction from *Spirula solidissima* contains only one polypeptide which has been putatively identified as lamin C.[20] The definitive assignment of these proteins with comparable lamins of somatic cells must wait until the peptide mapping is done. It has been shown that lamin C is a minor component in chicken erythrocyte.[18] However, the different composition between avian and mammalian lamins may be due to proteolysis, since lamin A is very sensitive to proteolytic attack and lamin C is probably a cleavage product of A. It appears that lamina is not an absolute nuclear component. Thus, meiotic spermatocytes and spermatids do not reveal the presence of lamina by immunofluorescence and electron microscopic techniques.[21]

Although the nuclear pore complex is a well-described ultrastructure, its components remain to be characterized. Attempts have been made to isolate the nuclear pores, but unfortunately, it is extremely difficult to obtain them free from both lamins and nuclear envelope proteins.[12] It is conceivable that the nuclear pore is either an integral part of the nuclear envelope or a distinct nuclear entity, but one which is connected to the proteins of the nuclear envelope. Thus, disintegration of the nuclear envelope reveals a network of fibrils which interconnects with the pore complexes and is different from the intrapore fibrils.[22] Since pore complexes can exist without lamina in meiotic spermatocytes and spermatids, it is evident that lamina and pore complexes are two distinct nuclear entities.[21] More recently, a protein of 68 kilodaltons has been observed as a major polypeptide of the nuclear pore complex.[23] The 68-kilodalton protein was localized by incubating isolated nuclear envelope with antibodies raised against undenatured proteins of *Xenopus* oocyte nuclear skeleton. This is the first demonstration of a polypeptide associated with the pore complex. Since lamina also contain a polypeptide of similar electrophoretic mobility, it remains to be determined whether the lamina and pore complex contain two different polypeptides but with identical electrophoretic mobilities.

Most studies of the nuclear envelope have been focused upon the characterization of lamina proteins, since lamins are the most predominant proteins of the nuclear envelope fraction.[12,17,18] It seems however, that nuclear envelope and lamina are two distinct entities since spermatocytes and spermatids contain nuclear envelope, but not lamina.[21] As is the case with the proteins of the nuclear pore complexes, little is known about the protein composition of the nuclear envelope and its functions are not at all clear, although there is a general consensus (but not proof) that all the nucleo-cytoplasmic traffic takes place at the level of the pore complex. Attempts to characterize the proteins of the nuclear envelope were made by iodinating purified rat liver nuclei with lactoperoxidase immobilized onto giant Sepharose® beads.[24] In this way, proteins of the outer nuclear membrane and the cytoplasmic surface of the nuclear pore complexes were selectively labeled. A complex pattern of polypeptides was observed in the range of 33 to 200 kilodaltons. In addition, polypeptides of 70 and 67 kilodaltons were heavily iodinated. The localization of these polypeptides, however, must be obtained by immunocytochemical techniques.

## IV. NUCLEAR MATRIX OR NUCLEAR SKELETON

An intranuclear network of fibrils and a residual nucleolus should be designated, in a strict sense, as nuclear matrix (Figure 1). As originally described by Berezney and Coffey[8] the network of tangled and ill-defined fibrils is formed by nonhistone proteins containing very little DNA. If nuclei are neither digested with RNAase A nor treated with nonionic detergent, lipids and RNA are regular components of the nuclear matrix. The lipids are localized in the nuclear envelope and are therefore not components of the nuclear interior. However, whether or not there are lipoproteins in the network remains to be demonstrated. After Triton® X-100 treatment, the lipids of both inner and outer membranes are removed.

Since the early studies of Mayer and Gulick[25] as well as those of Mirsky and Ris,[26] it was known that a high salt treatment and digestion of isolated nuclei with DNAase I yields a subset of insoluble nonhistone proteins. However, it was the work of Georgiev and

Chentsov[27] as well as that of Busch and co-workers[28] that demonstrated that a network of fibrils and a residual nucleolus exist within the nuclear residues after this treatment. The intranuclear network and residual nucleolus have been obtained from a variety of cells and tissues, including normal[6,8] and regenerating rat[29] and mouse liver cells,[30] prostate,[29] uterus,[31] 3T3 fibroblasts,[32] normal and viral infected HeLa cells,[33] Zajdela hepatoma cells,[34] Friend erythroleukemia cells,[35] *Tetrahymena pyriformis*,[36,37] slime mold *Physarum polycephalum*,[38] and more recently, *Drosophila melanogaster* embryos.[39] They have not been observed in avian erythrocytes[40] and *Xenopus laevis* oocytes.[41]

Several investigators have attempted to visualize a nonchromatinous network of fibrils in intact cells. Thus, using the regressive ethylenediaminetetraacetic acid (EDTA) staining technique which bleaches the chromatin, Monneron and Bernhard[42] were able to detect in whole cells a network of fibrils and granules connecting the nuclear pores with chromatin areas as well as with nucleoli. Ghosh et al.[43] and Brasch[44] have attempted to identify nuclear matrix fibrils in conditions where there is chromatin retraction, such as in α-amanitin-treated HeLa cells and rat kangaroo cells. Penman and co-workers[45] have identified nonchromatinous fibrils in poliovirus-infected human fibroblasts treated with detergent. Like the α-amanitin treatment, the poliovirus infection produces chromatin retraction. Although these observations support the concept that in vivo there is a nonchromatinous network, it is unknown whether this network and nuclear matrix are identical structures.

Both the presence of a network and a residual nucleolus have been attributed to the effects of cross-linking sulfhydryl-containing proteins and the aggregation of nonhistone proteins with undigested RNA.[46] The sulfhydryl-containing proteins can form disulfide bonds during nuclear isolation, DNA digestion, and high salt treatment. To assess the effect of sulfhydryl-containing proteins on both the morphology of the nuclear residues and the solubilities of its proteins, two sets of experiments were performed. In one, disulfide bonds were produced either during nuclear isolation in the absence of reducing reagents or by incubation of isolated nuclei with $H_2O_2$. In another set of experiments, the formation of disulfide bonds was prevented by the addition during nuclear isolation of *N*-ethylmaleimide, an irreversible and very effective sulfhydryl-blocking reagent. To remove nucleic acid, DNA and RNA were rendered nucleotide-soluble by a simultaneous digestion of isolated nuclei with high concentrations of DNAase I (500 μg/mℓ) and RNAase A (50 μg/mℓ). Afterwards, the lipids of the nuclear envelope were dissolved with 0.5% of Triton® X-100 and the proteins were solubilized either by high salt or by a mixture of dextran sulfate and heparin. It was demonstrated that the mixture of polyanions removes histones and many nonhistone proteins effectively.[47]

The morphology of the nuclear residues varies whether or not disulfide bonds are produced. When nuclear isolation is done without a sulfhydryl-blocking reagent, the residual structure contains a well-preserved nucleoli and an intranuclear network (Figure 2). However, the appearance of these structures differs whether the solubilization of histones and nonhistone proteins is done with salt or polyanions. Thus, a high salt treatment contracts the intranuclear network and nucleoli while the polyanions expand them (Figure 2). The nucleoli obtained by the dextran sulfate and heparin treatment appear to be formed by a very thin mesh and the intranuclear network is also formed by thin fibrils. The density of fibrils, however, is much higher in the nucleolus than in the internal network. By contrast, the high salt treatment yields highly compact structures of granular configuration, very striking in the nucleolus. The intranuclear network is highly aggregated, yielding large interfibrillar spaces. It appears that the high salt treatment has produced the agglutination of fibrils.

Nuclear residues obtained from nuclei isolated in the presence of *N*-ethylmaleimide contain distorted nucleolar residues and a sparse intranuclear network. As when nuclei are prepared without *N*-ethylmaleimide, the high salt treatment yields highly contracted structures while the dextran sulfate/heparin treatment yields highly expanded structures (Figure 3). The former treatment always produces densely aggregated and heavily stained material. These obser-

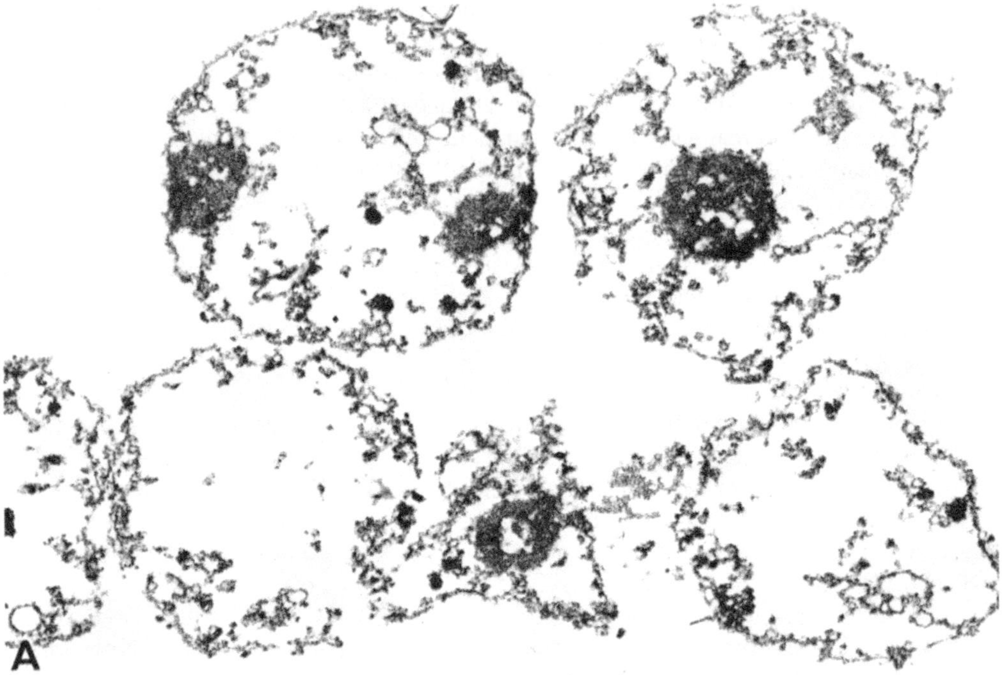


FIGURE 2. Electron microscopy of nuclear residues obtained from nuclei isolated without sulfhydryl-blocking reagents. Undifferentiated Friend erythroleukemia cells were resuspended at a density of 3 × $10^7$/mℓ in 10 m*M* Tris-acetate buffer (pH 7.4), 10 m*M* sodium acetate, 10 m*M* EDTA, 0.1% octyl-β-D-glucopyranoside, 0.15 m*M* spermine, 0.5 m*M* spermidine, 0.5 m*M* phenylmethysulfonyl fluoride (PMSF), 20% Percoll (Pharmacia Fine Chemicals) and disrupted with a 15-mℓ Dounce tissue homogenizer using a type-A pestle. Nuclei were pelleted, washed several times in 20 m*M* Tris-acetate (pH 7.7), 5 m*M* magnesium acetate, 0.5 m*M* calcium acetate, 100 m*M* sodium acetate, 146 m*M* sucrose, resuspended in this buffer at a density of 1.5 × $10^8$ nuclei per milliliter, and digested with DNAase I 500 μg/mℓ and RNAase A 50 μg/mℓ for 60 min at 8°C. Afterwards, the nuclear suspension was made in 0.5% Triton® X-100, washed in the same buffer, divided into two aliquots, and pelleted by centrifugation. One nuclear pellet was resuspended in 2 *M* NaCl, 10 m*M* EDTA, 10 m*M* sodium barbital (pH 9.0) and 0.1% octyl-β-D-glucopyranoside. The other was resuspended in 2 mg dextran sulfate per milliliter and 0.2 mg heparin per milliliter, 10 m*M* EDTA, 10 m*M* sodium barbital (pH 9.0), and 0.1% octyl-β-D-glucopyranoside. Both aliquots were incubated for 15 min in ice and then fixed with 1.5% glutaraldehyde in 0.1 *M* cacodylate buffer (pH 7.0). The fixed material was processed for electron microscopy as explained.[35] (A) Nuclear residues obtained in high salt; (B) nuclear residues obtained in dextran sulfate/heparin. (Magnification × 9000).

vations are consistent with the concept that nucleolar morphology and the solubilization of most of the internal network depend upon preventing the formation of disulfide bonds. Contrary to results obtained by Kaufman et al.[46] in rat liver nuclei, our observations in Friend erythroleukemia cells indicates that the morphologically identifiable nucleoli do not depend upon the presence of undigested RNA. In this regard, it is consistent with results obtained by Van Eekelen et al. in HeLa cells,[48] Fisher et al. in *Drosophila* embryos,[39] and Capco et al. in mouse 3T3 fibroblast.[45] At present, we cannot explain why rat liver nucleoli are solubilized after RNAase digestion and high salt treatment.

To study the effect of disulfide bonds on protein interactions, the pattern of cross-linked nuclear proteins in whole nuclei, high salt extracts, and high salt residues was examined. The formation of disulfide bonds in neighboring proteins was done either by the addition of $H_2O_2$ to isolated nuclei or by cell lysis without sulfhydryl-blocking reagents. The proteins were isolated and after a first electrophoresis step without reducing reagents, 2-mercapto-ethanol was added to cleave the disulfide bonds and obtain the separation of the polypeptides by electrophoresis in a second dimension (Figure 4). Polypeptides which were not cross-linked appeared along a diagonal line since they had the same mobility in both dimensions.

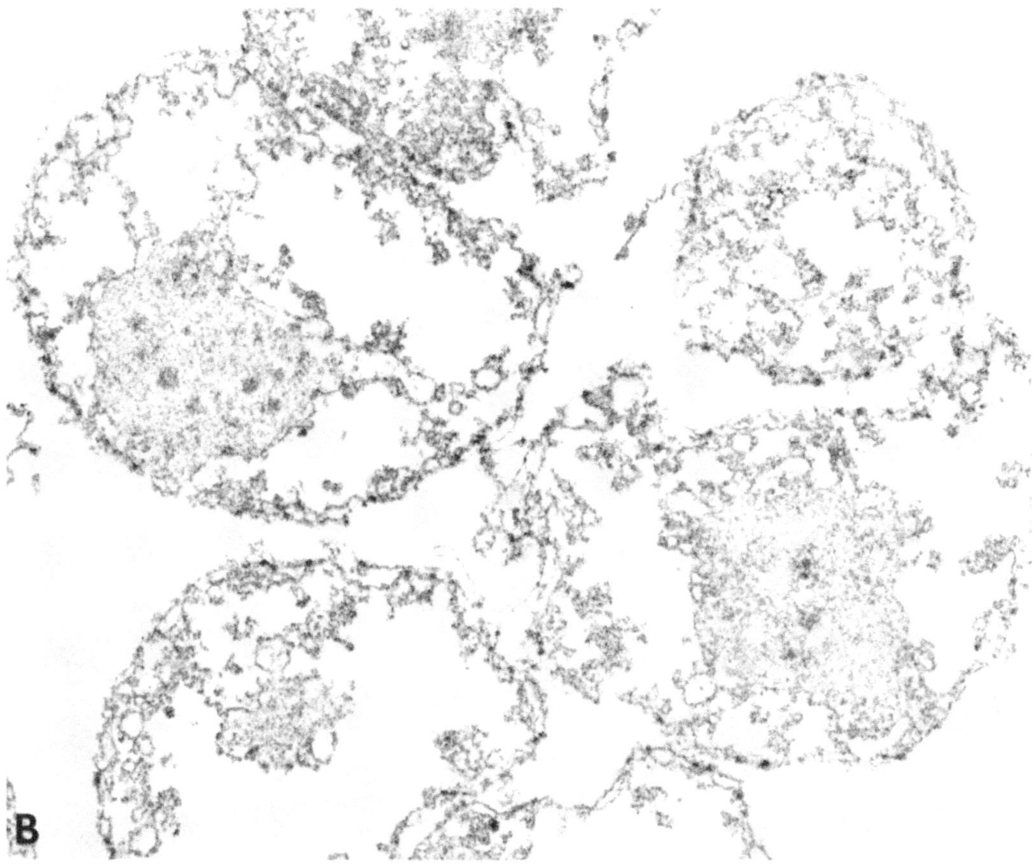


FIGURE 2B

Cross-linked proteins appeared below the diagonal, with the partners in a cross-linked heterotypic oligomer running along the same vertical line and homotypic oligomers along a horizontal line. Some proteins appeared above the diagonal; these probably had an intrachain cross-link which gave them a high electrophoretic mobility in the first dimension. Several nonhistone proteins are able to form disulfide bonds either in nuclei isolated without *N*-ethylmaleimide or upon the addition of $H_2O_2$. They must therefore contain accessible sulfhydryl groups in such a configuration that when they are oxidized, they can readily form disulfide bonds. Several of these proteins are present as homotypic oligomers like lamins A and B in the nuclear residues, but some of them appear to form heterotypic oligomers. The pattern of the off-diagonal spots is so complex, however, that in many cases it is difficult to distinguish between homotypic and heterotypic oligomers. The experiment clearly shows that the disulfide bonds are formed as an early step in the preparation of the nuclei before much opportunity arises for rearrangement of elements (Figure 4). The cross-links can be produced even in the absence of $H_2O_2$, i.e., by the $O_2$ that regularly exists in the buffers. However, no cross-links can be produced when nuclei are lysed.

It is interesting to note that the cross-linked protein complexes are found in both the high salt extract and the residuum. Further, the residuum consists of one subset of polypeptides that forms disulfide bonds and another subset which does not, as does the group of proteins which are soluble in high salt. Moreover, there is a third group of proteins which are high salt insoluble only when disulfide bonds are formed. For example, two bands which indicate proteins with molecular weights of less than 43K are high salt insoluble (i.e., remain in the

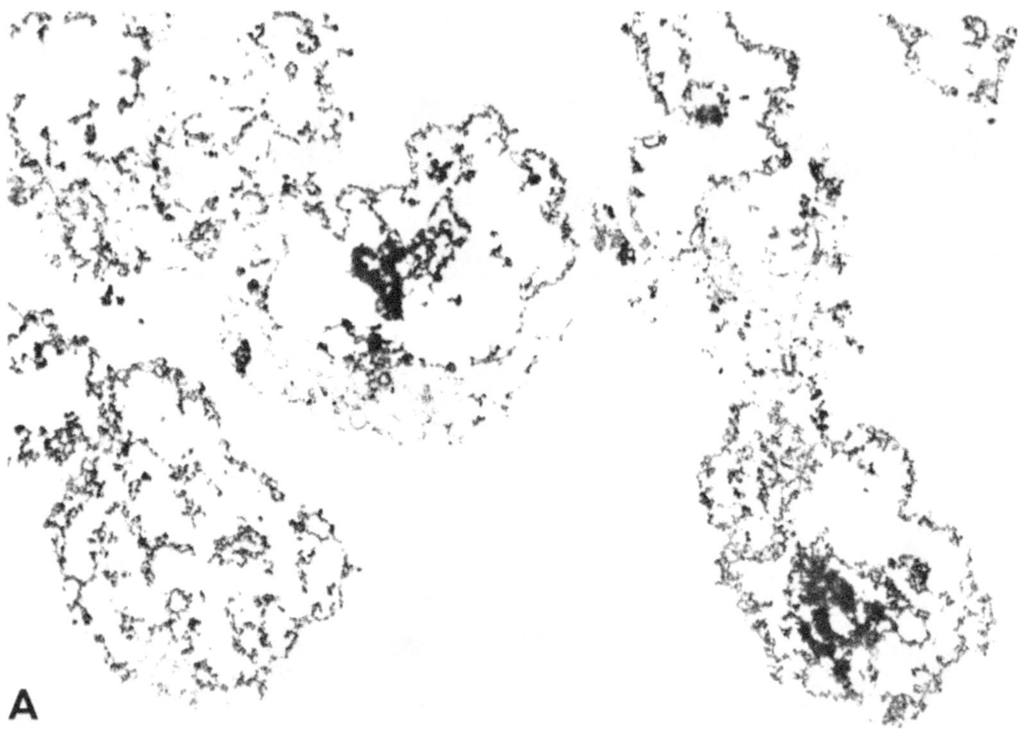

A

FIGURE 3. Electron microscopy of nuclear residues obtained from nuclei isolated in the presence of *N*-ethylmaleimide (NEM). Undifferentiated Friend erythroleukemia cells were disrupted as explained in Figure 2, but in the presence of 5 m*M* NEM. Nuclear isolation, digestion, treatments, and fixation were done as explained in Figure 2. (A) Nuclear residues obtained in high salt; (B) nuclear residues obtained in dextran sulfate/heparin. Notice the large increase in volume and deformation of the dextran sulfate/heparin residues. Since they appeared round and uniform when they were examined under a phase microscope, the deformation of the expanded structures is due to the centrifugal force used to pellet them. (Magnification × 9000).

nuclear residues) when disulfide bonds are formed (Figure 5). However, we do not know whether their insolubility in high salt occurs because they contain reactive sulfhydryl groups or because they interact with a group of proteins which form the disulfide bonds.

More recently, Laemmli and co-workers[49,50] postulated that some proteins of the nuclear residues are metalloproteins. They have observed that in the absence of a reducing reagent the addition of $Ca^{2+}$ during nuclear isolation or the incubation of nuclei with $Ca^{2+}$ at 37°C changes the configuration of these proteins which then become irreversibly insoluble in both high salt and dextran sulfate. No such effect is observed with $Mg^{2+}$, and a similar but reversible effect is observed with $Cu^{2+}$.[50] Franke and co-workers[41] have demonstrated that $Mg^{2+}$ at 2 m*M* and/or 10 m*M* affects the morphology of the extrachromosomal nucleoli of *Xenopus* oocytes.[41]

It is a difficult task to determine the effect of metals in the configuration of nucleoprotein complexes. With few exceptions, such as DNAase II, metals are necessary requirements for the enzymatic digestion of DNA. Since the quantities and types of metals within the nucleus are unknown, most probably the nuclei are exposed to nonphysiological metals for the enzymatic digestion of DNA. The procedure for complete removal of DNA requires DNAase I digestion in the presence of both $Mg^{2+}$ and $Ca^{2+}$. This condition renders 90% of the DNA to soluble nucleotides.[35] The presence of $Ca^{2+}$ is essential to cut both DNA strands simultaneously, since the cation keeps the enzyme bound to the DNA and the second strand is digested immediately after the first strand.[51] To assess the effect of $Mg^{2+}$ and $Ca^{2+}$ on the morphology of the nuclear residues, DNA can be chemically cleaved in the absence of both

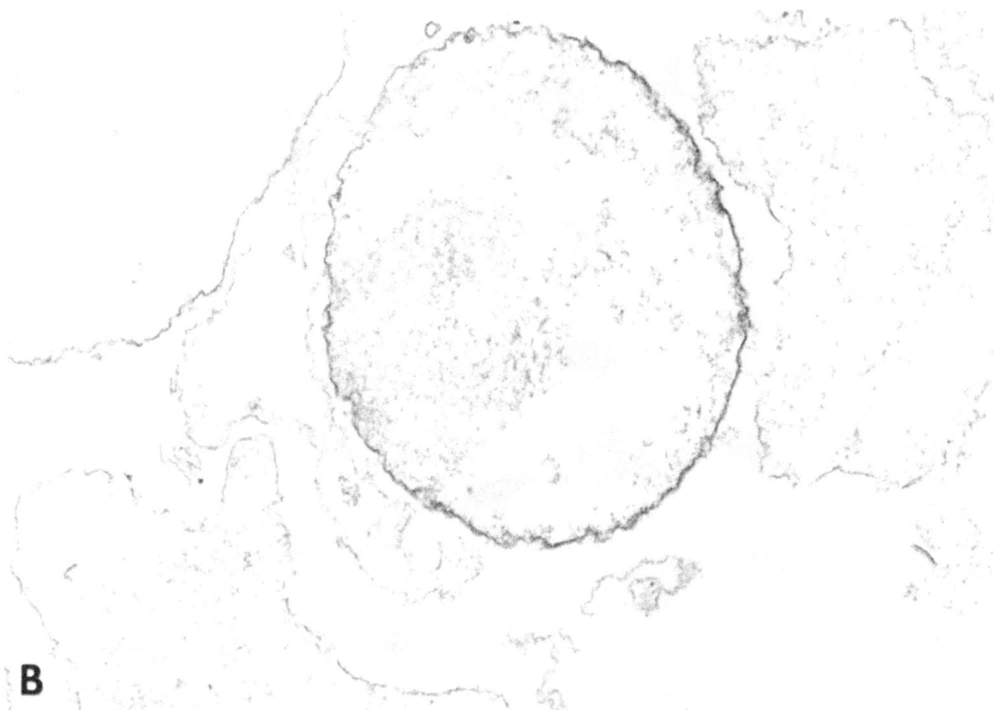


FIGURE 3B

metals by the 1,10-phenanthroline-cuprous coordination complex $(OP)_2Cu(I)$. Phenanthroline is a tricyclic compound which intercalates in double-stranded nucleic acids and produces the cleavage of phosphodiester bonds at pH 7.0.[52] Upon the addition of $(OP)_2Cu(I)$ and after a high salt treatment, most of the nuclear residues have no nucleoli and very little, if any, internal material. This method produces empty nuclear shells (Figure 6)! At this time, the effect of $(OP)_2Cu(I)$ in the dissolution of the internal network and nucleoli is not clear. The phenomenon can be attributed to the absence of metals, particularly $Ca^{2+}$.[49,50] However, it appears to be a more complex process. Thus, micrococcal nuclease digestion which requires $Ca^{2+}$ yields empty shells under certain conditions (to be published). This experimental evidence indicates that the interactions of whichever proteins may be responsible for what is known as nuclear matrix structures are drastically affected by the manner in which DNA is digested.

## V. DNA BINDING TO NUCLEAR MATRIX

The packing of DNA by histones into nucleosomal subunits and the packing of nucleosomes into a supranucleosomal structure are the two levels of DNA folding in the nucleus. The supranucleosomal structure forms a chromatin fiber having the configuration of either solenoid or superbeads.[53] There is evidence that the chromatin fiber is constrained by a proteinaceous network in both the metaphase and interphase.[54-58] The organization of the chromatin fiber can be described by a radial loop model in which chromatin loops 50 to 100 kb long emanate from a central core or axis. The chromosomal core which links the base of the DNA loop is assumed to be formed by a network of nonhistone proteins and was named the chromosomal scaffold by Laemmli and co-workers.[59] Therefore, there is at least one DNA binding site in chromosomes and two DNA binding sites in interphase nuclei. Scaffold-DNA is defined as the DNA fragment which binds to the chromosomal scaffold;

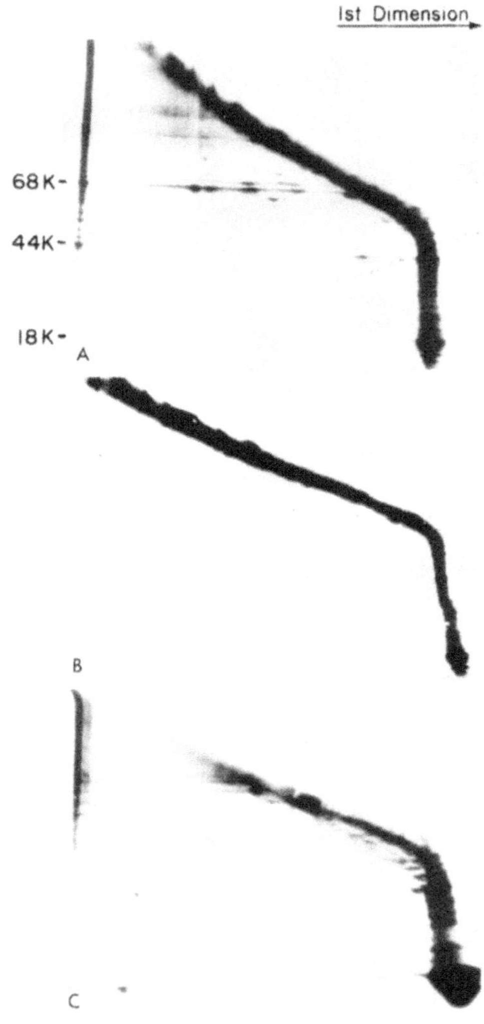


FIGURE 4. Effect of disulfide bonds on nucleoprotein interaction. The high salt residues and extracts were obtained from [$^{35}$S] methionine-labeled cells. Nuclei were isolated and digested as explained (Figure 2 and Reference 35). The nuclear residues were washed with 10 m*M* Tris-HCl (pH 7.7), 10 m*M* EDTA, 1 m*M* EGTA, dissolved with 2% lithium dodecyl sulfate (LDS) and clarified by centrifugation in the SW 50.1 Spinco rotor at 49 K for 60 min. The high salt extract was diluted with $H_2O$ and concentrated in an Amicon® ultrafiltration cell filter (PM30). O'Farrell buffer without reducing reagent was added to both the LDS lysate of the nuclear residues and the concentrated high salt extract. An appropriate amount of sample was applied onto the first acrylamide slab gel. It consisted of 5% acrylamide on top of a 5 to 6% acrylamide gradient and a 3% acrylamide stacking gel. After the run each lane was excised, soaked at room temperature for 30 min in a buffered 1% β-mercaptoethanol solution, and placed sideways on top of the second 5 to 15% acrylamide gradient gel. After an overnight run, the gels were fluorographed according to Bonner and Laskey.[92] (A) High salt residues obtained from nuclei isolated in the absence of NEM; (B) high salt residues obtained from nuclei isolated in the absence of NEM; (C) high salt extract obtained from nuclei isolated in the absence of NEM.

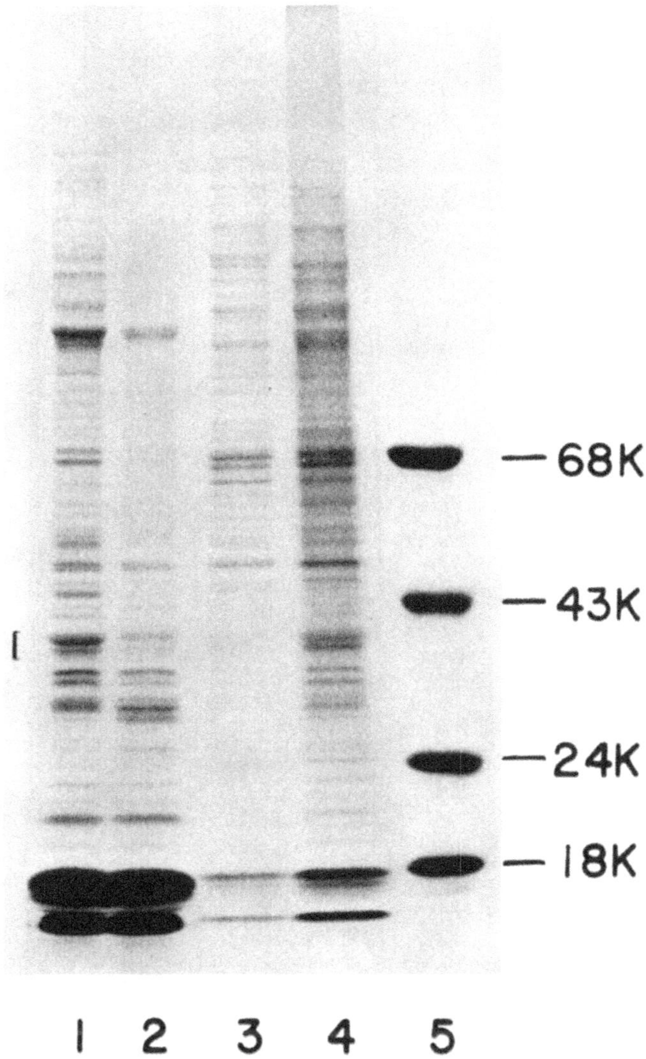


FIGURE 5. Effect of disulfide bond formation on nucleoprotein distribution between high salt residues and extracts. The high salt residues and extracts were obtained as explained in Figure 4 and Reference 35. The O'Farrell buffer with a reducing reagent (β-mercaptoethanol) was added, boiled for 5 min and aliquots were applied on a 5 to 15% acrylamide slab gel gradient. After an overnight run, the gels were stained and destained as explained.[35] (1) High salt extract obtained from nuclei isolated in the presence of NEM; (2) high salt extract obtained from nuclei isolated in the absence of NEM and treated with $H_2O_2$; (3) high salt residues obtained from nuclei isolated in the presence of NEM; (4) high salt residues obtained from nuclei isolated in the absence of NEM and treated with $H_2O_2$; (5) standards. The bracket indicates one set of two bands of molecular weight <43 K which are present in the high salt extract when a sulfhydryl-blocking reagent is added during nuclear isolation. Concurrently, these bands are present in the nuclear residues obtained from nuclei isolated in the absence of a sulfhydryl-blocking reagent.

matrix-DNA, the DNA fragment which binds to the intranuclear network, and lamina-DNA, the DNA fragment which binds to the nuclear periphery, i.e., the lamina-pore complexes. The proteins associated with scaffold-DNA and matrix-DNA are located at the base of the DNA-loop. Since lamina is known to be dissociated late in the G2 phase of the cell cycle

FIGURE 6. Electron microscopy of nuclear residues obtained from nuclei isolated in the presence of NEM and treated with $(OP)_2Cu(I)$ complex. Nuclei were isolated as explained in Figure 2, resuspended at a density of 7 × $10^7$ nuclei per milliliter in 20 m*M* Tris-acetate (pH 7.0), 15 m*M* sodium acetate, 60 m*M* potassium acetate, 250 m*M* sucrose, and 0.5 m*M* PMSF. A reaction mixture containing 100 m*M* Tris-acetate )pH 7.0), 1 m*M* 1,10-phenanthroline, 0.1 m*M* $CuSO_4$ and 14 m*M* 3-mercaptopropionic acid was freshly prepared. A 0.2 volume was added to the nuclear suspension and incubated at 8°C for 2 min. Ten microliters of $H_2O_2$ per milliliter (0.84% freshly prepared in 0.1 *M* Tris acetate pH 7.0) was added and incubated at 8°C for 13 min. The suspension was made in 0.5% Triton® X-100 and nuclei were pelleted by centrifugation. The pellet was resuspended in the same buffer and digested with RNAase 50 μg/mℓ at 8°C for 20 min. The nuclear suspension was pelleted by centrifugation, resuspended in a high salt buffer, and fixed as explained in Figure 2. Notice the absence of internal nuclear material (nuclear matrix) in the residues. (Magnification × 9000.)

as chromosome condensation proceeds, the binding of lamins A and B to DNA is lost in chromosomes.[60,61] Therefore it is reasonable to expect commonalities only between scaffold- and matrix-DNA as far as protein and DNA sequences are concerned. Also, it is reasonable to expect that some proteins that bind to the base of the loop may remain DNA-bound during the entire cell cycle. Laemmli and co-workers have found some similarities between the proteins that bind to scaffold-DNA and matrix-DNA.[49,50] It appears that $Ca^{2+}$ or $Cu^{2+}$ is necessary to maintain or stabilize these structures, and since $Ca^{2+}$ produces an irreversible toughening of the nuclei, Laemmli et al. propose that $Cu^{2+}$ may form a physiological coordination complex with the sulfhydryl-containing proteins.[50] The idea is supported by the fact that chelators like neocuproine but not ethyleneglycol-*bis*-(β-aminoethyl ether) *N,N'*-tetraacetic acid (EGTA) lead to the solubilization of many proteins and their disassembly from the base of the loop. Some polypeptides of scaffold- and matrix-DNA appear to have similar electrophoretic mobilities; their identities, however, remain to be determined by polypeptide mapping.[49,50]

The proteins that bind to matrix-DNA have been thought to play an important role in

DNA replication and gene expression. Coffey and co-workers[32] have observed that matrix-DNA obtained from regenerating rat liver or exponentially growing 3T3 fibroblasts pulse-labeled with $^3$H-thymidine is highly enriched in newly synthesized DNA. This is also true in several viruses such as polyoma, adenovirus, and SV40 which remain matrix-bound throughout the entire period of replication.[62-64] Several groups have postulated the existence of a fixed site of DNA synthesis in which replication complexes are anchored to the intranuclear matrix network, and DNA is reeled through these complexes as it is replicated.[32,65] Since the solubility of many of the so-called matrix proteins depends upon the formation of disulfide bonds and DNA digestion, care needs to be taken in drawing the conclusion that the replicative apparatus is a component of the matrix. This issue must be reexamined in nuclei prepared with sulfhydryl-blocking reagents. It would be important to know whether the proteins involved in replication are insoluble in high salt and tightly bound to DNA only when they are cross-linked by disulfide bonds. Another major point worth mentioning is that newly synthesized DNA like active genes are preferentially sensitive to DNAase I, but after a high salt treatment they become DNAase I resistant. There is something very peculiar in the effect of high salt on the proteins associated with the newly synthesized DNA.

Whether or not scaffold and matrix-DNA contain unique sequences and whether these sequences are similar in both structures remain to be demonstrated, since conflicting evidence has been reported. Georgiev and co-workers[66,67] using mouse L cells and Jeppesen and Bankier[68] using Don Chinese hamster cells found a slight, roughly twofold enrichment of highly repetitive sequences in scaffold-DNA. However, Bowen,[69] using HeLa S3 cells did not find a uniform class of sequence. The only DNA fragment that he detected was a component of satellite II and III DNA which is located primarily on human chromosome I. Kuo,[70] using the chromosomes of chicken lymphoblastoid cell MSB-I, did not find a specific sequence such as ovalbumin gene. However, he was able to detect a slight enrichment in repetitive sequences. The same discrepancy has been reported when experiments were designed to determine whether or not matrix DNA is enriched in active genes. Using chicken oviduct cells, Vogelstein and co-workers[71] have reported the preferential association of ovalbumin genes with nuclear matrix, and Cook and Brazell[72] using HeLa cells have found the preferential attachment of α-globin genes. Although Kuo[70] has made a similar observation, no depletion in globin genes is found in laying hen oviduct, and the copy number of the ovalbumin gene in matrix-DNA obtained from erythrocyte or liver nuclei is not significantly different from that of total unfractionated DNA.

Several shortcomings exist in the experimental design of these studies. To begin, in interphase nuclei, matrix- and lamina-DNA were analyzed together and a priori we expect that they must contain different DNA sequences. Although Kuo[70,73] has attempted to purify matrix-DNA complexes, his procedure does not discriminate between the DNA which is bound to lamina and the DNA which is bound to matrix. Assuming that matrix-DNA can be obtained free from lamina-DNA, it is possible that matrix-DNA contains at least two sets of DNA fragments, one belonging to scaffold-DNA and the other to active genes (unique sequences). Thus, it is known that the RNA-polymerase DNA complexes are not dissociated by high salt, and therefore factors for transcription and subunits of RNA-polymerase associated with DNA may become matrix components.[74] If highly repetitive sequences exist in scaffold- and matrix-DNA, they must be short — on the order of 9 bp as Jeppesen and Bankier[68] have suggested. In this case, recombinant DNA techniques must be used to detect them. As has been mentioned before, it is important to establish whether or not the binding of unique or repetitive DNA sequences to matrix proteins is affected either by the formation of disulfide bonds or by the manner in which matrix is prepared.

## VI. HnRNA, SnRNA, AND rRNA BINDING TO NUCLEAR MATRIX

Bound to the internal network are heterogeneous nuclear RNA (hnRNA), small nuclear

RNA (snRNA), and the 28S RNA which belong to the large ribosomal subunit in transit to the cytoplasm; bound to the nucleolus are ribosomal precursors. The presence of 18S RNA in nuclear matrix preparation is due to the presence of the small ribosomal subunit of the ribosomes that are attached to the outer nuclear membrane and to the cytoskeletal fibrils.[1,3] In rat liver nuclei these ribosomes remain attached even after dissolution of lipids by a nonionic detergent.[75] HnRNA and snRNA exist as nRNP* complexes associated with the intranuclear network (for review, see Pogo[76]). Recently it has been shown in HeLa cells infected with adenovirus type 2 that the virus-specific heterogeneous nuclear RNA is quantitatively associated with nuclear matrix.[77] Analysis of this matrix-associated RNA by S1 nuclease mapping revealed that precursors and processed messenger RNA (mRNA) remain associated with nuclear matrix. Chae and co-workers[78] using chicken erythroblasts and erythrocytes have observed that greater than 90% of the globin nuclear RNA remains bound to the nuclear matrix. More recently, Ciejek et al.[79] using chick oviduct nuclear matrix have observed that all of the precursors to rRNA, ovalbumin, and ovomucoid mRNAs (including various splicing intermediates) are associated with the nuclear matrix fraction. By contrast, mature ovalbumin and ovomucoid mRNAs are distributed between matrix and nonmatrix fractions. Using influenza-infected chick embryo fibroblasts, Cook and co-workers[80] have observed that most of the nascent viral transcripts are closely associated with nuclear matrix. They concluded that not only transcription and replication of viral RNA take place in the matrix, but also that they occur at fixed sites in the nucleus. The same group has challenged the widely accepted concept that it is the RNA polymerase which moves along the DNA; rather it is the DNA which passes through a transcription complex fixed to a subnuclear structure.[81] The model is similar to that postulated by Coffey and co-workers for replication.[32] There are two possible mechanisms that can explain the model. One is that there are no DNA fixed sites within the nucleus; the other is that these sites exist at the base of the DNA loop but during unfolding of the chromatin, the DNA moves through a transcription complex which in turn is attached to the nuclear matrix.

Van Venrooij and co-workers[82] have demonstrated that hnRNP (i.e., nRNP containing hnRNA) is quantitatively associated with matrix structures. They found that a 41.5 and a 43.0 K protein were not released by ribonuclease treatment, suggesting that these proteins are involved in hnRNA binding to nuclear matrix structures. However, the so-called hnRNA packaging proteins (32 to 38 K) were released after ribonuclease treatment of nuclear matrix. On the contrary, Maundrell et al.[83] have found that the bulk of the proteins of nRNP particles, except for 43 and 73 K, are extracted during matrix preparation. This discrepancy may be due to the different procedures of matrix isolation employed by these two groups. Since the procedures for nRNP isolation involved nuclear sonication, it is difficult to differentiate protein species which are associated with hnRNA from co-sedimenting matrix fragments.[76] Peters and Comings[84] have tried to examine this question by analyzing the proteins obtained from nuclear washes, nuclear matrix, and the 40S hnRNA particles in two-dimensional gel electrophoresis. In spite of the RNAase treatment, approximately one third of total matrix proteins have electrophoretic mobilities similar to that of hnRNP proteins. However, in this kind of analysis it is difficult to know whether the proteins that have similar mobilities in both structures are the same proteins and whether all the proteins of the 40S hnRNP particles are bona fide hnRNA-associated proteins. In view of the fact that matrix protein solubilities in high salt can depend upon the way in which DNA is digested as well as the formation of disulfide bonds, the solubility of hnRNP and snRNP (i.e., nRNP containing snRNA) has to be reexamined. For example, we have observed that in nuclei where disulfide bonds were produced, high salt as well as heparin did not release hnRNA.[76] On the contrary, if the sulfhydryl groups were blocked with *N*-ethylmaleimide, high salt as well as dextran sulfate/

* nRNP: All nuclear ribonucleoprotein particles excluding those containing preribosomal and ribosomal RNA species.

heparin were able to release native hnRNA, ribosomal precursor, and snRNA. The status of these RNA species is under investigation.

## VII. CONCLUSION

The major obstacle to the detection of an underlying nuclear structure is the presence of huge DNA molecules, highly folded and packed within a tiny nuclear space. In most eukaryotes (yeast is one exception), there are enough DNA and histones to fill all the nuclear space. Therefore, the space-filling nuclear network must a priori have a thin texture. Since DNA, RNA, and histone removal is a prerequisite for the biochemical and morphological analysis of nuclear matrix, in assessing the nature of this structure it is essential to know whether its occurrence is due to the method that solubilizes nucleic acids and histones. The important issue is to ascertain whether the removal of DNA, histones, and RNA produces such rearrangement of components that yields a nuclear matrix. The precise elucidation of nuclear matrix hinges upon this critical point. No internal network and residual nucleolus are obtained if DNA is chemically cleaved with $(OP)_2Cu(I)$ or digested with micrococcal nuclease. On the contrary, nuclear matrix is present when nuclei are digested with DNAase I. At the present time we do not know the reason for this phenomenon. Since DNAase I digests DNA into soluble fragments of 10 to 11 bases long, it is possible that positive-charged proteins may establish ionic bonds with negative-charged proteins. The electrostatic attractions may produce rearrangement of a subset of proteins and give rise to the opportunity for other protein-protein interactions such as the formation of strong hydrophobic and hydrogen bonds. In turn, these bonds may change the high salt solubility of these proteins. Perhaps some of these processes are mediated through cations or, as Laemmli and coworkers[49,50] suggest, through metals. In addition, the formation of disulfide bonds among several nonhistone proteins may help to stabilize protein-protein interactions. When DNA is chemically cleaved, these processes do not occur because the electrostatic balance is maintained by the release of large DNA fragments together with proteins, the reaction takes place in the presence of a sulfhydryl reducing reagent, and phenathroline is an effective chelator.

Does this mean that the presence of an intranuclear network and a residual nucleolus is an artifact? At present, we are in no position to resolve this important issue. However, we want to point out here that many of the proteins that are high salt-insoluble after DNAase I digestion are not true matrix proteins.

It is commonly assumed that nuclear matrix structures are the interchromatin fibrils and granules described by Monneron and Bernhard[42] in whole cells. However, only a tenuous morphological similarity exists to support this contention. Most probably, DNAase I digestion and high salt treatment alter, distort, and rearrange the components of the structures described by Monneron and Bernhard.[42] Therefore, DNAase I digestion and high salt treatment might not be the appropriate procedures for the isolation of native interchromatin fibrils and granules.

Lamina, pore complexes, and nuclear envelope are obtained irrespective of the method used to digest DNA, either by DNAase I or phenanthroline. They are therefore true nuclear entities. It is reasonable to assume that interchromatin fibrils and granules may be organized either through the pore or the lamina. Cavalier-Smith[85] has postulated the former rather than the latter as the origin of the nuclear filaments. We visualize the presence of a nonchromatinous fibrogranular structure as a dynamic nuclear entity establishing connections with the nuclear pore and chromatin areas as well as the nucleoli. Its components may consist exclusively of protein-protein interactions, or perhaps of ribonucleoprotein complexes. Such a structure may exist for DNA synthesis (base of the loop), for the formation of chromosomes (scaffold), for transcription, for hnRNA packing and processing, for rRNA processing and assembly with proteins (the nucleolus), and for the transport of ribosomal subunits as well as nRNP.

What is the need for a nuclear matrix? The answer to this question must lie in the origin of nucleated cells. As far as we can tell, all eukaryotes are derived from a common ancestor.[85] Perhaps to adapt to a predatory way of life, a species had to first evolve to a size that would enable it to engulf smaller organisms by a phagocytic mechanism similar to that by which amoeboid cells now feed.[86] Two important changes had to occur with phagocytosis: the formation of vacuoles and the internalization of semi-permeable membranes. The latter may be related to the formation of the nuclear membrane since the DNA in prokaryotes remains attached to the membrane. Perhaps the ancestral nuclear membrane is the mesosome which is an invagination of the prokaryotic cell membrane.[87,88] With the formation of the cell nucleus, DNA attachment remains, but to the nuclear membrane, in particular to the lamina.[49,50] With the appearance of the nuclear membrane, replication, transcription, RNA processing, and gene regulation were compartmentalized and physically separated from translation. Despite the similarity between the nuclear envelope and other membranes, the nuclear envelope contains the largest amount of pore complex. These pores often have eight-sided polygonal structures and thin fibrils.[12,22] It is tempting to ascribe the origin of the internal nuclear filaments to the intrapore fibrils, and perhaps both filaments are alike.[89] Gene products from the nucleus reach the cytoplasmic site for protein synthesis by passing through the nuclear pores. The nucleocytoplasmic traffic is mediated through channels that exist within the pores themselves and the nuclear interior. The nuclear network may function to route molecules and macromolecules. Thus, ribosomal proteins and the subunits of RNA polymerase I must be directed toward the nucleolus; the subunits of RNA polymerase II and III as well as hormone receptors[31] and other regulatory molecules[90,91] are directed toward their active chromatin sites. Finally, the components of the splicing machinery are directed to sites where the primary transcripts are synthesized.

## ACKNOWLEDGMENTS

The authors are grateful to Tellervo Huima for electron microscopic work, to Paula Krauss for editing and correction of the manuscript, and to Valerie Capers for word processing. This work was supported by NSF Grant #PCM-8110074.

## REFERENCES

1. **Brown, S., Levinson, W., and Spudich, J. A.,** Cytoskeletal elements of chick embryo fibroblasts revealed by detergent extraction, *J. Supramol. Struct.*, 5, 119, 1976.
2. **Wolosewick, J. J. and Porter, K. R.,** Microtrabecular lattice of the cytoplasmic ground substance: artifact or reality?, *J. Cell Biol.*, 82, 114, 1979.
3. **Fulton, A. B., Wan, K. M., and Penman, S.,** The spatial distribution of polyribosomes in 3T3 cells and the associated assembly of proteins into the skeletal framework, *Cell*, 20, 849, 1980.
4. **Heuser, J. E. and Kirschner, M. W.,** Filament organization revealed in platinum replicas of freeze-dried cytoskeletons, *J. Cell Biol.*, 86, 212, 1980.
5. **Cook, P. R. and Brazell, I. A.,** Mapping sequences in loops of nuclear DNA by their progressive detachment from the nuclear cage, *Nucleic Acids Res.*, 8, 2895, 1980.
6. **Miller, T. E., Huang, C.-Y., and Pogo, A. O.,** Rat liver nuclear skeleton and ribonucleoprotein complexes containing HnRNA, *J. Cell Biol.*, 76, 675, 1978.
7. **Paulson, J. R. and Laemmli, U. K.,** The structure of histone-depleted metaphase chromosomes, *Cell*, 12, 817, 1977.
8. **Berezney, R. and Coffey, D. S.,** Nuclear matrix. Isolation and characterization of a framework structure from rat liver nuclei, *Biochem. Biophys. Res. Commun.*, 60, 1410, 1974.

9. **Fawcett, D. W.,** On the occurrence of a fibrous lamina on the inner aspect of the nuclear envelope in certain cells of vertebrates, *Am. J. Anat.*, 119, 129, 1966.
10. **Patrizi, G. and Poger, M.,** The ultrastructure of the nuclear periphery: the zonula nucleum limitans, *J. Ultrastruct. Res.*, 17, 127, 1967.
11. **Stelly, V. B., Stevens, B. J., and Andre, J.,** Étude cytochimique de la lamelle dense de l'envelope nucléare, *J. Microsc. (Paris)*, 9, 1015, 1970.
12. **Aaronson, R. P. and Blobel, G.,** Isolation of nuclear pore complexes in association with a lamina, *Proc. Natl. Acad. Sci. U.S.A.*, 72, 1007, 1975.
13. **Gerace, L., Blum, A., and Blobel, G.,** Immunocytochemical localization of the major polypeptides of the nuclear pore complex lamina fraction: interphase and mitotic distribution, *J. Cell Biol.*, 79, 546, 1978.
14. **Krohne, G., Franke, W. W., Ely, S., D'Arcy, A., and Jost, E.,** Localization of a nuclear envelope-associated protein by indirect immunofluorescence microscopy using antibodies against a major polypeptide from rat liver fractions enriched in nuclear envelope-associated material, *Cytobiology*, 18, 22, 1978.
15. **Krohne, G., Dabauvalle, M. Ch., and Franke, W. W.,** Cell type-specific difference in protein composition of nuclear pore complex-lamina structures in oocytes and erythrocytes of *Xenopus laevis*, *J. Mol. Biol.*, 151, 121, 1981.
16. **Stick, R. and Hausen, P.,** Immunological analysis of nuclear lamina proteins, *Chromosoma*, 80, 219, 1980.
17. **Lam, K. S. and Kasper, C. B.,** Electrophoretic analysis of three major nuclear envelope polypeptides, *J. Biol. Chem.*, 254, 11713, 1979.
18. **Shelton, K. R., Higgins, L. L., Cochran, D. L., Ruffolo, J. J., and Egle, P. M.,** Nuclear lamins of erythrocyte and liver, *J. Biol. Chem.*, 255, 10978, 1980.
19. **Krohne, G., Franke, W., and Scheer, U.,** The major polypeptides of the nuclear pore complex, *Exp. Cell Res.*, 116, 85, 1978.
20. **Maul, G. G. and Audalovic, N.,** Nuclear envelope proteins from *Spirula solidissima* germinal vesicles, *Exp. Cell Res.*, 130, 229, 1980.
21. **Stick, R. and Schwarz, H.,** The disappearance of the nuclear lamina during spermatogenesis: an electron microscopic and immunofluorescence study, *Cell Differentiation*, in press.
22. **Scheer, U., Kartenbeck, J., Trendelenburg, M. F., Stadler, J., and Franke, W. W.,** Experimental disintegration of the nuclear envelope, *J. Cell Biol.*, 69, 465, 1976.
23. **Stick, R. and Krohne, G.,** Immunological localization of the major architectural protein associated with nuclear envelope of the *Xenopus laevis* oocyte, *Exp. Cell Res.*, 138, 319, 1982.
24. **Richardson, J. C. W. and Maddy, A. H.,** The polypeptides of rat liver nuclear envelope, *J. Cell Sci.*, 43, 253, 1980.
25. **Mayer, D. T. and Gulick, A.,** The nature of the protein of cellular nuclei, *J. Biol. Chem.*, 146, 433, 1942.
26. **Mirsky, A. E. and Ris, H.,** The composition and structure of isolated chromosomes, *J. Gen. Physiol.*, 34, 475, 1951.
27. **Georgiev, G. P. and Chentsov, Y. S.,** The structural organization of nucleochromosomal ribonucleoprotein, *Exp. Cell Res.*, 27, 570, 1962.
28. **Smetana, K., Steele, W. J., and Busch, H.,** A nuclear ribonucleoprotein network, *Exp. Cell Res.*, 31, 198, 1963.
29. **Shaper, J. H., Pardoll, D. M., Kaufmann, S. H., Barrack, E. R., Vogelstein, B., and Coffey, D. S.,** The relationship of the nuclear matrix cellular structure and function, *Adv. Enzyme Regul.*, 17, 213, 1979.
30. **Comings, D. E. and Okada, T. A.,** Nuclear proteins. III. The fibrillar nature of the nuclear matrix, *Exp. Cell Res.*, 103, 341, 1976.
31. **Barrack, E. R., Hawkins, E. F., Allen, S. L., Hicks, L. L., and Coffey, D. S.,** Concepts related to salt resistant estradiol receptors in rat uterine nuclei: nuclear matrix, *Biochem. Biophys. Res. Commun.*, 79, 829, 1977.
32. **Pardoll, D. M., Vogelstein, B., and Coffey, D. S.,** A fixed site of DNA replication in eukaryotic cells, *Cell*, 19, 527, 1980.
33. **Hodges, C. D., Mancini, P., Davis, F. M., and Heywood, P.,** Nuclear matrix of HeLa S3 cells. Polypeptide composition during adenovirus infection and in phases of the cell cycle, *J. Cell Biol.*, 72, 194, 1979.
34. **Berezney, R., Basler, J., Hughes, B. B., and Kaplan, S. C.,** Isolation and characterization of the nuclear matrix from Zajdela ascites hepatoma cells, *Cancer Res.*, 39, 3031, 1979.
35. **Long, B., Huang, C.-Y., and Pogo, A. O.,** Isolation and characterization of the nuclear matrix in Friend erythroleukemia cells: chromatin and heterogeneous RNA interactions with the nuclear matrix, *Cell*, 18, 1079, 1979.
36. **Wunderlich, F. and Herlan, G. A.,** A reversible contractile nuclear matrix, its isolation, structure and composition, *J. Cell Biol.*, 73, 271, 1977.

37. **Wolfe, J.,** A possible skeletal substructure of the macronucleus of *tetrahymena*, *J. Cell Biol.*, 84, 160, 1980.
38. **Michelson, K. R., Bekers, A. G. M., and Wank, F.,** Isolation of a residual protein structure from nuclei of the myxomycete *Physarum polycephalum*, *J. Cell Sci.*, 39, 247, 1979.
39. **Fisher, P. A., Berrios, M., and Blobel, G.,** Isolation and characterization of a proteinaceous subnuclear fraction composed of nuclear matrix, peripheral lamina and nuclear pore complexes from embryos of *Drosophila melanogaster*, *J. Cell Biol.*, 92, 674, 1982.
40. **La Fond, R. E. and Woodcock, C. L. F.,** The status of the nuclear matrix in adult chicken erythrocyte nuclei, *J. Cell Biol.*, 91, 57a, 1981.
41. **Franke, W. W., Kleinschmidt, J. A., Sprung, H., Krohne, G., Grund, C., Trendelenburg, M. F., Stohr, M., and Scheer, U.,** A nucleolar skeleton protein filament demonstrated in amplified nucleoli of *Xenopus Laevis*, *J. Cell Biol.*, 90, 289, 1981.
42. **Monneron, A. and Bernhard, W.,** Fine structural organization of the interphase nucleus in some mammalian cells, *J. Ultrastruc. Res.*, 27, 266, 1969.
43. **Ghosh, S., Paweletz, N., and Ghosh, I.,** Cytological identification and characterization of the nuclear matrix, *Exp. Cell Res.*, 111, 363, 1978.
44. **Brasch, K.,** Fine structure and localization of the nuclear matrix in situ, *Exp. Cell Res.*, 140, 161, 1982.
45. **Capco, D., Wan, K. M., and Penman, S.,** The nuclear matrix. Three dimensional architecture and protein composition, *Cell*, 29, 847, 1982.
46. **Kaufman, S. H., Coffey, D., and Shaper, J. H.,** Considerations in the isolation of rat liver nuclear matrix, nuclear envelope and pore-complex lamina, *Exp. Cell Res.*, 132, 105, 1981.
47. **Paulson, J. R. and Laemmli, U. K.,** The structure of histone-depleted metaphase chromosomes, *Cell*, 12, 817, 1977.
48. **Van Eekelen, C. A. G., Salden, M. H. L., Habets, W. J. A., Van de Putte, L. B. A., and Van Venrooij, W. J.,** On the existence of an internal nuclear protein structure in HeLa cells, *Exp. Cell Res.*, 141, 181, 1982.
49. **Lewis, C. D. and Laemmli, U. K.,** Higher order metaphase chromosome structure: evidence for metalloprotein interactions, *Cell*, 29, 171, 1982.
50. **Lebkowski, J. S. and Laemmli, U. K.,** Nonhistone proteins and long range organization of HeLa interphase DNA, *J. Mol. Biol.*, 156, 325, 1982.
51. **Price, P. A.,** The essential role of $Ca^{2+}$ in the activity of bovine pancreatic deoxyribonuclease, *J. Biol. Chem.*, 250, 1981, 1975.
52. **Marshall, L. E., Graham, D. R., Reich, K. A., and Sigman, D. S.,** Cleavage of deoxyribonucleic acid by the 1,10 phenanthroline-cuprous complex. Hydrogen peroxide requirement and primary and secondary structure specificity, *Biochemistry*, 20, 244, 1981.
53. **Kornberg, R. D.,** Chromatin structure: a repeating unit of histones and DNA, *Science*, 184, 868, 1974.
54. **Adolph, K. W., Cheng, S. M., Paulson, J. R., and Laemmli, U. K.,** Isolation of a protein scaffold from mitotic HeLa cell chromosomes, *Proc. Natl. Acad. Sci. U.S.A.*, 74, 4937, 1977.
55. **Benyajati, C. and Worcel, A.,** Isolation, characterization and structure of the folded interphase genome of *Drosophila melanogaster*, *Cell*, 9, 393, 1976.
56. **Cook, P. R. and Brazell, I. A.,** Characterization of nuclear structures containing superhelical DNA, *J. Cell Sci.*, 22, 287, 1976.
57. **Igo-Kemenes, T. and Zachau, H. G.,** Domains in chromatin structure, *Cold Spring Harbor Symp. Quant. Biol.*, 42, 109, 1978.
58. **Vogelstein, B., Pardoll, D. M., and Coffey, D. S.,** Supercoiled loops and eucaryotic DNA replication, *Cell*, 22, 79, 1980.
59. **Laemmli, U. K., Cheng, S. M., Adolph, K. W., Paulson, J. R., Brown, J. A., and Baumback, W. R.,** Metaphase chromosome structure: the role of nonhistone proteins, *Cold Spring Harbor Symp. Quant. Biol.*, 42, 351, 1978.
60. **Gerace, L. and Blobel, G.,** The nuclear envelope lamina is reversibly depolymerized during mitosis, *Cell*, 19, 277, 1980.
61. **Jost, E. and Johnson, R. T.,** The nuclear lamina assembly, synthesis and disaggregation during the cell cycle in synchronized HeLa cells, *J. Cell Sci.*, 47, 25, 1981.
62. **Buckler-White, A. J., Humphrey, G. W., and Pigiet, V.,** Association of polyoma T antigen and DNA with nuclear matrix from lytically infected 3T6 cells, *Cell*, 22, 37, 1980.
63. **Younghusband, H. B. and Maundrell, K.,** Adenovirus DNA is associated with the nuclear matrix of infected cells, *J. Virol.*, 43, 705, 1982.
64. **Deppert, W.,** Simian virus 40 (SV40)-specific proteins associated with the nuclear matrix isolated from adenovirus type 2-SV40 hybrid virus-infected HeLa cells carry SV40 U-antigen determinants, *J. Virol.*, 26, 165, 1978.
65. **Dijkwel, P. A., Mullenders, L. H. F., and Wanka, F.,** Analysis of the attachment of replicating DNA to a nuclear matrix in mammalian interphase nuclei, *Nucleic Acids Res.*, 6, 219, 1979.

66. **Razin, S. V., Mantieva, V. L., and Georgiev, G. P.,** DNA adjacent to attachment points of deoxyribonucleoprotein fibril to chromosomal axial structure is enriched in reiterated base sequences, *Nucleic Acids Res.*, 5, 4737, 1978.
67. **Razin, S. V., Mantieva, V. L., and Georgiev, G. P.,** The similarity of DNA sequences remaining bound to scaffold upon nuclease treatment of interphase nuclei and metaphase chromosomes, *Nucleic Acids Res.*, 7, 1713, 1979.
68. **Jeppesen, P. G. N. and Bankier, A. T.,** A partial characterization of DNA fragments protected from nuclease degradation in histone depleted metaphase chromosomes of the Chinese hamster, *Nucleic Acids Res.*, 7, 49, 1979.
69. **Bowen, B. C.,** DNA fragments associated with chromosome scaffolds, *Nucleic Acids Res.*, 9, 5093, 1981.
70. **Kuo, M. T.,** Analysis of DNA attached to the chromosome scaffold, *J. Cell Biol.*, 93, 278, 1982.
71. **Robinson, S. I., Nelkin, B. D., and Vogelstein, B.,** The ovalbumin gene is associated with the nuclear matrix of chicken oviduct cells, *Cell*, 28, 99, 1982.
72. **Cook, P. R. and Brazell, I. A.,** Mapping sequences in loops of nuclear DNA by their progressive detachment from nuclear cage, *Nucleic Acids Res.*, 8, 2895, 1980.
73. **Kuo, M. T.,** Distribution of tightly bound proteins in the chicken ovalbumin gene region, *Biochemistry*, 21, 321, 1982.
74. **Pogo, A. O.,** Modification of ribonucleic acid synthesis in isolated rat liver nuclei by low salt concentrations and specific divalent cations, *Biochem. Biophys. Acta*, 182, 57, 1969.
75. **Miller, T. E., Huang, C.-Y., and Pogo, A. O.,** Rat liver nuclear skeleton and small molecular weight RNA species, *J. Cell Biol.*, 76, 692, 1978.
76. **Pogo, A. O.,** Heterogeneous nuclear RNA-protein complexes and nuclear matrix in *The Cell Nucleus*, Vol. 8, Busch, H., Ed., Academic Press, New York, 1981, chap. 9.
77. **Mariman, E. C. M., Van Eekelen, C. A. G., Reinders, R. J., Berns, A. J. M., and van Venrooij, W. J.,** Adenoviral heterogeneous nuclear RNA is associated with the host nuclear matrix during splicing, *J. Mol. Biol.*, 154, 103, 1982.
78. **Ross, D. A., Yen, R.-W., and Chae, C.-B.,** Association of globin ribonucleic acid and its precursors with chicken erythroblast nuclear matrix, *Biochemistry*, 21, 764, 1982.
79. **Ciejek, E. M., Nordstrom, J. L., Tsai, M.-J., and O'Malley, B. W.,** Ribonucleic acid precursors are associated with the chick oviduct nuclear matrix, *Biochemistry*, 21, 4945, 1982.
80. **Jackson, D. A., Caton, A. J., McCready, S. J., and Cook, P. R.,** Influenza virus RNA is synthesized at fixed sites in the nucleus, *Nature (London)*, 296, 366, 1982.
81. **Jackson, D. A., McCready, S. J., and Cook, P. R.,** RNA is synthesized at the nuclear cage, *Nature (London)*, 292, 552, 1981.
82. **Van Eekelen, C. A. G. and van Venrooij, W. J.,** HnRNA and its attachment to a nuclear protein matrix, *J. Cell Biol.*, 88, 554, 1981.
83. **Maundrell, K., Maxwell, E. S., Puvion, E., and Scherrer, K.,** The nuclear matrix of duck erythroblasts is associated with globin mRNA coding sequences but not with the major proteins of 40S nuclear RNP, *Exp. Cell Res.*, 136, 435, 1981.
84. **Peters, K. E. and Comings, D. E.,** Two-dimensional gel electrophoresis of rat liver nuclear washes, nuclear matrix and hnRNA proteins, *J. Cell Biol.*, 86, 135, 1980.
85. **Cavalier-Smith, T.,** Evolution of the nuclear matrix and envelope, in *The Nuclear Envelope and Nuclear Matrix*, Vol. 2, The Wistar Symposium Series, Maul, G. G., Ed., Alan R. Liss, New York, 1982, 307.
86. **Watson, J. D.,** *Molecular Biology of the Gene*, W. A. Benjamin, Menlo Park, Calif., 1976.
87. **Maul, G. G.,** The nuclear and cytoplasmic pore complex: structure, dynamics, distribution and evolution, *Int. Rev. Cyto. Suppl.*, 6, 75, 1977.
88. **Jacob, F., Brenner, S., and Cuzin, F.,** On the regulation of DNA replication in bacteria, *Cold Spring Harbor Symp. Quant. Biol.*, 28, 329, 1963.
89. **Maul, G. G.,** Aspects of a hypothetical nucleocytoplasmic transport mechanism, in *The Nuclear Envelope and the Nuclear Matrix, Vol. 2, The Wistar Symposium Series*, Maul, G. G., Ed., Alan R. Liss, New York, 1982, 1.
90. **Levinger, L. and Varshavsky, A.,** Heat-shock proteins of *Drosophila* are associated with nuclease-resistant, high-salt-resistant nuclear structures, *J. Cell Biol.*, 90, 793, 1981.
91. **Sinibaldi, R. M. and Morris, P. W.,** Putative function of *Drosophila melanogaster* heat shock proteins in nucleoskeleton, *J. Biol. Chem.*, 256, 10735, 1981.
92. **Bonner, W. M. and Laskey, R. A.,** A film detection method for tritium-labelled proteins and nucleic acids in polyacrylamide gels, *Eur. J. Biochem.*, 46, 83, 1974.

Chapter 9

# NONHISTONE PROTEINS IN ANDROGEN ACTION

**Tung Y. Wang and Linda L. Chamberlin**

## TABLE OF CONTENTS

# I. INTRODUCTION

The male sex steroid hormones, or androgens, act on target organs via a general mechanism common to all steroid hormones. According to this mechanism,[1,2] the steroid hormone, upon entering its target cell, binds to a specific high affinity cytosol receptor. The steroid receptor complex thus formed undergoes a temperature-dependent activation process or transformation of the receptor and translocates to the nucleus. In the nucleus, the steroid-receptor complex binds readily to chromatin with ensuing unspecified events leading to an altered pattern of gene expression. Simply stated, the action of steroid hormones is primarily regulation of steroid-specific gene activities.

While the steroid binding by the receptor initiates specific hormonal responses, it is the nuclear events that define the molecular mechanism of steroid control of transcription. These events are largely unknown, but presumably commence with the interaction of the receptor complex with chromatin. This interaction is saturable, with high affinity and specificity with respect to both the steroid and chromosomal components to which the receptor complex binds. The chromosomal or nuclear components which contain the specific binding sites are called acceptors. Several androgen acceptor fractions have been reported, almost exclusively in rat prostate. Most, if not all, of these acceptor preparations are probably nonhistone chromosomal (NHC) proteins.

In this chapter, attention is focused on that aspect of androgen action subsequent to translocation of the androgen-receptor complex to the cell nucleus and on various acceptor preparations. With this survey, we then consider the possible functional significance of the acceptors which participate in processes resulting in activated transcription of androgen-specific genes.

# II. INTERACTION OF ANDROGEN-RECEPTOR COMPLEX WITH CHROMATIN

## A. Translocation of the Androgen-Receptor Complex to Target Cell Nucleus

Rat prostate receptor has generally been prepared by precipitation of cytosol with 40% ammonium sulfate or by DNA-affinity chromatography. In sucrose gradients, the receptor preparation sediments as an 8S[3-6] or a 3-4S[7-9] fraction, depending on its association-dissociation state.[5,10-12] In the presence of high salt, the receptor assumes the 3-4S form, but under low salt conditions, the 8S form is seen. These two forms are interconvertible.[5,10-12] A homogeneous androgen receptor has been purified from steer seminal vesicle by Chang et al.[13] It is a 3.8S acidic polypeptide with a molecular weight of 57,000 and an isoelectric point of 6.6. This may be compared with a pI of 5.8 for partially purified receptor preparations from rat prostate[14] and epididymis.[15] Whether the disparity represents organ differences is not known.

Before translocating into the nucleus, the androgen receptor undergoes a temperature-dependent (e.g., heating at 30°C for 10 min) transformation, manifested by a decrease in sedimentation coefficient from 8 to 4.5S.[16] Such a transformation, which reflects an activation of the receptor, can be mimicked by treatment with 0.5 *M* KCl.[5,12] Liao et al.,[17] based on evidence from prostate,[18,19] chick oviduct[20] and mouse L cells,[21,22] proposed that the cytosol receptor is inactive with respect to steroid binding and is activated to bind steroid ligand by an energy requiring process.

In rat prostate, testosterone is converted by the enzyme 5α-reductase to dihydrotestosterone (17β-hydroxy-5α-androstan-3-one) (DHT) which is the active androgen.[23,24] Translocation of DHT has been shown to involve both the steroid and the receptor molecule.[25] The translocated DHT-receptor complex, radioactively labeled in vivo or in vitro, may be partly recovered by extraction of the labeled prostate nuclei with 0.4 to 0.5 *M* KCl.[5,7,18] The nuclear receptor sediments in sucrose gradients as a 4.2 to 4.5S peak[11,16] with a pI of 6.5,[16] indicating a transformed receptor. Translocated DHT, rather than testosterone, is selectively retained in vivo and in vitro by nuclei,[16, 24-28] probably because cytosolic as well as nuclear testosterone-receptor complexes have a greater rate of dissociation than do cytosolic and nuclear DHT-receptor complexes.[28]

Intranuclear binding sites for administered [$^3$H]testosterone in vivo are predominantly in the chromatin, representing about 70% of the total nuclear retained radioactivity.[29] Binding of DHT-receptor complex to chromatin in vitro has also been shown to be target tissue-specific, requiring both the receptor and chromatin from prostate (or seminal vesicle). Spleen and kidney chromatins give only feeble binding, while liver chromatin gives low but significant binding.[16] The retained androgen remains at a high level up to 16 hr, similar to that observed in vivo. The long retention period of [$^3$H]DHT-receptor complex by chromatin of rat prostate may have some bearing on the biological response to testosterone treatment. In mouse kidney, for example, androgen-specific stimulation of ornithine decarboxylase depends on the length of time the receptor complex is retained in the nucleus, and not on the amount of translocated androgen-receptor complex.[30] This suggests temporal events of the androgen-receptor complex in target cell nucleus.

The chromatin binding sites for [$^3$H]DHT-receptor complex appear to be associated with transcriptionally active regions. When chromatin was fractionated into transcriptionally "active" and "inactive" chromatins by brief micrococcal nuclease digestion of rat prostate nuclei, active chromatin from rat castrated and injected with [$^3$H]testosterone bound more radioactivity than did inactive chromatin. This active chromatin also contained over 80% of the RNA polymerase II activity and was enriched fivefold in endogenous RNA.[31]

Analysis of micrococcal nuclease-digested nuclei by isokinetic sucrose gradient centrifugation reveals that the nuclear retained androgen is associated with nucleosome dimers and linker regions and not with mononucleosomes.[16] Binding of nuclear androgen receptor to interchromosomal linker regions has been further established by Rennie[32] and Bruchovsky et al.[33]

The foregoing brief summary points out some basic features of nuclear retention or chromatin binding of the androgen-receptor complex. First, translocation is steroid and target cell-specific regarding both the receptor and the binding chromosomal components. Second, a transformed receptor complex is involved. Third, it appears that chromatin binding sites of the receptor complex are involved in active transcription. As will become clear later, this last point is pertinent to possible functions of the androgen acceptors.

### B. Effect of Translocated Androgen-Receptor Complex on Chromatin

At least two effects are associated with the binding of androgen-receptor complex to rat prostate chromatin, namely, an increased endogenous RNA polymerase activity and an altered structure of chromatin to an "activated" conformation. Using the endogenous template and the RNA polymerase of intact castrated prostate nuclei, or supplemented with exogenous prostate RNA polymerase, Davies and Griffiths[34,35] reported that RNA synthesis was stimulated by addition of cytosolic or nuclear DHT-receptor complex. The stimulated RNA polymerase activity consisted of both α-amanitin-sensitive and -insensitive activities. The RNA polymerase activity stimulated by cytosolic DHT-receptor complex was, however, completely α-amanitin sensitive, indicating class-II type RNA polymerase activity. The stimulation was target tissue-specific, for it was undetectable with liver or spleen chromatin. RNA polymerase activity in chromatin[36] or nuclei[37] from castrated rat prostate has also been shown to be stimulated after 1 to 2 hr androgen treatment.

The observed early androgen stimulated RNA polymerase II activity is accompanied by a structural alteration of prostate chromatin. This conclusion is based on several lines of evidence.[36-38] The administration of testosterone in vivo to castrated rat results in enhanced binding of [$^3$H]actinomycin D to prostate chromatin.[36] The stimulated binding occurs within the first hour following testosterone treatment, and is prostate-specific. Circular dichroism spectra of prostate chromatin from castrated rat injected with testosterone for 2 hr reveal significant changes in amplitude near 220 nm due to DNA bound protein.[38] The CD spectral changes also vary during the 12-hr period after testosterone treatment, suggesting dynamic conformational changes in chromatin protein. Concomittant with these changes the chromatin exhibits an increase in template activity in transcription in vitro, in euchromatin (diffused chromatin) content, in melting of nonhistone protein (NHP) bound DNA, and in binding of polylysine.[38] In addition, testosterone treatment of castrated rat stimulates chromatin protein kinase activity in prostate within 1 to 2 hr following the steroid replacement, in parallel with the nuclear receptor level and RNA polymerase activity.[39,40] Androgen administration also enhances the rate of phosphorylation of prostate nuclear proteins, a majority of which are chromosomal NHP.[41] Phosphorylation of prostate chromatin has been shown[42] to stimulate its interaction with DHT-receptor complex. These observations taken together indicate that binding of the androgen-receptor complex to prostate chromatin stimulates RNA synthesis quantitatively by an enhanced RNA polymerase activity and qualitatively by altering chromatin structure to provide additional transcribing sequences.

The significance of the androgen-altered prostate chromatin structure depends on whether or not it is correlated with stimulated transcription of androgen-specific gene sequences. Using cDNA synthesized from prostate binding protein (PBP)-mRNA (an androgen-regulated mRNA)[43,44] as a hybridization probe, it has been shown[45] that micrococcal nuclease sensitive DNA in normal, but not in castrated nuclei is enriched sixfold in PBP coding sequences vis-a-vis total DNA. This shows that chromatin from normal prostate has a different structure (nuclease sensitive) than the castrated chromatin (nuclease insensitive). Since testosterone stimulates synthesis of PBP-mRNA, the observations correlate directly actively transcribed androgen regulated genes with the changed structure of rat prostate chromatin.

## III. THE CHROMOSOMAL OR NUCLEAR COMPONENTS THAT BIND ANDROGEN-RECEPTOR COMPLEX

### A. Binding of Androgen-Receptor Complex to DNA

It is apparent from the previous documentation that an early event of the translocated androgen-receptor complex is alteration of chromatin structure. The changed chromatin assumes an activated conformation which permits selective gene sequences (androgen-specific) to be transcribed. Since the effect is steroid- and tissue-specific, the chromosomal components with which the androgen-receptor complex interacts must also be similarly specific. Mammalian chromatin consists of DNA, proteins (histones and NHP), and RNA. Interaction of the androgen-receptor complex with chromatin may involve any one or all of the three kinds of macromolecules. The receptor complex is known to bind DNA. DNA binding has in fact been employed for partial purification of the receptor.[6,8,14] There is certainly an attractive incentive for considering DNA as an acceptor molecule. Binding of the receptor complex to a specific regulatory region of DNA, for example, could confer a direct role on the receptor molecule, analogous to a regulatory protein in bacterial systems, in promoting transcription of structural genes. However, prostate [$^3$H]DHT-receptor binding to DNA has been shown to be nonspecific, showing affinity for both calf thymus and rat prostate DNA,[14] although selective binding to DNA from these two tissues has also been used for partial purification of receptor from rat prostate.[6,8] While crude DNA fraction isolated from prostate chromatin from castrated rat receiving [$^3$H] DHT contains significant bound radioactive androgen,[46,47] phenol extracted DNA shows an insignificant level of radioactivity.[16] Studies of DNA binding of [$^3$H]DHT-receptor complex from kidney of castrated mouse[48,49] have demonstrated that the receptor complex binds nonselectively to DNA of different tissue or cell origin, with preference for denatured DNA over native DNA. Further, androgen receptor also binds to synthetic polynucleotides, especially poly(G), and all classes of RNA. Recent reports, however, show that the progesterone-receptor complex specifically binds to cloned cDNA coding for chicken egg-white protein[50] and that the glucocorticoid receptor recognizes murine mammary tumor virus sequences in DNA.[51] There is certainly a need to investigate further the DNA binding characteristics of prostate cytosolic receptor using specific rat DNA sequences.

Treatment of prostate chromatin with RNase has no effect on the binding of [$^3$H]DHT-receptor complex.[16] This observation would rule out RNA in chromatin binding of the receptor complex. This conclusion, however, is not entirely warranted as will be seen later in the case of nuclear ribonucleoprotein particles.[52]

### B. Binding of Androgen-Receptor Complex to Nuclear Proteins

#### *1. Histones*

The protease sensitivity of prostate nuclei in binding androgen-receptor complex[53] suggests that the acceptors are proteins or contain protein. Of the chromosomal proteins, the histones bind only 7 to 8% of total chromatin bound androgen in vivo.[46,47] Even this low level binding may not be attributable to histone binding because of contaminating NHP. For example, histones isolated by dissociation of prostate chromatin with 2 *M* NaCl/5 *M* urea at pH 6[46] includes NHP that are soluble at pH 6.[54] DNA-histone precipitate obtained from 2 *M* NaCl chromatin extract by lowering the salt concentration to 0.14 *M*[47] also contains some NHP.[55] The quantitative and qualitative constancy of the basic histones further make it questionable that histone binding, whatever its extent, has specificity.

#### *2. The Loosely Bound NHP Acceptors*

The NHP that are readily released from chromatin by 0.3 to 0.4 *M* NaCl are the loosely bound NHP which include several biochemically important proteins. Studies with tissues other than prostate have shown that the 0.35 *M* NaCl-soluble NHP contain a regulatory factor(s) that stimulates initiation of transcription.[56] This NHP fraction also contains nuclear RNA polymerase activity[57] as well as the high mobility group (HMG) proteins, especially the HMG-14 and -17 which are believed to be responsible for conferring DNase I sensitivity on chromatin and which are associated with transcriptionally active chromatin.[58,59] In castrated rat prostate, an early increase of HMG proteins has been observed in chromatin following injection of testosterone.[60] Chromatin reconstitution experiments have shown that the 0.35 *M* NaCl-soluble NHP are responsible for the androgen-induced alteration of chromatin structure[60] which could be due to the HMG proteins. Much of the prostate chromatin protein kinase activities is in this NHP fraction and several 0.35 *M* NaCl-soluble NHP are highly phosphorylated in a tissue-specific manner.[47]

The partial release of translocated [$^3$H]DHT by 0.4 *M* KCl or NaCl from prostate nuclei suggests that this solubilized fraction may contain acceptor molecules. Evidence of this was first shown by Tymoczko and Liao[61] who incubated 0.4 *M* KCl nuclear extract of castrated rat prostate with [$^3$H]DHT-receptor preparation and obtained an aggregated complex by centrifugation or by Millipore® filtration. The aggregate bound radioactivity was quantitatively recovered by extraction with 0.4 *M* KCl. Since liver nuclear extract was inactive, the reaction was prostate-specific. The ability of the prostate nuclear extract to interact with [$^3$H]DHT-receptor complex was heat labile and trypsin sensitive. The presence of acceptor in the nuclear extract was further supported by partial purification using ammonium sulfate and ethanol precipitation or precipitation at pH 4.0 to 4.5. The soluble acceptor is therefore an acidic NHP. The acceptor activity of the nuclear extract could be stimulated threefold by native calf thymus DNA as well as by synthetic purine polyribonucleotides. DNA probably provides nonspecific initial binding sites for the receptor complex to facilitate interaction with the soluble acceptor.

In this laboratory, we have made an attempt to purify the soluble acceptor from the loosely bound NHP of rat prostate chromatin. The chromatin was extracted with 0.35 *M* NaCl from which the soluble acceptor was isolated by rat DNA-cellulose chromatography and Bio-Rex®-70 treatment.[62] The purified acceptor is a 14,000 mol wt polypeptide and according to its amino acid composition, appears to be an acidic protein. Like the acceptor fraction of Tymoczko and Liao,[61] interaction of the purified acceptor with [$^3$H] DHT-receptor complex is stimulated by native rat DNA. The purified acceptor can stimulate transcription of DNA in vitro.

Saline-soluble nuclear proteins also contain a basic protein acceptor(s). Mainwaring et al.[53] prepared nuclear proteins from castrated rat prostate by extraction with 2 *M* NaCl and fractionation by chromatography on Bio-Rex®-70 into a nonadsorbed fraction and fractions eluted successively with 0.4 *M* NaCl and 10% guanidinium chloride. These nuclear fractions were each coupled to CNBr-activated Sepharose® 2B and assayed for binding of [$^3$H]DHT-receptor preparation in the presence and absence of 0.5 *M* KCl. Only the fraction eluted by guanidinium chloride bound specifically to the receptor complex. Liquid isoelectric focusing of the nuclear extract showed that the high-affinity binding nuclear proteins had pI of 8.6 to 9.3, indicating they are basic proteins. The low-affinity binding nuclear proteins were acidic proteins with pI 3 to 7. Fractionation of the nuclear extract either by sucrose gradient centrifugation or by gel filtration on Sephadex® G-75 showed that the high affinity acceptor proteins had an average sedimentation coefficient of 6.5 S and a molecular weight of 70,000. By comparison, the low-affinity nuclear binding proteins were 1.5 to 5.5 S proteins. These results are at variance with the soluble acceptor preparations described previously which apparently are acidic proteins and, at least in one case, of low molecular weight. The basic acceptor could be localized in the 0.4 to 2.0 *M* KCl fraction. These results show that the saline-soluble nuclear proteins contain more than one acceptor.

*3. The Tightly Bound NHP Acceptors*

In addition to the saline-soluble acceptors described above, the rat prostate chromatin or nuclear fraction that is resistant to high salt (1 to 3 *M* NaCl) extractions contains specific binding sites for DHT or the androgen-receptor complex. Androgen acceptor activity has been associated with several components in the insoluble fraction, namely, the nuclear matrix,[63] the nuclear ribonucleoprotein (RNP),[61] the NP fraction[42] and the residual chromatin.[47,64] To these may be added the nuclear envelope which has been shown to contain high-affinity androgen binding sites.[65] The nuclear envelope apparently is involved in translocation of androgen-receptor and presumably has a different function than the chromatin acceptors. These are grouped together as the tightly bound chromosomal or nuclear NHP.

***a. The Residual Chromatin***

The residual chromatin is the fraction of chromatin that resists extraction by 2 *M* NaCl. It contains mainly NHP tightly bound to about 5 to 8% of the DNA. Metaphase HeLa chromosomes, briefly digested with micrococcal nuclease and depleted of histone, bulk NHP, and DNA by extraction with 2 *M* NaCl, have been shown to retain a highly organized and compact DNA consisting of 10- to 30-μm loops attached to NHP in a scaffold structure.[66,67] The backbone NHP of the HeLa scaffold structure contain DNA-dependent cell-specific antigen activity.[68] Protein scaffold has also been isolated from interphase chromosomes.[69-72] In the mouse L cell both metaphase and interphase chromosomes show similar nonrandom conserved DNA enriched in abundant middle repetitive sequences.[72] The residual chromatin thus plays an important role in the organization of chromosomes.

Evidence that the residual chromatin contains androgen acceptor sites was first observed in the distribution of in vitro translocated androgens in prostate chromatin.[47] In this study, the chromatin from castrated rat prostate nuclei incubated with [$^3$H] androgen-receptor complex was salt fractionated into 0.4 and 2 *M* NaCl soluble, DNA-histone, and 2 *M* NaCl insoluble fractions. Of the total chromatin-bound radioactive androgens, about 45% was associated with the residual chromatin, as compared with 14% in the 0.4 *M* NaCl soluble (the soluble acceptor) fraction. Prostate chromatin fractions from castrated rat injected with [$^3$H] testosterone gave essentially similar androgen distribution. Since the residual chromatin contains only a small amount of the total NHP, it is a more effective acceptor fraction than the soluble acceptor. This is also true in the estrogen systems where 34 to 50% of nuclear retained [$^3$H] estradiol, translocated either in vivo or in vitro, is resistant to high salt extraction.[73-76] The high salt resistant binding sites have been suggested to represent long-term high affinity acceptor sites responsible for steroid-induced growth.[76]

When [$^3$H] DHT-labeled nuclei from castrated rat prostate were digested with micrococcal nuclease, the fractionated residual protein maintained a high level of bound [$^3$H]DHT throughout the nuclease digestion period (0 to 30% acid soluble).[60] Similar digestion with pancreatic DNase resulted in a slow but steady depletion of residual protein bound radioactivity.[60] The residual protein acceptor(s) is probably associated with the nucleosomal core.

Reconstitution of chromatins with residual proteins from normal and castrated rat prostates and hybridization of RNA transcribed from the reconstituted chromatins with rat DNA show that the residual protein is required for full and tissue-specific androgen-receptor binding to, and transcription of, prostate chromatin.[47,64]

### *b. The Nuclear RNP*

Extraction of isolated cell nuclei with 1 to 3 *M* NaCl yields an insoluble residual fraction[77-79] which can be partially solubilized by brief DNase I digestion and subsequent treatment with sodium deoxycholate.[78] In rat liver and calf thymus, the solubilized residual proteins contain 33S ribonucleoprotein particles comprised of 5S RNA and heterogenous acidic and phosphorylated proteins.[78] Liao et al.[52] incubated the ribonucleoprotein particles prepared from castrated rat prostate nuclei with [$^3$H]DHT-receptor and demonstrated the formation of a ternary complex by sucrose gradient centrifugation. No such complex formation was observed between labeled androgen-receptor complex and RNP particles derived from rat liver. The binding activity of the nuclear RNP toward [$^3$H]DHT-receptor complex is sensitive to RNase and protease indicating that both RNA and the protein moieties are required for the acceptor activity. Insensitivity of prostate chromatin to RNase treatment in binding to [$^3$H]DHT-receptor complex,[16] as documented in Section II, is probably due to inaccessibility of the RNP to the nuclease, or protection by the RNA-bound proteins. It is likely that the nuclear RNP particles are of chromosomal origin.

### *c. The Nuclear Matrix*

Barrack and Coffey[63] have described the presence of androgen-binding sites in another nuclear fraction of rat prostate, the nuclear matrix. The isolation of the nuclear matrix includes an initial digestion of nuclei with DNase I, then extraction with 2 *M* NaCl and finally washing the insoluble residue with Tris EDTA. Morphologically the nuclear matrix consists of a peripheral lamina, the residual nucleoli and a fibrogranular network of RNP.[63] In a detailed study of total occupied and unoccupied DHT binding sites on nuclear matrix prepared from rat prostate, Barrack and Coffey[63] have shown by steroid exchange assay that the nuclear matrix binding sites for DHT account for 45% of the total nuclear-specific binding sites. By judicious treatment of nuclei with RNase which selectively removes the internal RNP network of the nuclear matrix, the internal RNP network was found to contribute 75% of the high affinity matrix binding sites. The results are harmonious with the observation by Liao et al.[52] on the nuclear RNP particles. The extent of specific androgen binding of nuclear matrix is also similar to the 45% androgen binding by the residual chromatin[47] cited earlier.

The specific DHT binding sites on rat prostate nuclear matrix were reduced to 2% of the intact control level after 2 days of castration.[63] The reduced binding sites were completely restored 1 hr after an s.c. injection of DHT. Steroid competition binding assays further showed that [$^3$H] binding to prostate nuclear matrix was 100% competed by DHT and R1881 (methyltrienolone), a specific binding ligand for rat prostate cytosolic receptors,[80] 20% by testosterone, and about 1% by estrogens and cortisol. These results indicate that the prostate nuclear matrix binding sites are steroid and tissue-specific, and are under the control of androgens. Similar observations have also been obtained with rat uterine nuclear matrix in estradiol binding.[63,81]

### *d. The Chromatin NP Fraction*

Klyzsejka-Stefanowicz et al.[42] have used chromatins reconstituted from chromosomal fractions for the identification of androgen acceptors in rat prostate NHP. Chromatin was sequentially fractionated into 5 *M* urea-soluble NHP (UP fraction), 2.5 *M* NaCl/urea (pH 5) dissociated histones (HP fraction), 2.5 *M* NaCl/urea (pH 8) released NHP (NP fraction), and DNA (NC fraction).[82] Hybrid chromatins were reconstituted from these fractions, or some of these fractions, prepared from rat prostate, testis, and liver. These complete and incomplete hybrid chromatins were incubated with [$^3$H]DHT-receptor preparation and the complexed chromatins were collected by centrifugation and Millipore® filtration. Complex formation was found to depend on the NP fraction. In fact, prostate or testis NHP fraction alone, complexed with rat DNA exhibited the same binding affinity toward [$^3$H]DHT-receptor as did native chromatin.

Since the NP fraction represents the remaining NHP after removal of most of the NHP (>90%) and histones, it is tightly bound in chromatin. In rat liver and Novikoff hepatoma, the NP fraction comprises less than 5% of the total chromatin protein and about 2% of total DNA.[83,84] It binds specifically to homologous DNA and forms immunospecific complexes.[84] Chromatin reconstitution and RNA-DNA hybridization studies showed, moreover, that transcription of globin mRNA from chicken reticulocyte chromatin in vitro depends on a tissue-specific NP fraction.[85] These characteristics indicate that the NP fraction determines the tissue specificity of a chromatin and is involved in the regulation of transcription.

The importance of the tightly bound NHP fraction as acceptors is also manifested in chromatins from other steroid target cell systems. The chick oviduct $AP_3$ fraction of Spelsberg et al.,[86] which contains specific acceptor sites for progesterone-receptor, is isolated by dissociation of chromosomal proteins in 3 *M* NaCl/7.5 *M* urea at pH 8.5 after removal of histones and much of the NHP by 2 *M* NaCl/5 *M* urea at pH 6, and is thus a tightly bound NHP fraction, analogous to the NP fraction. This fractionation procedure was used by Nozu et al.[26] who, by determining intrachromatin distribution of translocated DHT-[$^{131}$I] receptor, obtained a 55% binding by chromosomal NHP which is actually the $AP_3$ fraction.

Tightly bound NHP have been fractionated by Bekhor et al.[87] and Gates and Bekhor[88] using 5 *M* urea containing increasing concentrations of NaCl up to 2.0 *M*. The NHP released by 1.5 to 2.0 *M* NaCl (DNA-P fraction) contains 6 and 4% of the total chromatin protein and DNA, respectively. A simplified procedure for the preparation of DNA-P fraction consists of direct extraction of chromatin with buffered 2 *M* NaCl and removal of free DNA cosedimented with the 2 *M* insoluble DNA-P by extensive buffer wash.[89] Significantly, the small amount of DNA associated with the tightly bound NHP in the DNA-P fraction represents a functional subset of DNA sequences enriched in transcribing sequences. Using cDNA synthesized from chicken liver poly(A)RNA as a probe in a DNA-DNA reassociation experiment, Gates and Bekhor[88] observed a rate of reannealing of the cDNA with DNA-P faster than that with total DNA or bulk DNA lacking DNA-P, equivalent to a 25% enrichment in transcribing sequences in DNA-P. In human placental DNA-P fraction, similar reassociation kinetic studies using a cDNA probe prepared from lactogen mRNA showed a 22-fold enrichment in lactogen gene sequence as compared to total placental DNA.[89] The DNA-

P NHP apparently are a different set of proteins since they have a different electrophoretic profile from the bulk NHP. These results have led to the suggestion[89] that the DNA-P proteins may select, or preferentially bind to, specific sequences for transcription. The inferred significance of the tightly bound NHP in regulation of gene activity is apparent.

The nuclear RNP particles, the nuclear matrix, and the residual chromatin that contain androgen-acceptor sites are very similar preparations, and the former two may be considered as solubilized residual proteins. Considering that the majority of the nuclear matrix binding sites are attributable to its internal RNP network, the comparison between the nuclear matrix and the nuclear RNP is especially striking. Together with the NP fraction (and also the DNA-P and $AP_3$ fractions), these preparations, in a general way, all have a common denominator in that they are insoluble in high salt solutions under various conditions. These tightly bound NHP preparations, of course, could have different acceptors. In view of the presence of the soluble acceptors, androgens apparently operate in their target cells via a multiple acceptor system.

## IV. CONCLUSIONS

The evidence documented so far indicates that androgen acceptors are nuclear or chromosomal NHP. Whether the nuclear acceptors are also chromosomal proteins cannot be discerned at present. Here, on the basis of isolation procedures, we have classified the androgen acceptors into (saline) soluble and insoluble groups, corresponding to loosely and tightly bound NHP. The necessity of dissociating DNA-histone in order to obtain the tightly bound NHP, and the different characteristics of the loosely and tightly bound NHP indicate that such a classification is more than operational convenience. The soluble and insoluble acceptors are certainly distinctly different proteins and may have different functional roles as acceptors. Although intragroup preparations may be similar, there are notable differences among some. The acidic and basic soluble acceptor preparations are an example. The nuclear matrix and the nuclear RNP are similar, but they probably are different from the NP fraction, despite their 2 *M* NaCl insolubility under various conditions. The tightly bound NHP are apparently more prominent acceptors than are the loosely bound NHP.

It is interesting to note that the acceptor preparations have various but important properties other than acceptor activity. For example, the NP fraction binds specifically to homologous DNA,[84] and the resulting complexes are immunologically tissue-specific.[83,90] The NP[85] and the residual chromatin[47,64] are required for specific transcription of chromatin. The soluble acceptor binds to DNA and stimulates initiation of RNA synthesis in vitro.[65] Since specific transcription is a consequence of the receptor complex binding to acceptor, the properties related to transcription are likely those of the acceptor molecules.

The multiple acceptor system in rat prostate is apparent. What is not obvious is how the multiple acceptors interact with the androgen-receptor complex resulting in altered transcription. For convenience, the multiple acceptors may be grossly simplified as a soluble acceptor (group) and an insoluble acceptor (group). Based on available information, it is possible to postulate the mode of action for the acceptors in relation to the translocated androgen-receptor complex. We propose herein a tentative model as follows.

The acceptors, being DNA binding, are located in DNA regions which are assumed to be the transcribable sequences that may or may not be available for transcription. Interaction with the androgen-receptor complex, simultaneously involving both the soluble and insoluble acceptors, results in a conformational change in chromatin structure which makes accessible, or facilitates transcription of, specific gene sequences. In this case, the acceptors would be markers tagging specific DNA sequences for recognition by the receptor complex. An alternative is that interaction of the receptor complex with the two acceptors could be temporal. In this case, the androgen-receptor complex would bind to one acceptor, for example, the soluble acceptor, effecting an alteration in chromatin structure in such a way that the second insoluble acceptor now becomes accessible. Binding to the second acceptor would then result in a further conformational change of the chromatin to an activated state which allows transcription of the steroid regulated genes.

With such a model, it is possible to allow the involvement of direct binding of the receptor complex to DNA. For example, initial binding of the receptor complex to DNA would be immediate and nonspecific. The DNA-bound receptor complex then travels along the exposed DNA length and locates the acceptors, a process analogous to the nonspecific binding of *Escherichia coli* RNA polymerase to DNA template, followed by either of the two alternatives described previously. However, if DNA binding of the receptor complex is sequence-specific, then the specific DNA binding region must be included as an additional acceptor site. Interaction of the receptor complex with DNA would directly initiate synthesis of early androgen specific mRNA or regulatory molecules for further control steps. This would be in addition to nonspecific DNA binding of the receptor complex via acceptors as suggested above.

To make the above statement, we have considered and attempted to correlate the presence of more than one acceptor, the DNA binding of the acceptors, and the structural alteration in chromatin resulting from binding of the receptor complex to chromatin. We have also drawn upon information obtained from nonandrogen systems such as DNA-P and $AP_3$ fractions. It is clear that further work on purification, identification, and characterization of the various acceptor fractions are desirable for the verification of the hypothesized scheme presented here.

## ACKNOWLEDGMENTS

Work in our laboratory has been supported by a grant (HD-09443) from the National Institutes of Health.

## REFERENCES

1. **Gorski, J., Toft, D., Shyamala, G., Smith, D., and Notides, A.,** Hormone receptors: studies on the interaction of estrogen with the uterus, *Recent Prog. Horm. Res.*, 24, 45, 1968.
2. **Jensen, E. V., Suzuki, T., Kawashima, T., Stumpf, W. E., Jungblut, P. W., and DeSombre, E. R.,** A two-step mechanism for the interaction of estradiol with rat uterus, *Proc. Natl. Acad. Sci. U.S.A.*, 59, 632, 1968.
3. **Unhjem, O., Tveter, K. J., and Aakvaag, A.,** Preliminary characterization of an androgen macromolecular complex from the rat ventral prostate, *Acta Endocrinol.*, 62, 153, 1969.
4. **Mainwaring, W. I. P.,** A soluble androgen receptor in the cytoplasm of rat prostate, *J. Endocrinol.*, 45, 531, 1969.
5. **Bauliu, E. E. and Jung, I.,** A prostatic cytosol receptor, *Biochem. Biophys. Res. Commun.*, 38, 599, 1970.
6. **Hu, A. L., Loor, R. M., Chamberlin, L. L., and Wang, T. Y.,** Study of androgen-binding cytosol proteins from rat prostate purification of androgen receptor, *Arch. Biochem. Biophys.*, 185, 134, 1978.

7. **Fang, S., Anderson, K. M., and Liao, S.,** Receptor proteins for androgens. On the role of specific proteins in selective retention of 17 β-hydroxy-5α-androstan-3-one by rat ventral prostate *in vivo* and *in vitro*, *J. Biol. Chem.*, 244, 6584, 1969.
8. **Hu, A. L., Loor, R. M., and Wang, T. Y.,** Purification of a 3S cytosol androgen receptor from rat prostate that stimulates DNA-dependent RNA synthesis *in vitro*, *Biochem. Biophys. Res. Commun.*, 65, 1327, 1975.
9. **Wilson, E. M. and French, F. S.,** Binding properties of androgen receptors. Evidence for identical receptors in rat testis, epididymis and prostate, *J. Biol. Chem.*, 251, 5620, 1976.
10. **Liao, S.,** Cellular receptors and mechanisms of action of steroid hormones, *Int. Rev. Cytol.*, 41, 87, 1975.
11. **Wilson, E. M. and French, F. S.,** Effect of proteases and protease inhibitors on the 4.5S and 8S androgen receptor, *J. Biol. Chem.*, 254, 6310, 1979.
12. **Colvard, D. S. and Wilson, E. M.,** Identification of an 8S androgen receptor-promoting factor that converts the 4.5S form of the androgen receptor to 8S, *Endocrinology*, 109, 496, 1981.
13. **Chang, C. H., Rowley, D. R., Lobl, T. J., and Tindell, D. J.,** Purification and characterization of androgen receptor from steer seminal vesicle, *Biochemistry*, 21, 4102, 1982.
14. **Mainwaring, W. I. P. and Irving, R.,** The use of deoxyribonucleic acid-cellulose chromatography and isoelectric focusing for the characterization and partial purification of steroid-receptor complexes, *Biochem. J.*, 134, 113, 1973.
15. **Tindell, D. J., Hansson, V., McLean, W. S., Ritzen, E. M., Nayfeh, S. N., and French, F. S.,** Androgen-binding proteins in rat epididymis: properties of a cytoplasmic receptor for androgen similar to the androgen receptor in ventral prostate and different from androgen binding protein (ABP), *Mol. Cell. Endocrinol.*, 3, 83, 1975.
16. **Mainwaring, W. I. P. and Peterken, B. M.,** A reconstituted cell-free system for the specific transfer of steroid-receptor complexes into nuclear chromatin isolated from rat ventral prostate gland, *Biochem. J.*, 125, 285, 1971.
17. **Liao, S., Hiipakka, R. A., Schilling, K., Oberhausor, A. K., Chang, C., and Judge, S.,** Cellular dynamics of androgen receptor and protein induction in rat ventral prostate, in *The Prostate Cell: Structure and Function*, Part A, Alan R. Liss, New York, 1981, 381.
18. **Liao, S. and Fang, S.,** Receptor proteins for androgens and the mode of action of androgens on gene transcription in ventral prostate, *Vitam. Horm. N.Y.*, 27, 18, 1969.
19. **Liao, S., Rossini, G. P., Hiipakka, R. A., and Chen, C.,** Factors that control the interaction of the androgen-receptor complex with the genomic structure in the rat prostate, in *Perspectives in Steroid Receptor*, Bresciani, F., Ed., Raven Press, New York, 1980, 99.
20. **Nishigori, H. and Toft, D. O.,** Inhibition of progesterone receptor activation by sodium molybdate, *Biochemistry*, 19, 77, 1980.
21. **Sando, J., Laforest, A., and Pratt, W.,** ATP-dependent activation of L cell glucocorticoid receptors to the steroid binding form, *J. Biol. Chem.*, 254, 4772, 1979.
22. **Sando, J., Hammond, N., Stratford, C., and Pratt, W.,** Activation of thymocyte glucocorticoid receptors to the steroid binding form, *J. Biol. Chem.*, 254, 4779, 1979.
23. **Liao, S., Tymoczko, J. L., Castaneda, E., and Liang, T.,** Androgen receptors and androgen-dependent initiation of protein synthesis in the prostate, *Vitam. Horm. N.Y.*, 33, 297, 1975.
24. **Bruchovsky, N. and Wilson, J. D.,** The conversion of testosterone to 5α-androstan-17β-ol-3-one by rat prostate *in vivo* and *in vitro*, *J. Biol. Chem..*, 243, 2012, 1968.
25. **Nozu, K., Yitok, H., and Tamaoki, B. I.,** Direct evidence on incorporation of the receptor into the nuclei of rat ventral prostate in the form of the complex with dihydrotestosterone, *Endocrinology (Japan)*, 22, 537, 1975.
26. **Nozu, K. and Tamaoki, B. I.,** On the role of the cytosol receptors in the incorporation of androgens into the prostatic nuclei of rat, *J. Steroid Biochem.*, 6, 57, 1975.
27. **Tveter, K. J. and Attramadal, A.,** Selective uptake of radioactivity in rat ventral prostate following administration of testosterone-1,2-$^3$H, *Acta Endocrinol. (Copenhagen)*, 59, 218, 1968.
28. **Liang, T. and Heiss, C. E.,** Inhibition of 5α-reductase, receptor binding and nuclear uptake of androgens in the prostate by a 4-methyl-4-aza-steroid, *J. Biol. Chem.*, 256, 7998, 1981.
29. **Mainwaring, W. I. P.,** The binding of [1,2-$^3$H] testosterone within the nucleus of rat ventral prostate, *J. Endocrinol.*, 44, 323, 1969.
30. **Pajunen, A. E. I., Isomaa, V. V., Jänne, O. A., and Bardin, C. W.,** Androgenic regulation of ornithine decarboxylase activity in mouse kidney and its relationship to changes in cytosol and nuclear androgen receptor concentrations, *J. Biol. Chem.*, 257, 8190, 1982.
31. **Davies, P., Thomas, P., Borthwick, N. M., and Giles, M. G.,** Distribution of acceptor sites for androgen-receptor complexes between transcriptionally active and inactive fractions of rat ventral prostate chromatin, *J. Endocrinol.*, 87, 225, 1980.

32. **Rennie, P. S.,** Binding of androgen receptor to prostatic chromatin requires intact linker DNA, *J. Biol. Chem.*, 254, 3947, 1979.
33. **Bruchovsky, N., Rennie, P. S., and Comeau, T.,** Partial purification of nuclear androgen receptor by micrococcal nuclease digestion of chromatin and hydrophobic interaction chromatography, *Eur. J. Biochem.*, 120, 399, 1981.
34. **Davies, P. and Griffiths, K.,** Stimulation of RNA polymerase activity *in vitro* by prostatic steroid-protein receptor complexes, *Biochem. J.*, 136, 611, 1973.
35. **Davis, P. and Griffiths, K.,** Further studies on the stimulation of prostate RNA polymerase by 5α-dihydrotestosterone-receptor complexes, *J. Endocrinol.*, 62, 385, 1974.
36. **Mainwaring, W. I. P. and Jones, D. M.,** Influence of receptor complexes on the properties of prostate chromatin, including its transcription by RNA polymerase, *J. Steroid Biochem.*, 6, 475, 1975.
37. **Wang, T. Y. and Loor, R. M.,** Testosterone-activated RNA synthesis in isolated prostate nuclei, *J. Steroid Biochem.*, 10, 299, 1979.
38. **Loor, R. M., Hu, A. L., and Wang, T. Y.,** Structurally altered and transcriptionally activated rat prostate chromatin induced by androgens, *Biochim. Biophys. Acta*, 477, 312, 1977.
39. **Ahmed, K. and Wilson, M. J.,** Chromatin-associated protein phosphokinases of rat ventral prostate: characteristics and effects of androgenic status, *J. Biol. Chem.*, 250, 2370, 1975.
40. **Ahmed, K., Wilson, M. J., Gouell, S. A., and Williams-Ashman, H. G.,** Effects of polyamines on prostatic chromatin- and nonhistone-protein associated protein kinase reactions, *Biochem. J.*, 176, 739, 1978.
41. **Ahmed, K., Davis, A. T., Gouell, S. A., and Wilson, M. J.,** Phosphorylation of nonhistone protein fraction which coextracts with the HMG proteins of chromatin, *Biochem. Biophys. Res. Commun.*, 96, 326, 1980.
42. **Klyzsejko-Stefanowicz, L., Chiu, J. F., Tsai, Y. H., and Hnilica, L. S.,** Acceptor proteins in rat androgenic tissue chromatin, *Proc. Natl. Acad. Sci. U.S.A.*, 73, 1954, 1976.
43. **Parker, M. G., White, R., and Williams, J. C.,** Cloning and characterization of androgen-dependent mRNA from rat ventral prostate, *J. Biol. Chem.*, 255, 6996, 1980.
44. **Peeters, B. L., Mous, J. M., Rombauts, W. A., and Heyns, W. J.,** Androgen-induced mRNA in rat ventral prostate. Translation, partial purification, and preliminary characterization of the mRNAs encoding the components of prostatic binding protein, *J. Biol. Chem.*, 255, 7017, 1980.
45. **Hiremath, S. T. and Wang, T. Y.,** Nuclease-sensitivity of prostatic binding protein gene sequences in rat ventral prostate, *Cell Biol. Int. Rep.*, 6, 279, 1982.
46. **Nozu, K. and Tamaoki, B. I.,** Incorporation of [$^{131}$I]-labeled androgen-receptor into nuclei of rat prostates, *Biochem. Biophys. Res. Commun.*, 38, 145, 1974.
47. **Nyberg, L. M. and Wang, T. Y.,** The role of the androgen-binding nonhistone proteins in the transcription of prostatic chromatin, *J. Steroid Biochem.*, 7, 267, 1976.
48. **Lin, S.-Y. and Ohno, S.,** The binding of androgen receptor to DNA and RNA, *Biochim. Biophys. Acta*, 654, 181, 1981.
49. **Lin, S.-Y. and Ohno, S.,** The interactions of androgen receptor with poly(A)-containing RNA and polyribonucleotides, *Eur. J. Biochem.*, 124, 283, 1982.
50. **Mulvihill, E. R., LePennec, J.-P., and Chambon, P.,** Chicken oviduct progesterone receptor; location of specific regions of high-affinity binding in cloned DNA fragments of hormone-responsive genes, *Cell*, 28, 621, 1982.
51. **Payvar, F., Wrange, O., Carlstedt-Duke, J., Okret, S., Gustafsson, J. A., and Yamamoto, K. R.,** Glucocorticoid receptors bind selectively *in vitro* to a cloned DNA fragment whose transcription is regulated by glucocorticoids *in vivo*, *Proc. Natl. Acad. Sci. U.S.A.*, 78, 6628, 1981.
52. **Liao, S., Liang, T., and Tymoczko, J. L.,** Ribonucleoprotein binding of steroid-"receptor" complexes, *Nature (London) New Biol.*, 241, 211, 1973.
53. **Mainwaring, W. I. P., Syms, E. K., and Higgins, S. J.,** Nuclear components responsible for the retention of steroid-receptor complexes, especially from the standpoint of the specificity of hormonal responses, *Biochem. J.*, 156, 129, 1976.
54. **Wang, T. Y.,** The isolation, properties, and possible functions of chromatin acidic proteins, *J. Biol. Chem.*, 242, 1220, 1967.
55. **Kostraba, N. C. and Wang, T. Y.,** Inhibition of transcription *in vitro* by a nonhistone protein isolated from Ehrlich ascites tumor chromatin, *J. Biol. Chem.*, 250, 8938, 1975.
56. **Kostraba, N. C., Montagna, R. A., and Wang, T. Y.,** Study of the loosely bound nonhistone chromatin proteins. Stimulation of DNA-templated RNA synthesis by a specific DNA-binding phosphoprotein fraction, *J. Biol. Chem.*, 250, 1548, 1975.
57. **Montagna, R. A. and Wang, T. Y.,** A comparison of the loosely and tightly bound nonhistone proteins from Ehrlich ascites tumor chromatin, *Cancer Res.*, 36, 3138, 1976.
58. **Weisbrod, S. and Weintraub, H.,** Isolation of a subclass of nuclear proteins responsible for conferring a DNase I-sensitive structure on globin chromatin, *Proc. Natl. Acad. Sci. U.S.A.*, 76, 630, 1979.

59. **Gazit, B., Panet, A., and Cedar, H.,** Reconstitution of DNase I-sensitive structure on active genes, *Proc. Natl. Acad. Sci. U.S.A.*, 77, 1787, 1980.
60. **Hiremath, S. T., Maciewicz, R. A., and Wang, T. Y.,** The loosely bound nonhistone chromosomal proteins of rat prostate in androgen action, *Biocheim. Biophys. Acta*, 653, 130, 1981.
61. **Tymoczko, J. L. and Liao, S.,** Retention of an androgen-protein complex by nuclear chromatin aggregates: heat-labile factors, *Biochim. Biophys. Acta*, 252, 607, 1971.
62. **Hiremath, S. T., Loor, R. M., and Wang, T. Y.,** Isolation of an androgen acceptor from salt extract of rat prostatic chromatin, *Biochem. Biophys. Res. Commun.*, 97, 981, 1980.
63. **Barrack, E. R. and Coffey, D. S.,** The specific binding of estrogens and androgens to the nuclear matrix of sex hormone responsive tissues, *J. Biol. Chem.*, 255, 7265, 1980.
64. **Wang, T. Y.,** The role of nonhistone chromosomal proteins in the interaction of prostate chromatin with androgen-receptor complex, *Biochim. Biophys. Acta*, 518, 81, 1978.
65. **Lefebre, Y. A. and Novosad, Z.,** Binding of androgens to a nuclear-envelope fraction from the rat ventral prostate, *Biochem. J.*, 186, 641, 1980.
66. **Adolph, K. W., Cheng, S. M., Paulson, J. R., and Laemmli, U. K.,** Isolation of a protein scaffold from mitotic HeLa cell chromosomes, *Proc. Natl. Acad. Sci. U.S.A.*, 74, 4937, 1977.
67. **Pauslon, J. R. and Laemmli, U. K.,** The structure of histone-depeleted metaphase chromosomes, *Cell*, 12, 817, 1977.
68. **Campbell, A. M., Briggs, R. C., and Hnilica, L. S.,** Cell specific antiserum to chromosome scaffold proteins, *Nucleic Acids Res.*, 6, 205, 1979.
69. **Brasch, K. and Setterfields, G.,** Structural organization of chromosomes in interphase nuclei, *Exp. Cell Res.*, 83, 175, 1974.
70. **Cook, P. R. and Brazell, I. A.,** Supercoils in human DNA, *J. Cell Sci.*, 19, 261, 1975.
71. **Adolph, K. W.,** Organization of chromosomes in HeLa cells: isolation of histone-depleted nuclei and nuclear scaffolds, *J. Cell Sci.*, 42, 291, 1980.
72. **Razin, S. V., Mantieva, V. L., and Georgiev, G. P.,** The similarity of DNA sequences reamining bound to scaffold upon nuclease treatment of interphase nuclei and metaphase chromosomes, *Nucleic Acids Res.*, 7, 1713, 1979.
73. **Labeau, M. C., Massol, N., and Baulieu, E. E.,** An insoluble receptor for oestrogens in the "residual" nuclear proteins of chick liver, *Eur. J. Biochem.*, 36, 294, 1973.
74. **Mester, J. and Baulieu, E. E.,** Dynamics of oestrogen-receptor distribution between the cytosol and nuclear fractions of immature rat uterus after oestradiol administration, *Biochem. J.*, 146, 617, 1975.
75. **Baudendistil, L. J. and Ruh, T. S.,** Antiestrogen action: differential nuclear retention and extractability of the estrogen receptor, *Steroids*, 28, 223, 1976.
76. **Clark, J. H. and Peck, E. J., Jr.,** Nuclear retention of receptor-estrogen complex and nuclear acceptor sites, *Nature (London)*, 260, 635, 1976.
77. **Wang, T. Y.,** Solubilization and characterization of the residual proteins of the cell nucleus, *J. Biol. Chem.*, 241, 2913, 1966.
78. **Hu, A. L. and Wang, T. Y.,** The ribonucleoprotein particles from nuclear residual proteins of calf thymus, *Arch. Biochem. Biophys.*, 144, 549, 1971.
79. **Patel, G., Patel, V., Wang, T. Y., and Zobel, C. R.,** Properties of the residual proteins of rat liver, *Arch. Biochem. Biophys.*, 128, 654, 1968.
80. **Boune, C. and Raynaud, J. P.,** Methyltrienolone, a specific ligand for cellular androgen receptors, *Steroids*, 26, 227, 1975.
81. **Barrack, E. R., Hawkins, E. F., Allen, S. L., Hicks, L. L., and Coffey, D. S.,** Concepts related to salt resistant estradiol receptors in rat uterine nuclei: nuclear matrix, *Biochem. Biophys. Res. Commun.*, 79, 829, 1977.
82. **Chiu, J.-F., Hunt, M., and Hnilica, L. S.,** Tissue-specific DNA-protein complexes during azo dye hepatocarcinogenesis, *Cancer Res.*, 35, 913, 1975.
83. **Chiu, J.-F., Wang, S., Fujitani, H., and Hnilica, L. S.,** DNA-binding chromosomal nonhistone proteins. Isolation, characterization and tissue specificity, *Biochemistry*, 14, 4552, 1975.
84. **Wang, S., Chiu, J.-F., Klyszejko-Stefanowicz, L., Fujitani, H., and Hnilica, L. S.,** Tissue-specific chromosomal nonhistone protein interactions with DNA, *J. Biol. Chem.*, 251, 1471, 1976.
85. **Chiu, J.-F., Tsai, Y. H., Sakuma, K., and Hnilica, L. S.,** Regulation of *in vitro* mRNA transcription by a fraction of chromosomal proteins, *J. Biol. Chem.*, 250, 9431, 1975.
86. **Spelsberg, T. C., Steggles, A. W., Chytil, F., and O'Malley, B. W.,** Progesterone binding components of chick oviduct. V. Exchange of progesterone-binding capacity from target to nontarget tissue chromatins, *J. Biol. Chem.*, 247, 1368, 1972.
87. **Bekhor, I., Lapeyre, J. N., and Kim, J.,** Fractionation of nonhistone chromosomal proteins isolated from rabbit liver and submandibular salivary glands, *Arch. Biochem. Biophys.*, 161, 1, 1974.
88. **Gates, D. M. and Bekhor, I.,** Distribution of active gene sequences: a subset associated with tightly bound chromosomal proteins, *Science*, 207, 661, 1980.

89. **Norman, G. L. and Bekhor, I.,** Enrichment of selected active human gene sequences in in the placental DNA fraction associated with tightly bound nonhistone chromosomal proteins, *Biochemistry*, 20, 3568, 1981.
90. **Krejewska, W. M., Briggs, R. C., Chiu, J.-F., and Hnilica, L. S.,** Immunologically specific complexes of chromosomal nonhistone proteins with DNA in chicken erythroid nuclei, *Biochemistry*, 19, 4667, 1980.

Chapter 10

# ANALYSIS OF NORMAL AND NEOPLASTIC TISSUE NHC PROTEINS BY HIGH-RESOLUTION TWO-DIMENSIONAL GRADIENT ELECTROPHORESIS AND SILVER STAINING

**Lewis V. Rodriguez, Louis S. Ramagli, and Dennis A. Johnston**

## TABLE OF CONTENTS

## I. INTRODUCTION

"Cancer" represents behavior of an abnormal cell phenotype characterized by uncontrolled cell division, invasiveness, metastasis, and inheritance in daughter cells. Transmissibility and stability of cancer most likely reflects presence of an inheritable DNA alteration (mutation) and/or expression or derepression of single genes, related groups, or both. A large number of recent studies have focused upon a proposed link between alterations of DNA (broadly termed mutagenesis) and eventual appearance of cancers. This "genetic" basis for carcinogenesis has been proposed to possibly result from a broad spectrum of alterations ranging from error-accumulation through gene redundancy and/or expression of oncogenes. No question remains as to the relation between certain types of DNA damage and induction of tumors.[44] However, significant gaps still exist between demonstration of DNA damage and manifestation of the malignant phenotype. For example, not all carcinogens have been shown to interact with DNA.[29] Also, the target size for neoplastic transformation appears to be 20- to 25-fold larger than that for mutation.[50,100] Further, selected cancers are able to express normal phenotypes under special conditions.[62,73] Although evidence indicates that alteration of DNA plays a major role in carcinogenesis, it is important to establish that part played by non-DNA factors, i.e., "epigenetic" effects. This is particularly appropriate since intriguing, although controversial evidence indicates that enhanced expression of normal cell proteins, or derepression of others may play a role in establishment and maintenance of malignant phenotype.[28,60] Epigenetic theories of carcinogenesis have derived from studies of normal differentiation and a major characteristic of most cancers — abnormality in control of normal gene expression. Generally, mutational events are not involved in differentiation; sequential expression of different genes effected by chromatin organization rearrangements. Examples of derepression and repression occur in human cervical and gastric carcinomas[39] where the normally repressed condensed X chromosome decondenses and reactivates, and occur during hepatocarcinogenesis by loss of expression of genes for $\alpha$-2-euglobulin production[91] and tryptophan oxygenase.[33] Reexpression of "oncofetal" proteins by malignant cells and during carcinogen exposure represents another example of such phenomenon. Should disruption of normal gene expression be a critical, initial event in malignancy, then detection of changes in macromolecules regulating specific genes (chromosomal proteins) might provide insight into malignant phenotype-inductive sequence. Chromosomal proteins (histone and nonhistone [NHC]) control both structural and functional properties of eukaryotic genomes.[31,56] Histones are involved in genome packaging and are nonspecific repressors of DNA-dependent RNA synthesis.[32,49] Components of NHC proteins also control genomic maintenance but more importantly, catalyze defined gene sequence transcription.[35,48,57,70] Changes, therefore, in NHC proteins or NHC protein subgroups during carcinogenesis, could be "hallmarks" for malignant conversion and might play key roles in the process. The potential for chromosomal proteins as neoplastic transformation markers has been recently reviewed.[92] Functional and structural properties of chromosomal proteins in cells and changes occurring in malignancy have been recently reviewed,[40,79,80] and were subjects of previous reviews[11,94] and monographs.[24,25,93]

While a number of studies have reported differences in NHC proteins in normal and malignant cells,[27,34,40] few exist concerning in vivo changes in chromatin proteins during chemically induced or spontaneous carcinogenesis.[1,10,17,19,20,21,23,37,41,42,51,87,102] Of these, most concern chemical hepatocarcinogenesis. For example, rats treated with thioacetamide showed NHC protein alterations that varied with age of animals,[10,37] while comparison of normal liver nuclei with those of diethylnitrosamine (DEN) treated rats indicated a selective loss of an NHC protein polypeptide in treated rats.[41,42] In chromatin of ethylnitrosourea-induced neuroectodermal tumors, NHC protein characterizing both fetal and adult tissue have appeared in electrophoretic patterns.[17] Tumor-specific changes in 2 $M$ salt soluble and residual protein electrophoretic patterns of rat liver treated with 3-methylcholanthrene have also been reported.[23] Accumulation of two specific NHC proteins during induction of colonic cancer with dimethylhydrazine has been reported by Allfrey and co-workers.[1,19-21] During nitrosomorpholine-induced hepatocarcinogenesis, only slight quantitative changes in two NHC proteins were observed in primary hepatomas, whereas, in their metastases, both quantitative and qualitative changes were detected.[102] Quantitative changes in NHC proteins were reported in rats exposed to DEN in drinking water.[86] Lastly, qualitative changes in NHC proteins resolved by bidirectional electrophoresis have been reported in hepatocellular carcinomas induced with aflatoxin $B_1$.[51]

Major questions that these and other studies concerning chromosomal protein change and neoplasia have left unanswered are as follows: (1) are there chromosomal NHC proteins that can be definitively associated with obligate cellular alterations induced during carcinogenic exposure or spontaneous carcinogenesis or in resulting tumors; (2) can these tumor-associated proteins (TAP) be used as identifying markers for carcinogenesis of cell alterations; and (3) does a specific identifiable NHC protein exist that regulates the gene(s) responsible for malignant phenotype or aberrant control? Identification of such protein(s) should facilitate our search for clues to understanding underlying mechanisms of carcinogenesis as well as for possible points of attack.

In this chapter, we report on our studies of changes induced in NHC proteins during chemically induced and spontaneous murine hepatocarcinogenesis. Particular attention is directed at pointing out potential problems during isolation of chromatin and chromosomal proteins. Our use of high-resolution multiple linear gradient two-dimensional electrophoresis, silver staining of resolved polypeptides, and computer-assisted image processing in the study of NHC proteins in normal and neoplastic tissue will similarly be detailed.

## II. CHEMICALLY INDUCED AND SPONTANEOUS CARCINOGENIC SYSTEMS

### A. Rat Chemically Induced Hepatocarcinogenesis

We have utilized two basic hepatocarcinogenic regimen in Sprague-Dawley (Charles River Farms) and ACI (The University of Texas Environmental Science Park) male rats to produce discrete carcinogenic stages and primary hepatocellular carcinomas (PHCs) for analysis. Two prototypic hepatic carcinogens, *N*-2-acetylaminofluorene (AAF) and DEN were administered (Table 1).[12,95,97] While DEN in rats does not result in production of discrete nodular stages that characterize AAF intermittent feeding, at termination of a fully carcinogenic DEN exposure histological examination of such livers indicated that >90% of the hepatocytes exhibit histochemically characteristics of preneoplastic cells. Normal liver from age-matched rats and 24- and 48-hr regenerating rat liver provide control tissues for protein pattern analyses.

### B. Spontaneous Mouse Hepatic Carcinomas

The inbred mouse strain C3H/HeN (C3H) exhibits a genetic tendency for high rates of spontaneous hepatocellular tumorigenesis. While a number of extraneous factors (diet, animal density, type of bedding, etc.) influence tumor incidence in C3H mice,[30,47,85] no known cause for PHCs exist. In C3H males, incidence of spontaneous liver tumors is age related, achieving rates of 65 to 100% within 18 to 24 months of birth.[13,14,16,45,76,105] Possible induction of PHCs by a variety of viral species present in these mice as a PHC cause has been pursued in a number of laboratories, but no evidence presently links PHC occurrence to a viral agent(s).[80] Mice harboring spontaneous carcinomas are detected by use of a radioimmunoassay for α-fetoprotein serum level. In previous studies, we have determined the existence of an absolute relationship in C3H mice between an elevated α-fetoprotein level (AFPL)

**Table 1**
**RAT HEPATOCARCINOGENIC REGIMEN AND TISSUES ANALYZED**

A. AAF feeding: four consecutive 3-week feedings of AAF at 0.06% in basal diet #101 with 1 week of normal diet #101 between each 3-week carcinogen feeding
   1. Cycles and tissues analyzed
      a. 1 × 3: liver at the termination of one 3-week feeding
      b. 2 × 3: liver after the termination of the second 3-week feeding
      c. 3 × 3 nodules: at the termination of three 3-week feeding cycles, livers exhibit small hyperplastic nodules which eventually disappear if dietary carcinogen exposure ceases
      d. 4 × 3 nodules: at the termination of four 3-week cycles of feeding, livers have been >90% converted to hyperplastic nodules which are considered to be putative premalignant lesions; carcinogen exposure ceases after this feeding cycle
      e. Late nodules: after carcinogen feeding cessation, most putative premalignant nodules regress, leaving a subpopulation that continue proliferation
      f. Primary hepatocellular carcinoma (PHC): cancers which appear 8 to 12 weeks after completion of four cycles of 3 weeks each of AAF feeding despite cessation of carcinogen exposure
B. DEN exposure: DEN is administered continuously in drinking water for various periods of time in weeks at 60 ppm
   1. Exposure Level
      a. 60 ppm for 8 weeks: represents acute carcinogen effects; livers in such animals histologically exhibit foci of altered cells but not gross or microscopic malignancy
      b. 60 ppm for 8 weeks, normal diet for 8 weeks, sacrifice
      c. 60 ppm for 10 weeks, normal diet for 11 weeks
      d. Livers harvested at 8 to 11 weeks after cessation of a fully carcinogen DEN regimen (60 ppm for 8 to 10 weeks) exhibit well-defined primary hepatocellular carcinomas

and the presence of hepatic tumors. All mice examined to date having AFPL levels of >0.5 μg/mℓ of blood have exhibited one to four well-defined, spontaneous, single hepatocellular carinomas of 0.7 to 2.2 g wet weight. Normal age matched and regenerating livers provide control samples.

## III. ISOLATION OF CHROMATIN AND CHROMOSOMAL PROTEINS

### A. Chromatin Preparation

Chromatin represents the interphase state of chromosomes in eukaryotic cell nuclei consisting of a macromolecular complex composed of DNA, RNA, and protein. Chromatin is generally operationally defined; the proteins associated with this isolated genome reflect methodology used in their isolation. In chromatin isolation schema, both loss of true chromosomal proteins and possible contamination by nucleoplasmic, membrane, and cytoplasmic components during isolation require consideration. Purified nuclei prepared from freshly isolated tissue is an absolute requirement since chromatin prepared from whole tissue or from frozen tissue is severely contaminated with cytoplasmic proteins.[52,106] Potential for cytoplasmic and membrane contamination of chromatin preparations can be minimized and in most cases eliminated by utilizing solutions containing nonionic detergents such as Triton® X-100 during nuclei isolation and centrifugation of isolated crude chromatin through heavy buffered sucrose.[9,22] During nuclei preparation, at each preparative step, monitoring of aliquots by phase contrast microscopy and electron microscopic analysis can ensure isolation of nuclei without cytoplasmic "tags". Further, incorporation of aliquots of radioactively labeled cytoplasm into all isolation and buffer solutions allows for determination of potential cytoplasmic contamination in isolated chromatins.[77] It should be noted, however, that since chromosomal proteins are synthesized in the cytoplasm of the cell, presence of a "cytoplasmic" protein in a chromatin preparation need not automatically indicate a preparation artifact but, rather, may indicate existence of a cytoplasmic pool in dynamic equilibrium with chromatin. The many examples of movement of proteins in and out of the nucleus of the cell renders existence of such pools plausible.

Availability of purified chromatin from liver tissues and tumors is critical to our studies. For the past few years, clean nuclei have been prepared in our laboratory in high yields from excised rat and mouse liver and tumor tissues according to a modified version of Reeder's procedure, which was based on original methods for rat liver by Chaveau as modified by Blobel and Potter.[18,63,64,66,75,77-81,104] Briefly, the isolation scheme is as follows:

Normal tissues and/or tumors are removed from fasted animals, weighed and washed in ice-cold physiologic saline containing phenylmethylsulfonyl fluoride (PMSF). Washed tissues are minced and homogenized in 10 volumes of 0.25 *M* STM (0.25 *M* sucrose-0.01 *M* Tris-HCl (pH 7.4), 0.025 *M* $MgCl_2$) in 1.0 m*M* PMSF. This and all subsequent solutions and buffers contain 1.0 m*M* PMSF and all steps are performed at 4°C. After filtering homogenates through four layers of cheesecloth, crude nuclei are pelleted by centrifugation at 5000 rpm for 10 min (Sorval SS34 head or comparable rotor). Supernatants are collected and set aside for later cytosolic protein isolation. Pelleted nuclei are resuspended in 0.25 *M* STM, an aliquot examined by light microscopy to determine that all whole cells have been disrupted, and then two volumes of 2.3 *M* sucrose, 0.5 *M* KCl, 0.025 *M* $MgCl_2$ in 0.01 *M* Tris-HCl (pH 7.4) containing 0.1% Triton® X-100 are added. Nuclei are isolated by centrifugation at 30,000 rpm in a fixed angle 30 rotor (Beckman) for 1 hr. Nuclei are resuspended and washed three times in 10 m*M* Tris (pH 8.0)-1.0 m*M* PMSF. Phase contrast and electron microscopic analysis of Triton® X-100-prepared nuclei confirm the absence of cytoplasmic tags and the removal of the outer nuclear membrane. Histones are extracted with dilute mineral acids, recovered and acid-urea gels run to analyze for presence of electrophoretic sub-bands, a highly sensitive assay for chromosomal protein proteolysis.[81] Nuclei suspended in Tris-buffer are lysed by 8 strokes in a tight-fitting Dounce homogenizer ("B" pestle). Then, 7 mℓ of lysed chromatin solution are layered over sucrose step gradients of 15 mℓ each of 1.6 *M* sucrose and 15 mℓ of 1.3 *M* sucrose (both in 10 m*M* Tris-PMSF at pH 8.0) and sedimented at 27,000 rpm for 90 min (SW27 rotor, Beckman). Chromatin pellets are resuspended in 0.01 *M* Tris HCl pH 8.0, lightly homogenized and layered a second time over step gradients composed of 3 mℓ each of the 1.6 and 1.3 *M* sucrose. Gradients are centrifuged at 41,000 rpm for 60 min (SW41, Beckman). Pellets are washed gently three times in 0.01 *M* Tris-HCl (pH 8.0) and then dialyzed overnight against 10 m*M* Tris-HCl (pH 8.0)-PMSF, quick-frozen in a dry ice-acetone bath, and stored at −80°C. To ensure that chromatin preparations are free of cytosol contamination, aliquots of [$^3H$]-labeled cytoplasm prepared from livers of male 175-g Sprague-Dawley rats exposed to 0.1 Ci/g body weight of [$^3H$]-labeled amino acid mix are used as part of the homogenization medium. Post-nuclear supernatants (crude nuclear isolation step) are combined and centrifuged at 60,000 rpm in a 60Ti rotor (Beckman) for 1 hr. This procedure pellets mitochondria, ribosomes, and particulate components. Cytosolic supernatants are combined, dialyzed against $H_2O$ containing 1.0 m*M* PMSF, shell frozen, lyophilized, and stored at −80°C. Yields of DNA-chromatin with these procedures typically are 0.8 to 1.0 mg/g of starting wet weight, yield from PHCs reflecting that from necrosis-free tissue.

Isolated chromatins are characterized by composition and UV spectrum;[77] DNA concentration is determined by absorbance at 260 nm and by fluorescence when bisbenzimidazole (Hoechst 33258) binds to DNA.[26] Chromatin and cytosolic protein contents are determined according to the procedure of Lowry et al.[59]

### B. Isolation of Chromosomal Proteins

While the histone chromosomal proteins can easily be dissociated from chromatin-DNA and isolated by a variety of techniques,[69] a substantial amount of the NHC proteins are difficult to isolate due to tight binding to DNA and aggregation with histones and each other. A great number of procedures have been utilized to extract histones and NHC proteins from chromatin and further separate histone and NHC protein components.[69] In pilot studies, we evaluated protein extraction and yields from whole and dehistonized chromatins by procedures utilizing: (1) 4 *M* CsCl (pH 11.6) or 4 *M* CsCl (pH 14.0)[15,54] (2) sodium dodecyl sulfate (SDS),[71] (3) hot (>90°C) SDS,[105] (4) phenol,[98,109] (5) 0.4 *M* guanidine hydrochloride-6 *M* urea,[58] (6) BAL 31 nuclease,[38] (7) DNase-RNase,[67,108] (8) $S_1$ nuclease,[72] (9) 3 *M* NaCl-5 *M* urea,[51] (10) 3 *M* NaCl-7 *M* urea,[84,89] and, (11) direct solubilization in isoelectric focusing electrophoretic sample solubilization urea mix or "urea mix" containing NaCl-protamine sulfate.[3,87] Chromatin extraction in high salt-urea buffers proved most reliable in terms of yield of extracted chromosomal proteins, recovery, and satisfactory electrophoretic resolution.[84,89] Briefly, chromosomal proteins are extracted as follows.

Respective chromatin samples at a concentration of 10 $A_{260}$ units/mℓ (500 μg/mℓ) are extracted in a Dounce homogenizer with spin bar in 3 *M* NaCl-7 *M* urea in 10 m*M* Tris-HCl pH 8.0, 1 m*M* PMSF, and 5 m*M* β-mercaptoethanol at 4°C. During extraction, the chromatin solubilization mix is "Dounced" (B pestle) at hourly intervals and then left to extract overnight. Proteins were separated from nucleic acids by centrifugation at 204,000 g for 20 hr in either an SW60TI or SW41 (Beckman) rotor. Solubilized proteins are collected, DNA pellets washed with extraction buffer, and collected proteins dialyzed against four 4-ℓ changes of distilled water containing 1 m*M* PMSF and two changes of distilled water minus PMSF. During dialysis, the chromatin proteins precipitate and are then recovered by centrifugation (20,000 rpm Sorvall SS34 rotor, 10 min). Dialysis bag supernatants were also lyophilized and the recovered protein combined with centrifugally recovered proteins. These procedures resulted in extraction of >95% of the chromatin-associated proteins.

## IV. TWO-DIMENSIONAL GEL ELECTROPHORESIS

Two-dimensional analysis of normal and neoplastic tissue chromosomal proteins are performed in the so-called ISO-DALT system developed by the Andersons.[2,5] The system evolved from extensive modification and/or improvements of the original high-resolution methods of O'Farrell and Klose.[55,67] As utilized in our laboratory, 40 first-dimension isoelectric focusing gels and 20 second-dimension slab gels may be simultaneously cast and run at a given time. Isofocusing is carried out in 1.2 mm ID × 11 cm gels and second-dimension slab electrophoresis on 17× 17 cm 10 to 20% linear polyacrylamide SDS slabs. The large gradient slabs give better resolution than small homogeneous slabs, and co-casting as well as co-electrophoresis of samples ensures high reproducibility. Since highly detailed descriptions of the ISO-DALT system have appeared,[2,5] its description as follows is brief.

Each ISO apparatus allows for simultaneous casting of 20 isoelectric focusing gels. Electrofocusing gels are composed of 8.25 g of urea, 0.75 mℓ of pH 3.5 to 9.5 preblended ampholytes (LKB), 2.0 mℓ of 30% acrylamide plus 1.8% bisacrylamide, and 6.0 mℓ of glass distilled $H_2O$. After dissolution under warm tap $H_2O$, the mix is degassed briefly (lyopholizer), 300 μℓ of NP-40, 50 μℓ of 10% ammonium persulfate and 7 μℓ of TEMED added, and then the mix poured into an ISO loading trough. Gels are cast by displacement of gel solution with $H_2O$ and polymerize in 1 hr. During electrophoresis, the upper reservoir buffer is 20 m*M* NaOH while 10 m*M* $H_3PO_4$ is used in the lower (anode) reservoir. Gels are prefocused for 1 hr at 200 V, samples are then applied with a microliter syringe and focused to equilibrium for for 12 to 20 hr at a voltage level so that 10,000 V hr is reached (ISO system). In order to adequately resolve basic chromosomal proteins, position of the ISO anode ($H_3PO_4$) and cathode (NaOH) buffers are reversed and focusing is performed for varying V hr total times (3400 to 4500 V hr). In such a system (NEPHGE), proteins migrate in relation to each other based on their pI but proteins do not reach equilibrium. Focused gels are extruded with $H_2O$ from a syringe and each individually rocked for 20 to 30 min in SDS equilibration buffer (10% glycerol, 5% β-mercaptoethanol, 2.1% SDS in 0.125 *M* Tris-HCl pH 6.8 containing 0.001% bromophenol blue dye marker. Equilibrated gels are either frozen (−80°C) or applied immediately to second-dimension DALT gels.

In the DALT system, 20 10 to 20% polyacrylamide gradient SDS gels are cast simultaneously, thereby ensuring reproducible electrophoretic displays of protein in the second dimension. Gels are cast in 7-in. square glass plates held together with strips of silicone rubber that contain glass spacers of 1.5-mm thickness glued to the inner surface of one of the squares. The gel-casting apparatus consists of a Reeve Angel gradient former with small magnetic mixer attached to a Tygon manifold connected to a gradient line, a vacuum line, a dense displaying fluid line, and a line to a distribution manifold. A second Tygon manifold connects with the gradient former to a gel-casting chamber containing 20 of the glass gel plates which can be rotated about one axis. Gradient gels are cast as described,[6,7] and polymerize in 1 hr. First-dimension SDS-equilibrated gels are stretched out, excess buffer removed, slid into second-dimension gel plate slots, and sealed to second-dimension slabs with an overlay of 0.5% molten agarose which then allows for differential orientation of slabs during electrophoresis. Gels are electrically isolated and plates cooled along their entire length. Gel slabs are slid into position in rubber septa that divide the second-dimension DALT tank lengthwise, and plates are uniformly cooled during electrophoretic runs by circulation of chilled (4°C) electrophoresis buffer. Electrophoresis is performed at 0.8 to 1.0 A until marker dye reaches the end of each gel plate (~3.5 to 4 hr). Gels are removed and placed in groups of 10 to 20 in 4-q plastic food keepers for fixation and staining.

Since, in two-dimensional electrophoretic analysis of chromosomal proteins, no advantage is gained by fractionation of histones and nonhistones prior to electrophoresis, lyophilized protein extracts are directly solubilized for analysis. Several sample solubilization mixes, each designed to favor resolution of acidic, neutral and/or basic chromosomal proteins are utilized. Lyophilized extracted proteins are weighed out, aliquots taken for protein determinations, and the remainder placed in a tight-fitting glass-on-glass Dual homogenizer (1-mℓ capacity) and sample solubilization mix added. Experimentation with each sample to determine optimal solubilization volume is necessary. Samples are solubilized by vigorous homogenization and then are ready for electrophoretic analysis. In general, solubilized protein concentrations of 4 to 6 μg/μℓ of sample mix are routinely obtained for the chromosomal proteins we have examined. Cytosolic protein sample mixes of 10 μg and higher per microliter concentration are easily obtained with these procedures. These procedures subject the sample to be analyzed to minimum manipulation, exposure to heating, and based on extensive control electrophoretic runs of proteins, with known isoelectric points no introduction of charge change. For optimal resolution of acidic-neutral NHC proteins under conditions of equilibrium isoelectric focusing (10,000 V hr), the solubilization mix is 9.0 *M* urea, 4% NP-40, 2% preblended ampholines (LKB, 80% pH range 3.5 to 9.5, 20% pH range 5.0 to 8.0) and 2% of β-mercaptoethanol (BME). Under these conditions, it is difficult to reproducibly produce stable pH gradients much above pH 7, thus excluding from gels NHC proteins with basic isoelectric points. Similarly, since histones have pIs >11, they are also excluded. In analyses designed to resolve basic as well as acid-neutral chromosomal proteins (NEPHGE), we utilize a solubilization mix consisting of 9.0 *M* urea, 2% NP-40, 2% preblended ampholines, 2% BME, 0.3 *M* NaCl and 1.6 mg/mℓ of protamine sulfate,[87] load samples on the acid side and focus for 4000 V hr. To protect samples from acid degradation, they are loaded under an overlay of 4 *M* urea. While this strategy facilitates resolution of basic NHC proteins, after 3000 elapsed V hr histones have migrated through and out of the individual focusing gels. In general, these parallel procedures have allowed us to resolve NHC proteins with isoelectric points ranging from 3.8 to 10.5. Additionally, for further stabilization of the basic end of pI gradients, 0.9 mg of lysine and 1.5 mg of arginine per milliliter can be added to focusing gel and sample solubilization mixes.[101]

To ensure that samples are free of particulate material, due to protein aggregation or precipitates due to basic ampholyte-residual nucleic acid complexes, chromosomal proteins are subjected to high-speed centrifugation at room temperature (50,000 rpm, Beckman VT150 rotor) in cut-down Beckman 400-μℓ polyethylene microcentrifuge tubes oriented horizontally in specially constructed polycarbonate or Delrin adapters (Figure 1). Cytosol samples are

FIGURE 1. Apparatus for removal of basic ampholyte-nucleic acid complexes. Lyophilized chromosomal protein samples resuspended in solubilization mix are centrifuged at 50,000 rpm at room temperature in a vertical VT150 rotor (Beckman) for 1 hr after respective samples have been loaded into cut-down 400-μl polyethylene microfuge tubes which have been horizontally oriented in polycarbonate (pictured) or Delrin fabricated adapters. The design for a prototypic polycarbonate adaptor was the kind gift of Dr. Norman Anderson (Argonne National Laboratory, Argonne, Ill.), its inventor. A rotor for this purpose of somewhat modified design and greater sample capacity (LP42-TI) is now available from Beckman Instruments.

clarified prior to loading in 1.5-mℓ microfuge tubes at 12,000 *g* for 10 min (Beckman Microfuge 12).

## V. DETECTION OF PROTEINS IN DALT GELS

During initial preliminary studies, we utilized Coomassie brilliant blue R-250 for the visualization of proteins. This dye can detectably stain 0.1 to 3.0 μg of protein in a single band in slab polyacrylamide SDS gels. Since in single-dimensional protein gel displays a band in a rod or slab may in fact represent several "spots" in a two-dimensional gel profile, many NHC proteins present in low concentration in samples are not detected by Coomassie. We now routinely use silver stain and in vitro labeling of proteins by reductive methylation for protein visualization.[43,96] In our hands, silver "stain" detects proteins some 370-fold more sensitively than Coomassie, this sensitivity being in the same range as that for $^{14}C$ and $^{3}H$-methylated proteins after a 5-day autoradiographic exposure. Silver staining utilizes a modification of de Olmos' cupric-silver method and possesses limits of sensitivity in our gel system of less than 3 pg/mm$^3$.[43]

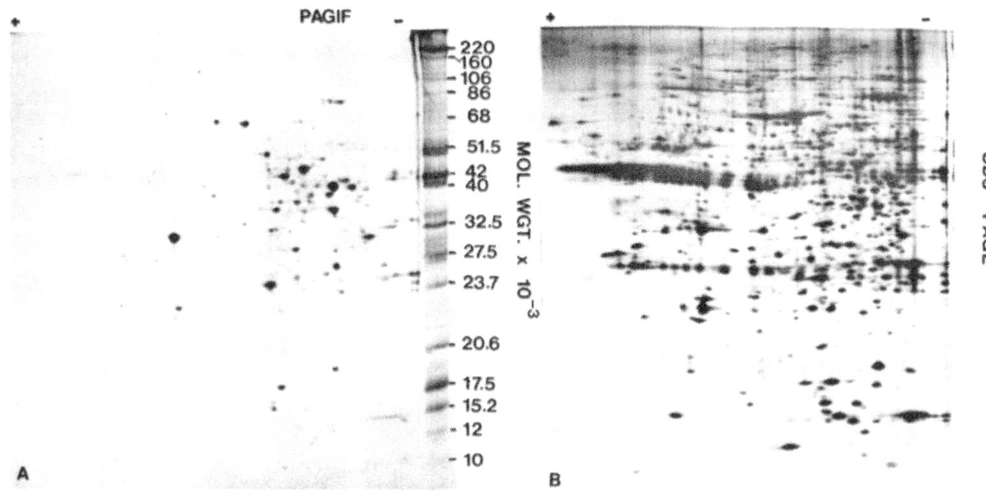


FIGURE 2. Comparison of Coomassie brilliant blue R-250 and silver staining in gradient two-dimensional PAGE. Cytosolic liver proteins (430 μg) isolated from a 1-year-old male C3H/HeN mouse were separated by equilibrium-two-dimensional-gradient PAGE and the second-dimension DALT-gel (A) first stained with Coomassie, destained, and then (B) silver stained.[43] Diluted whole rat heart homogenate (1:4) and protein carbamylation trains (at 42 and 25 × 10$^{-3}$ daltons in (B) were included for molecular weight and pI determinations. In this analysis, 187 polypeptides ranging from 20 to 150 Kd were evident when staining was with Coomassie, but over 1800 ranging from 10 to over 250 Kd were detected with silver. (From Guevara, J. Jr., Johnston, D. A., Ramagli, L. S., Martin, B. A., Capetillo, S., and Rodriguez, L. V., *Electrophoresis*, 3 197, 1982. With permission.)

### A. Silver Staining

Second dimension 10 to 20% linear gradient polyacrylamide gels are agitated overnight in fixing solution composed of 20% (v/v) ethanol, 5% (v/v) acetic acid, and 2.5% (w/v) sulfosalicyclic acid at a concentration of 200 mℓ of fixing solution per 50 mℓ of gel. Gel(s) are then washed in glass-distilled $H_2O$ (200 mℓ/50 mℓ of gel) with constant agitation for 1 hr with one change, followed by another 1-hr wash in $H_2O$ and a last 1-hr wash in 20% (v/v) ethanol. The gels are then placed in 50% (v/v) ethanol for a minimum of 2 hr or overnight with constant agitation. Individual ethanol-washed gels are placed in freshly prepared ammonicial silver solution (for 100 mℓ of solution, 1.4 mℓ of 14.8 N $NH_4OH$ is added to 21 mℓ of 0.36% freshly degassed NaOH and then accompanying vigorous vortexing, 4 mℓ of 20% $AgNO_3$ solution added). This solution is diluted with glass distilled $H_2O$ to 90 mℓ and absolute ethanol added to a total of 100 mℓ. A standard gradient gel requires 150 mℓ of this solution. Each gel is equilibrated in this solution for 1 hr. The silver solution is then decanted and the gel rinsed with distilled $H_2O$ once and then three times for 20-min periods each in 20% (v/v) ethanol. These washing steps remove most of the unbound silver; incorporation of ethanol retains the size integrity of gels. After decanting the last ethanol wash, the gel is again placed in freshly prepared 20% ethanol containing 0.005% citric acid and 0.0135% formaldehyde (developer). Proteins become visible during this development step, all proteins appearing by 20 min elapsed time. To stop development, the citric acid-formaldehyde solution is decanted and 20% (v/v) ethanol containing 1% acetic acid added. After 5 min in this solution, $NH_4OH$ is added for 10 min to neutralize the acid and the gel then stored in 20% ethanol. Neutralization of acetic acid with ammonium hydroxide prevents protein silver spot fading and a permanent record results (Figure 2).

### B. Radiolabeling of Proteins

Due to difficulty in obtaining sufficiently high specific activity of isotypically labeled proteins from tissues labeled in vivo, we utilize a method in which high specific activity protein labeling is accomplished in vitro. Display of radioactively labeled proteins serves as an adjunct to silver staining since it is not known whether silver stains or detects all types of proteins stoichiometrically. These two diverse methods allow for detection and quantitation of nearly all proteins of interest to us. Known amounts of lyophilized samples are reductively methylated with [$^{14}$C] or [$^{3}$H]-formaldehyde (New England Nuclear, specific activity 42 to 48 mCi/mmol) which has been purified by passage through a small Dowex® 1-X8 column. Reductive methylations are carried out in a 1-mℓ reaction volume containing 500 to 1000 μg of protein,[96] 2 m*M* [$^{14}$C]-formaldehyde (100,000 dpm/μmol), 20 m*M* $NaCNBH_3$ in 100 m*M* Hepes® [4-(2-hydroxyethyl)-1-piperazine ethane sulfonic acid] buffer and 8 *M* urea. After incubation at room temperature (22 to 25°C) overnight, 3 mℓ of 10% trichloroacetic acid (TCA) is added. Precipitated samples are isolated by centrifugation, a portion aliquoted, dissolved in 0.2 mℓ of 0.2 *M* NaOH and counted in Scintiverse® (Fisher). In test methylations of bovine serum albumin (BSA) and whole rat heart homogenates, we routinely obtained specific activities approaching 80 to 150,000 dpm/μg of protein. Gels containing $^{14}$C or $^{3}$H labeled proteins are processed by scintillation autography (fluorography) using X-ray film (Figure 3).[94] By co-electrophoresis of $^{14}$C and $^{3}$H in vitro labeled NHC proteins extracted from different test chromatins and a combination of fluorography and autoradiography,[61,88] comparisons with high confidence in a single matrix are possible. In addition, recent tests indicate that gels containing $^{35}$S, $^{14}$C, or $^{3}$H labeled proteins may be, with low quenching effect, first stained with silver and then processed by autoradiography and fluorography.[65,103]

### C. Photography

Silver stained gels are removed from 20% ETOH and placed directly on a light box (GE circline, Aristo PG-20C or Aristo MS-1417) to which is fixed a photographic step tablet calibrated in 16 steps in a density range of 0.04 to 1.54. Photographs are taken using a 215-mm f6.3 lens with Eastman Kodak® professional copy film 4125, 4 × 5 in. This high-contrast film facilitates visualization of very small faintly stained proteins. Films are developed in an S.S. White-Aluveleper automatic developer eliminating variability associated with manual development. Prints are made from negatives with the aid of an automatic-dodging reducer Log E scanning enlarger. Metering accuracy is checked with Eastman Kodak® Sensitometer strips; prints are made on Agfa® Geveart Pl-3 paper, 5 × 7-in. and 8 × 10-in. and processed through an Eastman Kodak Royal Rapid® print processor.

The use of a photographic step wedge, sensitometer strips, and automatic negative and print processors allows a high degree of reproducibility and standardization enabling our comparison of prints from many different gels.

### D. Analysis of Gel Protein Maps

Comparisons between gel patterns of cell fractions are relatively simple when gels are superimposable. However, should slight variations due to gel casting, protein load, migration of proteins, etc., occur, the task becomes more difficult. Quantitation of gels and intercomparison of hundreds of spots on hundreds of gels are handled by computer-assisted image analyses.

Each suitably stained gel, 4 × 5 negative, or fluorograph is analyzed by image processing. Once stained gels or fluorographs have been photographed, two methods of digitization and standardization of individual gels are used. In the first method, the original stained gel is scanned directly with a high-quality Cohu model 7120, 525, line black and white camera (512 × 480 pixels resolution, 256 levels of gray) and digitized by a Colorado Video model 270A digitizer. The system is equipped with a joystick and single-frame video disc. Utilizing the same imaging system, in the second method of analysis, a negative of the original gel or fluorograph is digitized with the video system. The digitizer is interfaced to a Modcomp® II/220 minicomputer having 96Kbytes memory, 25 Mbytes ISS disc, and a Tektronix® 4013

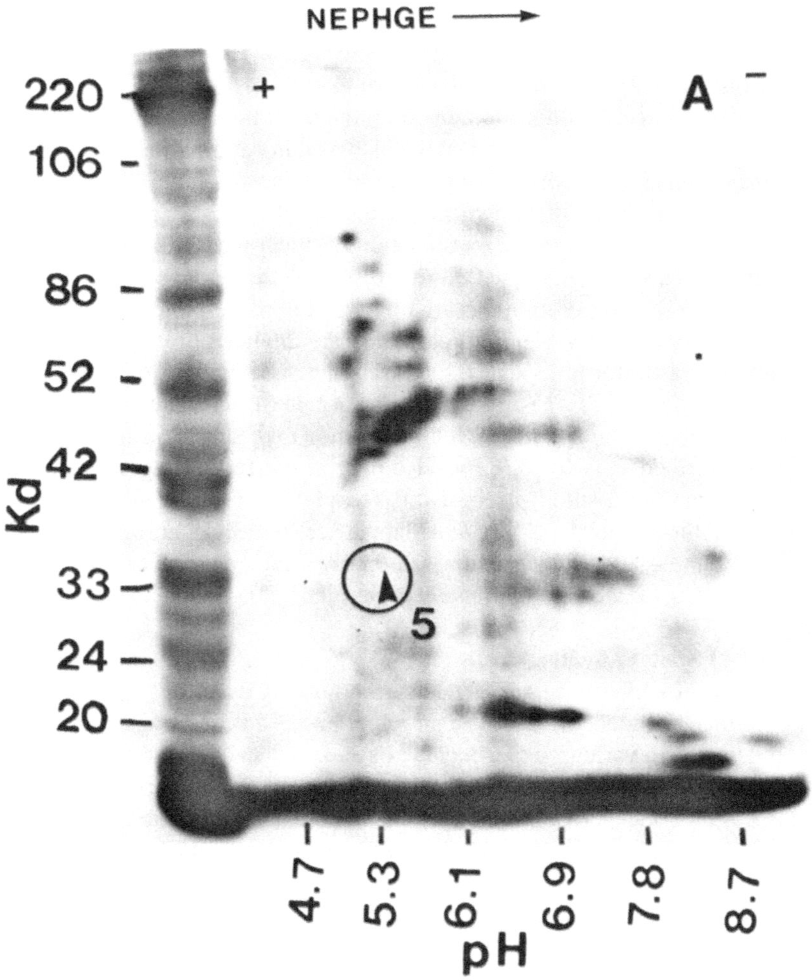


FIGURE 3. In vitro reductive methylation of phenol soluble chromosomal proteins. Total phenol soluble normal rat liver chromatin chromosomal proteins were solubilized in 8 *M* urea, radiolabeled in vitro with [$^{14}$C]-formaldehyde, free formaldehyde eliminated by dialysis and recovered $^{14}$C-labeled chromosomal proteins analyzed by two-dimensional nonequilibrium PAGE under conditions where histones migrate out of the first-dimension focusing gel.[94] Approximately $6 \times 10^5$ cpm as protein were loaded on the focusing gel and fluorography performed after 10 days exposure. Whole rat homogenate proteins radioactively labeled as described with $^{14}$C-formaldehyde were co-electrophoresed to serve as molecular weight standards and pH determined in ISO gels with a micro pH electrode. In vitro labeling provides an added complimentary protein detection method to that using silver. Presence of a polypeptide in the circled portion of the gel labeled #5 denotes presence of a phosphoprotein in AAF and DEN PHC NHCP populations which is not present in normal and regenerating rat liver NHCP.[94]

graphics terminal as a program control monitor. Data recovery ("spot maps") is provided as pseudogray level plots on a Versatec® 2160 plotter or by use of a Centronics® printer in association with the Modcomp® II/220 minicomputer. Additionally, a Grinnel® 2735 pseudocolor raster, graphics system with model 2000 Matrix Color Camera (35 mm and 8- × 10-in. formats) is interfaced to the Modcomp® II/220 to serve for recovery, display, and construction of standardized comparison maps of various sample protein displays.[53]

## VI. STANDARDIZATION OF GELS

### A. pH and Molecular Weight of Resolved Proteins

Comparison of gels containing hundreds of protein spots to determine spot identity (X, Y coordinates corresponding to pI and molecular weight) and to standardize gels requires external pH measurement or incorporation of proteins of known pI into samples to be analyzed as well as co-electrophoresis of known molecular-weight standards. To determine the isoelectric point, we have found that a combination of direct measurement of the pH gradient within individual focused ISO gels with a micro pH electrode just prior to loading onto the second dimension and incorporation of carbamylated reference proteins in samples result in reliable pH measurement.[8] By blending partially carbamylated protein samples and uncarbamylated protein, a uniform series of spots ranging from native to completely modified molecules (carbamylation train) is obtained. For a pattern of evenly spaced protein bands as molecular-weight standards in the SDS dimension, a rat heart whole homogenate prepared as previously described is utilized.[36] For electrophoresis, rat heart proteins in agarose are cast in rod gel form and a small piece of this rod placed on either side of the ISO gel on top of the second dimensions SDS-gradient slab. After co-electrophoresis with the particular first-dimension sample and silver staining, the heart standard resolves into some 80 bands ranging from 10,000 to 220,000 daltons mol wt (Figure 2).

### B. Gels and Gel Spot Patterns

Standardization of gels is necessary since variations may occur due to degree of migration and amount of material loaded. Since the principal purpose is to "fingerprint" and compare protein populations of various experimental tissues and cell fractions, standardization is necessary to ensure geometric compatibility. For geometric standardization, a heart homogenate at the edge of the gel, and at least ten internal marker proteins well separated on experimental gels were utilized. Use of ten markers and heart display permits full dimensional correction over the entire surface of gels. Position rotation and scaling to dalton and pH scale was developed and a multivariate regression model (MANOVA) used to effect standardization utilizing marker proteins occurring within respective test protein samples. In this manner, cytosolic proteins isolated from nine individual spontaneous PHCs occurring in 1-year-old C3H male mice were analyzed by equilibrium isoelectric focusing and second-dimension SDS-PAGE (2D-PAGE) (Figure 4). Ten proteins resolved in individual tumor cytosols and in normal cytosol of age-matched male mice representing all areas of possible gel resolution, were picked for position comparison and scaling by computer image processing. Chosen marker proteins (spots) occurring in the four PHCs were located using the joystick on each respective 4- × 5-in. negative and again on each negative of equilibrium-2D PAGE of separately "urea mix" constituted protein samples from the four PHCs which had been analyzed 2 weeks after the original electrophoretic run. The position of each of the respective 10 marker protein spots was standardized to the position for that same spot in the second-dimension pattern for PHC #1, the PHC #1 pattern similarly standardized against itself, and positions for the markers in age-matched normal C3H males also determined. Standardized positions for the marker proteins analyzed in this manner are presented in Figure 5. Marker positions in five additional C3H-PHCs have also been analyzed by this method. Of the greater than 470 polypeptides detected in the nine PHCs by equilibrium-2D PAGE and in control runs, the standardized pixel locations for the 10 markers in all gels only varied between 2 to 4 pixels in a pixel window comprised of 500 × 350 discrete vertical and horizontal pixels. These data graphically demonstrate the ability to isolate and segregate individual proteins or groups and perform comparative analyses having high confidence limits within complex protein spot patterns, representing complex protein sample populations.[53]

After standardization, two methods of gel comparison are used: (1) simple matching or

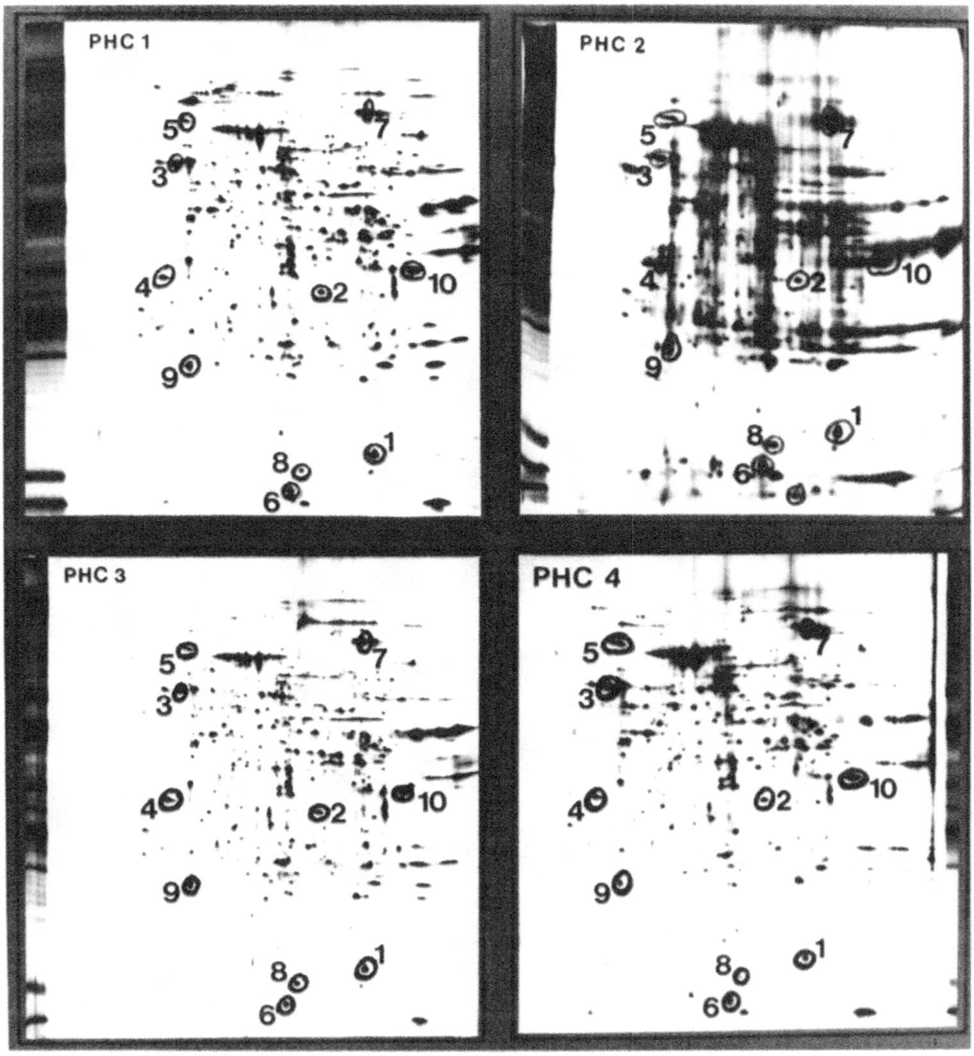


FIGURE 4. Second-dimension gel standardization using internal sample protein markers. 50 μg of cytosol proteins isolated from individual spontaneous PHCs occurring in 1-year-old C3H/HeN male mice and cytosol from age-matched males were isoelectrically focused for 10,000 V hr (equilibrium), separated in second-dimension 10 to 20% linear gradient SDS-polyacrylamide slabs, and the slabs silver stained. As noted in each of the gels representing the four PHCs, ten (circled and numbered) proteins occurring in the respective cytosols were picked to perform position, scaling, and rotational standardization (Section VI.B).

correlation of one gel with another over the entire surface, and (2) separation of each spot or spot-group and calculation of parameters of each which then are combined in a match type or correlative measure to obtain scores for each protein spot and combined to give overall scores for gel comparisons. The advantage of the first method is its simplicity and speed of operation. The potential disadvantage is that small errors in standardization can shift gels enough to effect results and while an overall score is obtained, spot-by-spot contributions to the score would be, in general, unknown. The disadvantage of the second technique is time. Such an algorithm (computer subprogram) requires additional time over the first method although the increase can be minimized by proper programming; i.e., the use of enhancement processing such as ramping background adjustment and differential

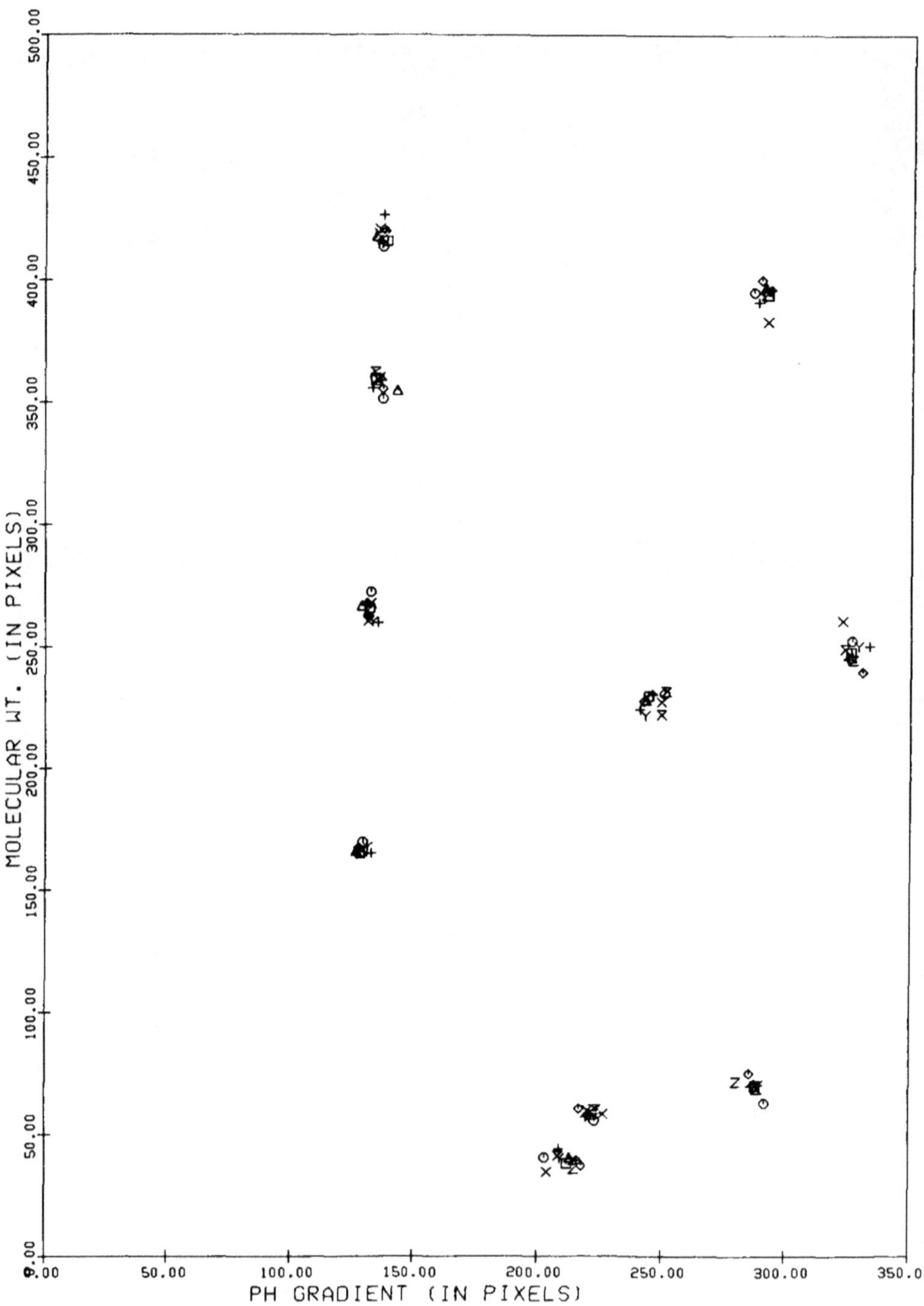


FIGURE 5. Standardized position of marker proteins occurring in individual mouse PHC cytosols. Standardized internal gel marker cytosolic protein estimated positions for four C3H-PHCs and age-matched normal males as well as positions for a second run of freshly prepared individual C3H-PHC cytosol proteins run 2 weeks after the first electrophoretic run are presented. In general, for a pixel window (picture points) of 500 × 350 pixels, positions for markers varied by only 2 to 4 pixels with multiple correlation coefficients greater than 0.999 and F ratios above 9000.[53] (□), Positions used as the standard; (○), age matched control liver; (△), C3H-PHC #1, 1st run; (+), C3H-PHC #1, repeat, new sample; (×), C3H-PHC #2, 1st run; (◇), C3H-PHC #2, repeat, new sample; (↑), C3H-PHC #3, 1st run; (x), C3H-PHC #3, repeat, new sample; (z), C3H-PHC #4, 1st run; (Y) C3H-PHC #4, repeat, new sample.

operators, and estimation of area and integrated optical density, shape, parameters, etc., while separating and identifying the spot. To date, comparisons have not been complicated by variations in spot location and therefore do not require spot-by-spot comparisons.[53] To date, processing includes (1) use of average background subtraction techniques to reduce background smears (ramping process); (2) use of simple differential operations, such as the Robert's cross-operator to identify points on each spot;[82] (3) localized threshold spot separation techniques; (4) use of standards on each run as well as marker proteins in each gel to calibrate gray densities and to simplify rotation and scaling to a standard scale; and (5) use of the original gel as well as photographically achieved images for immediate processing.

## VII. NHC PROTEIN CHANGE IN HEPATOCARCINOGENESIS

### A. Chemically Induced Cancer

Chromosomal proteins from chromatin of adult rat liver, 24- and 48-hr regenerating rat liver, AAF- and DEN-induced PHCs as well as 4 × 3 AAF nodules were extracted with 3 *M* NaCl-7 *M* urea Tris buffer, recovered proteins resolved by ISO or NEPHGE two-dimensional linear gradient PAGE and second-dimension gel protein patterns visualized by silver staining. Photographs, 4 × 5 negatives, and transparencies, produced by drying stained gels between porous cellophane, representing each chromatin analyzed were examined and NHC protein occurrence maps constructed. After examination of individual chromatin maps, composite two-dimensional NHC protein maps were constructed.

#### *1. Equilibrium Focusing Studies (ISO)*

When 80 μg of NHC protein from each respective chromatin was focused to equilibrium (10,000 V hr), separated in the second dimension by molecular weight, and silver stained, 350 to 370 NHC proteins ranging from over 220,000 to 10,000 in molecular weight and pIs from 4.4 to 8.5 could be resolved. Figure 6 shows representative two-dimensional gels of normal rat liver and AAF-induced PHC NHC protein preparations as well as a composite second-dimension map depicting rat liver NHC protein occurrence and changes during hepatocarcinogenesis induced by two diverse-acting carcinogens. A number of new NHC proteins appeared as a result of induced carcinogenesis; occurrence of some of these, however, reflected cycling tissue and, therefore, were not related directly to malignant induction. Potentially, the most significant result was appearance in NHC protein populations of AAF-nodules and -PHCs, and DEN-PHCs of three NHC proteins that were not present in normal tissues. Additionally, a number of NHC protein losses (some common to both carcinogen regimen, some unique to each respective carcinogen) occurred. A full presentation of results of this study as well as discussion of the potential significance of common and unique acquisition of and/or loss of specific NHC proteins during induced chemical carcinogenesis will appear elsewhere.[74]

#### *2. Nonequilibrium pH Gradient Analysis (NEPHGE)*

Since isofocusing (loading on the basic side and 10,000 V hr duration) favored resolution of acidic and neutrally charged NHC proteins, resolution above pH 7.0 being variable due to cathodic drift, we also examined chromatin NHC protein change during chemical hepatocarcinogenesis under NEPHGE conditions. 80 μg of salt-urea extracted NHC protein was suspended in "urea mix" containing 0.3 *M* NaCl and 1.6 mg/mℓ of protamine sulfate, loaded on the acid side (phosphoric acid-anodal buffer), and focusing performed for 4000 V hr. Second-dimension gradient SDS slab electrophoresis, fixing, staining, and processing were as described above. A number of qualitative changes and one quantitative change were evident in the 390+ NHC proteins resolved in normal and carcinogen chromatins by this methodology (Figure 7). A major basic protein present in normal rat liver (36,000 mol wt/9.75 pI) was greatly reduced in all other chromatins analyzed. Similarly, AAF and DEN

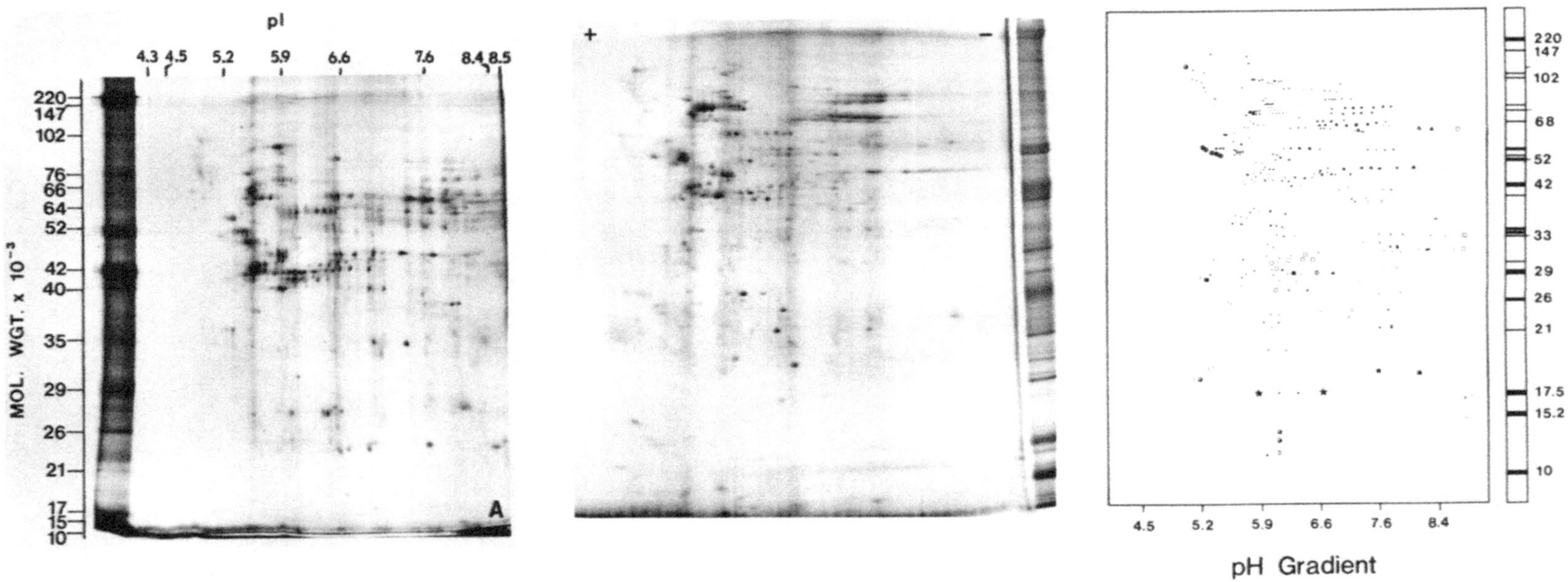


A B C

FIGURE 6. Equilibrium two-dimension gel analysis of NHC proteins in normal rat liver (A) and AAF-PHCs (B) and composite cartoon of NHC protein (C) change during diverse chemically induced rat hepatocarcinogenesis. Representative silver stained second dimension electrophoretic patterns for 80 μg of NHC protein isolated by salt-urea extraction of chromatins from (A) normal rat liver and (B) AAF-PHC are presented. Molecular weights for heart standard and range of pIs resolved appear in A. A composite map cartoon denoting unique or common NHC protein occurrence was constructed: (●), detected in normal rat liver; (■), DEN-PHC only; (△), AAF 4 × 3 nodules or AAF-PHC; (⊗), AAF-PHC only; (★), only in regenerating liver tissues; (○), in normal rat liver, but not DEN-PHC; (□), in normal, but not in AAF 4 × 3 nodules or PHCs; (△) in normal, but not in regenerating rat liver; (◐), in normal, but not in DEN-PHCs, AAF 4 × 3 nodules or AAF PHCs; (◪), in normal, but in no other chromatin; (▲), only in DEN-PHCs, AAF 4 × 3 nodules and AAF-PHCs; and (Φ), in all test chromatins, but not in normal.

PHCs exhibited loss of two constellations of NHC protein (72,000/8.2 to 8.7 and 32,000/6.2 to 7.2, respectively) from their maps. Some 16 NHC proteins were found to be uniquely induced by AAF, 8 by DEN, and 2 new NHC proteins were induced by both (Figure 8).

### *a. Marker NHC Protein Position Standardization*

Since a primary goal of our studies is the detection of possible "hallmark" malignant changes in potential gene transcription controlling NHC proteins that might occur in complex second-dimension protein patterns, the ability to pinpoint the position of such proteins (X,Y coordinates) when comparing many gels was assessed by utilizing scaling and rotation methodology (Section VI.B) developed for use with internally occurring selected sample markers. Since in general NEPHGE resolution of chromosomal proteins (in our experience) results in broader or "less defined" spot topography (basic side of gels) when compared to identical completely focused spot patterns (equilibrium) and the possibility that this type of electrophoretic rate analyses might lead to nonlinear spot-to-spot position relationships, selected marker proteins in NEPHGE-NHC protein gels (Figure 7) were examined. Test results of analysis of standardization of marker silver-stained NEPHGE NHC proteins in normal rat liver (NRL), regenerating rat liver (RRL), and AAF-induced (AAF) as well as DEN-induced (DEN) PHCs appear in Figure 9. In this experiment, spot positions for respective markers in two separate electrophoretic runs (designated Series 1 and 2) of freshly constituted samples were analyzed with a 60-day period between runs (Figure 9). Marker protein positions in the series 1 NRL profile were digitized and their respective positions used as the "standard positions" to which others were compared in both runs comprising series 1 and 2. Lastly, positions of the markers in series 1 NRL were redigitized and standardized to serve as an added control. The median pixel range, defined as the minimum to maximum difference in standardized spot position, for each of the nine analyses of each of the ten markers was 6.6 pixels in the horizontal (pI) and 8.2 pixels in the vertical (molecular weight) dimension. Marker spot positions in the AAF Series 2 gel varied the most when compared to standardized marker positions. Spot positions for NHC proteins #9 and #3 in RRL, AAF and DEN in series 1 also showed considerable variation. Spot #3 occurs in a horizontal constellation in the series AAF-2 gel; its correct scoring position within such constellations was found to be a function of correct "landmarking" and consideration of differences between standard position and standardized spots. Visual inspection of gels indicated that variation in spot #9 standardization was the result of the variability in position of spot #10 within the respective gels. Similarly, distances between spots #9 and #10 as well as between spots #9 and #7 were not proportional. These data point out that to assure complete standardization, some nonlinear adjustments will be required for chosen markers that occur in low molecular weight basic regions of NEPHGE-gradient gels.

## B. Spontaneous Mouse Hepatocarcinogenesis

In studies that parallel those in rat chemically induced hepatocarcinogenesis, NHC protein content in 12-, 18-, and 24-month-old C3H male mice exhibiting spontaneously occurring PHCs have been compared to age-matched control normal and regenerating liver as well as to background liver. Background liver in this carcinogenesis system represents nontumor-involved liver within the same livers of animals exhibiting spontaneous PHCs. A typical second-dimension equilibrium (ISO) silver stained NHC protein profile for PHCs in 1-year-old male C3H mice is presented in Figure 10. Upwards of 365 NHC proteins are easily discernible in such tumors when 80 μg of NHC proteins are analyzed. Comparative studies in these mice indicate that a number of new NHC proteins and some NHC protein loss occur during progression to and in resultant spontaneous liver cancers. Further, our initial analyses indicate that some changes seen in NHC proteins in mouse spontaneous liver cancer parallel and, in some cases, have been identical to those detected in chemically induced rat hepatocarcinogenesis.[74]

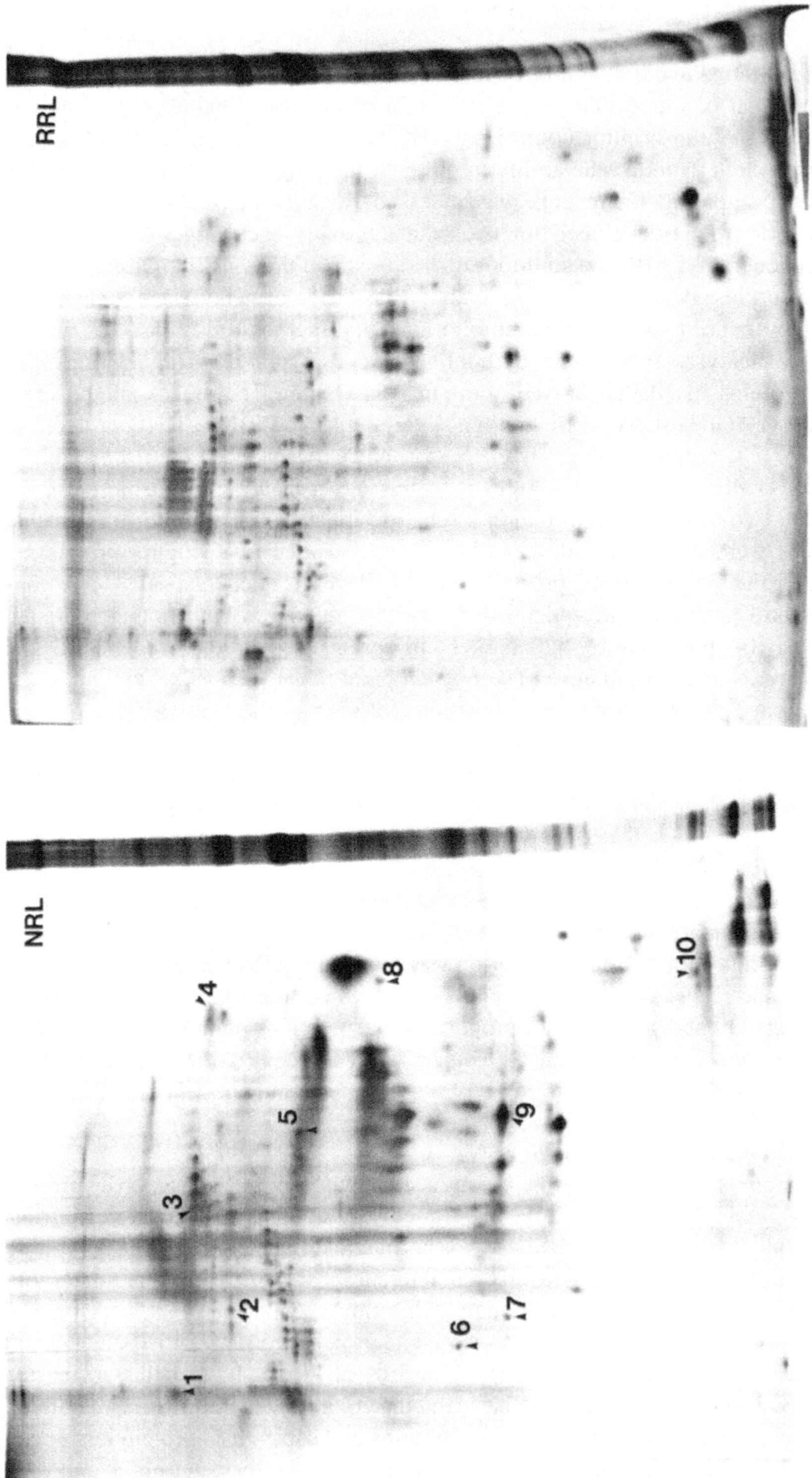


B

A

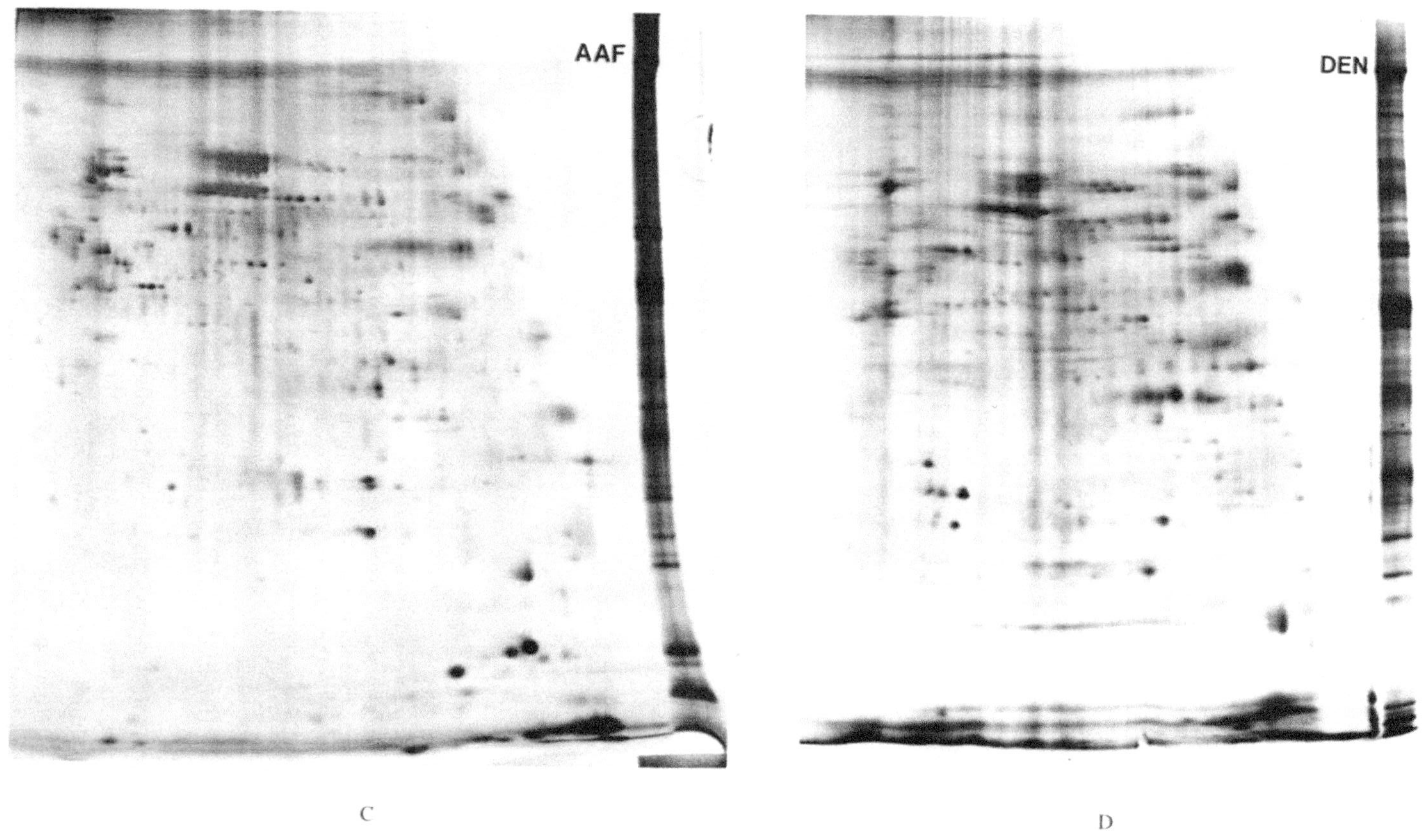


FIGURE 7. Nonequilibrium two-dimensional analysis of AAF- and DEN-induced NHC protein change. 80 μg of salt-urea extracted NHC protein from normal rat liver (NRL); regenerating rat liver (RRL), AAF-PHCs (AAF), and DEN-PHCs (DEN) were separated by NEPHGE for 4000 V hr and in second-dimension gradient SDS slabs. Second-dimension slabs were silver stained. Some 390+ NHC protein polypeptides were visualized in the normal and cancer chromatins and a number of qualitative and one striking quantitative change noted (see text and Figure 8).

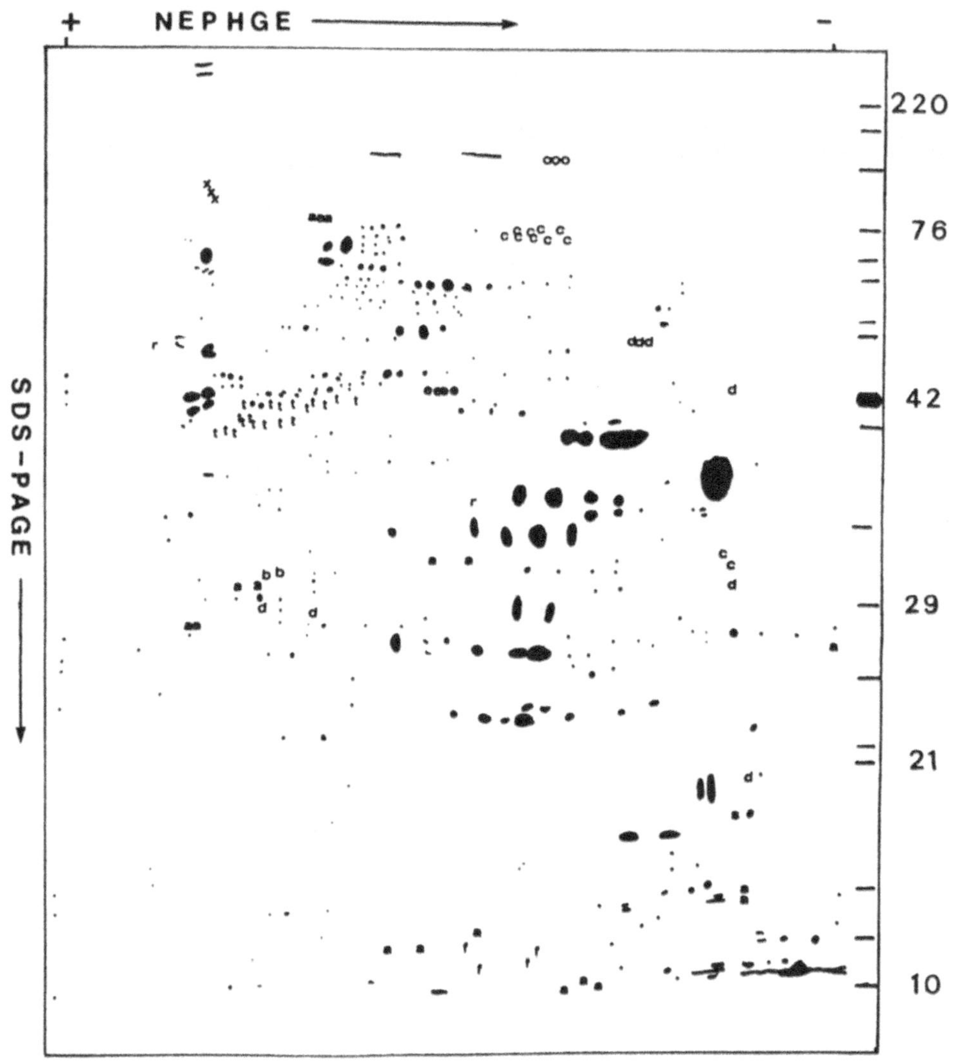


FIGURE 8. Comparison of NEPHGE-resolved NHC proteins. Four separate runs of NRL, RRL, AAF, and DEN NHCP were performed, the occurrence of NHC protein in respective silver stained second-dimension gels determined, similarities and differences noted, and a composite cartoon constructed. Occurrence code: (●), detected in normal rat liver; (r), regenerating liver tissue; (a), AAF-PHCs only; (d), DEN-PHCs only; (e), RRL, AAF, and DEN PHCs; (o), RRL and DEN-PHCs (b), AAF and DEN PHCs; (f), in normal, missing in RRL; (c), in normal, missing in AAF-PHCs; (t), in normal, missing in DEN-PHCs; (x), in normal, missing in all other chromatins; and (s), in normal, missing in RRL and AAF-PHCs.

## VIII. SUMMARY AND DISCUSSION

The NHC protein content of normal and neoplastic liver tissue during diverse chemical induction of liver cancer in rats and in spontaneous hepatocarcinogenesis in mice was examined by high resolution repetitive two-dimensional gel electrophoresis to assess if common and potentially obligatory changes in NHC proteins accompany etiologically diverse murine hepatocarcinogenic induction. Equilibrium and nonequilibrium first-dimension focused NHC proteins were analyzed on gradient second-dimension slabs allowing resolution of proteins with pIs ranging from 3.8 to 10.5 and molecular weights from 10,000 to 300,000 daltons. Standardization of gels by inclusion of carbamylated aliquots of known protein in the focusing dimension, direct pH gradient measurement in focused sample-experimental

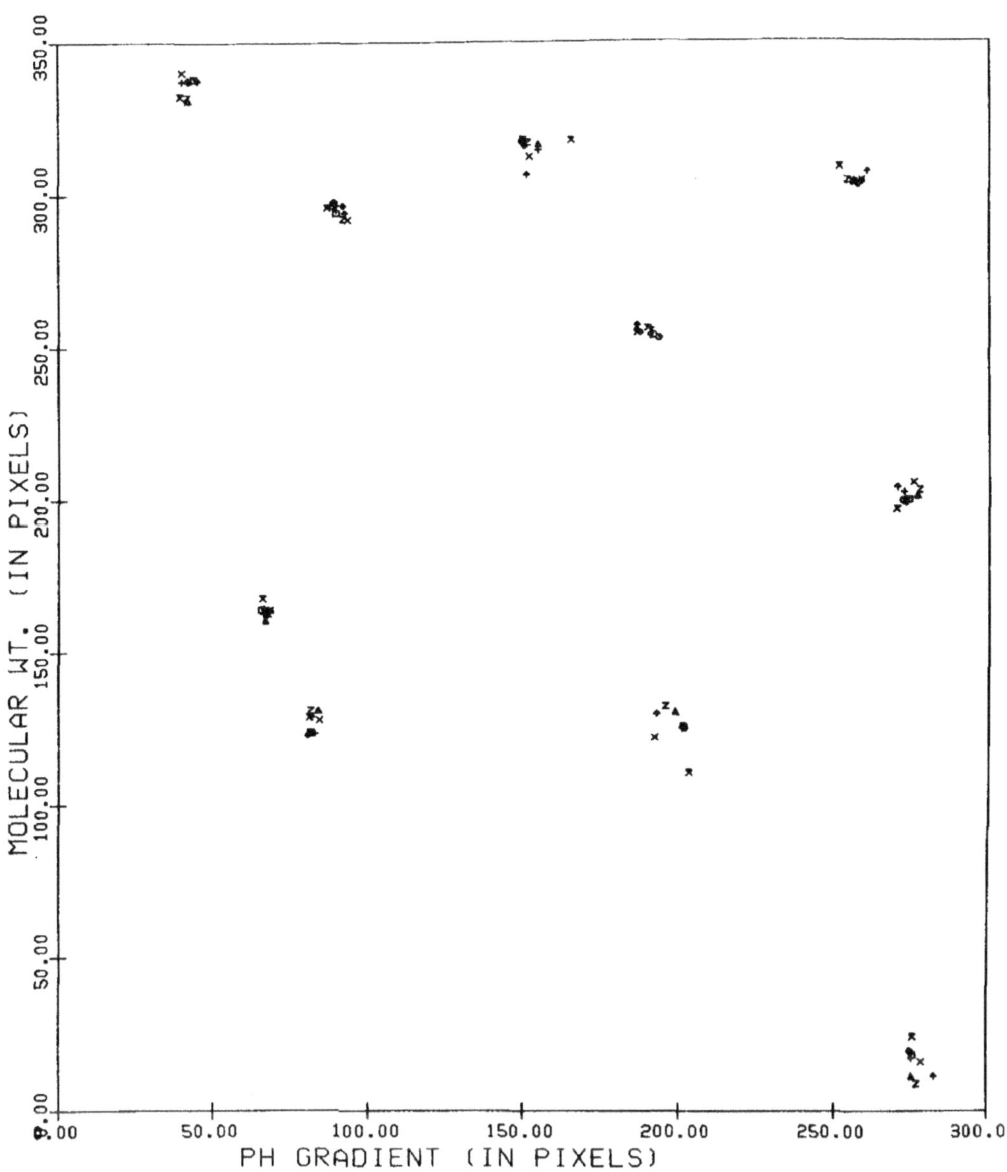


FIGURE 9. Standardization of NEPHGE-NHC protein position. Ten internally occurring NHC proteins in gel patterns of NRL (numbered 1 to 10 and positions noted in NRL gel, Figure 7), RRL, AAF-PHCs (AAF), and DEN-PHCs (DEN) were analyzed by scaling and rotation methodology (Section VI.B). Estimates of standardized gel marker positions for two separate electrophoretic runs of fresh sample, Series 1 and 2, performed some 60 days apart are plotted in a pixel window of 350 × 300 elements (picture points). The position of respective markers in the Series 1 NRL gel were used as the standard position for comparisons. (□), Denotes positions used as standards; (○), NRL series #1; (△), NRL series #2; (+), RRL series #1; (×), RRL series #2; (◇), AAF series #1; (↑), AAF series #2; (x), DEN series #1, and (z), DEN series #2.

gels, and co-electrophoresis of known molecular weight proteins in the second dimension indicated that patterns of resolved polypeptides were with few exceptions, very reproducible, in most cases approaching identity. Our development and use of silver-staining technology for thick gradient second-dimension gels resulted in detection of a minimum of 368 to a maximum of 400 NHC protein components in the respective chromatins examined. These resolved component levels exceed by approximately fivefold the number of components previously resolved in Coomassie blue R-250 stained gels that contained 15-fold the chro-

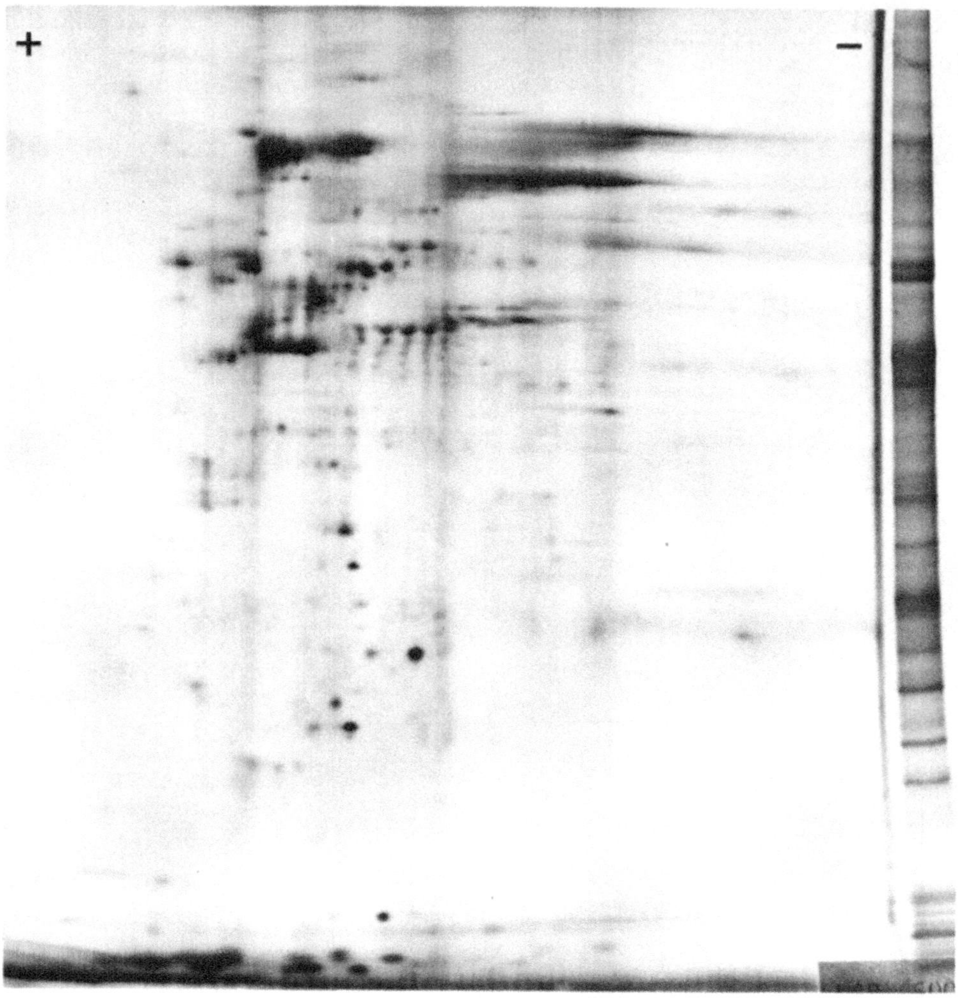


FIGURE 10. Equilibrium two-dimensional gel of NHC protein polypeptides in spontaneous hepatocarcinogenesis in C3H/HeN mice. Typical second-dimension silver stained patterns for spontaneous PHCs occurring in 1-year-old C3H/HeN male mice when 80 μg of salt-urea NHC proteins were analyzed. Over 365 polypeptides are evident; the overall occurrence pattern differed qualitatively from that for age-matched control liver and background liver.

mosomal protein load and are very close to the number of nucleoprotein components previously detected in two-dimensional PAGE analysis of cell cultures radioactively labeled in vitro.[71,72] The heightened detection afforded by silver use is graphically shown in Figure 2 where over ten times the number of polypeptides are seen in a second-dimension protein pattern when silver staining followed Coomassie staining in a single gel run of mouse cytosolic proteins (Section V.A).

By paying particular attention to avoiding nuclear, chromatin, and chromosomal protein isolation methods (proven to have a high probability for artifact and selection of tissues for study from normal, varying normal physiology, discrete stages in diverse agent-induced carcinogenesis as well as in spontaneously occurring liver cancers), we have found a number of potentially significant alterations in the NHC protein population during carcinogenesis. While a number of the NHC protein losses and gains we have detected seem to represent the particular carcinogen used (rats), cell cycling, or changes due to aging (mice), and therefore are not likely candidates for "hallmark changes", those NHC proteins that appear or disappear in common during diverse carcinogenesis may eventually prove to be such changes. Although it is difficult to compare our findings with previous studies in experimental hepatocarcinogenesis due to the variety of chromosomal protein isolation methods, electrophoretic systems utilized and low sensitivity of conventional protein stain comparisons of spot patterns indicate that most if not all proteins previously described in rat liver[37,71] also appear in our second-dimension silver stained electropherograms. To confidently establish whether common NHC protein losses or gains represent critical causal events during carcinogenesis or alternatively, merely represent tumor homeostasis and progression properties requires further study. In the present studies, the appearance of three NHC proteins as well as loss of two NHC proteins in maps of AAF-induced 4 × 3 nodules and occurrence of these same NHC protein losses and gains in both AAF and DEN PHCs indicate that some NHC protein changes cannot simply be explained as characterizing resultant tumors. Study of earlier stages in AAF and DEN carcinogenesis as well as in aging mouse populations should help clarify this picture.

The rigid standardization that has been possible in these studies by use of an analytical system comprised of multiple gel casting and run capacity (ISO-DALT), adaptation of highly sensitive staining to thick gradient gels, and multiple capacity video image computer-assisted processing has facilitated our progress to date. However, as previous authors have noted, a need still exists for heightened standardization and quality control in acrylamide, bisacrylamide, ampholytes, and chemical reagents available to investigators.[4,68,90] Although high-resolution two-dimensional polyacrylamide electrophoresis is evolving as a standard tool in many laboratories, a wide disparity still exists in the quality of electropherographs between laboratories and further standardization is necessary. In in vivo NHC protein studies, inability to obtain adequate radiolabeling of all NHC protein remains a problem that may be resolved by development of even more sensitive stains and in vitro radiolabeling techniques.

Production of standard two-dimensional PAGE maps of normal and normal physiologically altered tissue NHC protein will augment estimation of the statistical parameters required to accurately compare normal and experimental tissues. In continuing studies, we see the development of this capacity as follows. With this gel technique, images (spots) on gels are reduced primarily to a set of ordered "spots" or small collections of "spots" related directly to the characteristics of the proteins which generate them. Each spot will be characterized by the following parameters: (1) position — normalized coordinate position of the center of the spot; (2) area — measured as the number of normalized pixels within its boundary; (3) intensity — measured as the sum of gray values (normalized to adjust for amount of material, stain, etc.) within the spot (otherwise known as integrated optical density or "mass" of the spot); and (4) shape — measured as sums of ellipsoids or bivariate Gaussians. Shape is also used to identify and separate doublets and triplets. If staining proves to be essentially stoichiometric and spots unsaturated, intensity and area can be used as comparison to test whether spots reflect the presence of one or more proteins. This facilitates estimates of relative proportions of proteins in a given sample. Variation of each parameter in each spot will be compared to that utilizing normal sample electrophoretic runs that exhibit varying degrees of migration. A stock of normal tissue sample(s) will run at the same time as each experimental sample and this permits a factorial design yielding replication errors, time varying systematic errors, variation in distribution of proteins in test samples as well as correlations between spots of experimental and of normal tissues. After a suitable number of normal sample electrophoretic runs and analyses, a normal protein distribution spot pattern (NPSP) will be generated and stored in computer files. Patterns of test tissue samples will be compared to NPSP using a modified Mahalanobis $D^2$ and/or matching type algorithm with comparisons displayed visually as gray level pseudo-color "maps" on the Grinnel®/Matrix display system.[39,83] Results of each gel analysis will be provided on 3-ft square maps, each spot denoted as a circle or ellipse where size or gray value denotes the amount of protein present. In addition, X-Y coordinates for each spot and quantitation will be stored, since such maps are invaluable for comparative studies. In addition, retrieval and comparison programs will be developed to display spots occurring when normal tissue spots are subtracted from those of test tissues. Catalogs of spots for each sample will be stored for future use.

While considerable difficulty still exists in analysis of NHC proteins due to their heterogeneity, aggregative properties once isolated, lack of existence of comprehensive NHC protein maps for any given cell or tissue, and functional correlations, it is apparent they participate in differential genome function and undergo qualitative change during etiologically diverse carcinogenesis. Further, since cancers exhibit aberrant gene regulation, delineation of those changes in NHC proteins that can be shown to be obligatory to generation of the "prototypic" malignant phenotype will greatly enhance our understanding and eventually control of malignant neoplasia.

## ACKNOWLEDGMENTS

These studies were supported in part by Grants CA 28263, CA 11430, RR5511-18, and RR5511-19 from the National Cancer Institute, National Institutes of Health, Bethesda, Md. The authors wish to thank Dr. Frederick F. Becker and staff for their help in acquiring the various tissues utilized in these studies. Also we wish to thank Mrs. Sylvia Capetillo for excellent biochemical assistance and Mrs. Catherine Johns for manuscript preparation.

## REFERENCES

1. **Allfrey, V. G.,** in *Carcinogens: Identification and Mechanisms of Action,* Clark Griffin, A. and Shaw, C. R., Eds., Williams & Wilkins, Baltimore, 1978.
2. **Anderson, N. G. and Anderson, N. L.,** Analytical techniques for cell fractions. XXI. Two-dimensional analysis of serum and tissue proteins: multiple isoelectric focusing, *Anal. Biochem.*, 85, 331, 1978.
3. **Anderson, N. G., Anderson, N. L., and Tollaksen, S. L.,** in *Operation of the Iso-Dalt System,* ANL-BIM-79-2, Argonne National Laboratory, Argonne, Ill., 1979.
4. **Anderson, N. G. and Anderson, L.,** The human protein index, *Clin. Chem.*, 28, 739. 1982.
5. **Anderson, N. L. and Anderson, N. G.,** Analytical techniques for cell fractions. XXII. Two-dimensional analysis of serum and tissue proteins: multiple gradient-slab gel electrophoresis, *Anal. Biochem.*, 85, 341, 1978.
6. **Anderson, N. G., Price, C. A., Risher, W. D., Canning, R. E., and Borger, C. L.,** Analytical techniques for cell fractions. IV. Reorienting gradient rotors for zonal centrifugation, *Anal. Biochem.*, 7, 1, 1964.
7. **Anderson, N. G., Waters, D. A., Nunely, C. E., Gibson, R. F., Schilling, R. M., Denney, E. C., Caine, G. B., Babely, E. F., and Peradi, T. D.,** K-series centrifuges. I. Development of K-II continuous-sample flow with banding centrifuge system for vaccine purification, *Anal. Biochem.*, 32, 460, 1969.
8. **Anderson, N. L. and Hickman, B. J.,** Analytical techniques for cell fractions. XXIV. Isoelectric point standards for two-dimensional electrophoresis, *Anal. Biochem.*, 93, 312, 1979.
9. **Augenlicht, L. H.,** Nuclease digestion of RNA in chromatin, in *Methods in Cell Biology,* Vol. 16, Stein, G., Stein, J., and Kleinsmith, L. J., Eds., Academic Press, New York, 1977, 467.
10. **Ballal, N. R., Goldknopf, I. L., Goldberg, D. A., and Busch, H.,** The dynamic state of liver nucleolar proteins as reflected by their changes during administration of thioacetamide, *Life Sci.*, 14, 1835, 1974.
11. **Baserga, R.,** Nonhistone chromosomal proteins in normal and abnormal growth, *Life Sci.*, 15, 1057, 1974.
12. **Becker, F. F. and Sell, S.,** Differences in serum α-fetoprotein concentrations during the carcinogenic sequences resulting from exposure to diethylnitrosoamine or acetylaminofluorene, *Cancer Res.*, 39, 1437, 1979.
13. **Becker, F. F. and Sell, S.,** α-Fetoprotein levels and hepatic alterations during chemical carcinogenesis in C57BL/6N mice, *Cancer Res.*, 39, 3491, 1979.
14. **Becker, F. F., Stillman, D., and Sell, S.,** Serum α-fetoprotein in a mouse strain (C3H-AvyfB) with spontaneous hepatocellular carcinoma, *Cancer Res.*, 37, 870, 1977.
15. **Benjamin, W. B. and Gelhorn, A.,** Acidic proteins of mammalian nuclei: isolation and characterization, *Proc. Natl. Acad. Sci. U.S.A.*, 59, 262, 1968.
16. **Bickerton, R. K.,** in *Carcinogenesis Testing in Chemicals,* Goldberg, L., Ed., CRC Press, Cleveland, 1974, 55.

17. **Biessman, H. and Rajewsky, M. F.,** Nuclear protein patterns in developing and adult brain and in ethylnitrosourea-induced neuroectodermal tumors of the rat, *J. Neurochem.*, 24, 387, 1975.
18. **Blobel, G. and Potter, V. R.,** Nuclei from rat liver: isolation method that combines purity with high yield, *Science*, 154, 1663, 1966.
19. **Boffa, L. C. and Allfrey, V. G.,** Changes in chromosomal proteins in colon cancer, *Cancer*, 40, 2584, 1977.
20. **Boffa, L. C., Vidali, G., and Allfrey, V. G.,** Selective synthesis and accumulation of nuclear non-histone proteins during carcinogenesis of the colon induced by 1,2-dimethylhydrazine, *Cancer*, 36, 2356, 1975.
21. **Boffa, L. C., Vidali, G., and Allfrey, V. G.,** Changes in nuclear nonhistone protein composition during normal differentiation and carcinogenesis of intestinal epithelial cells, *Exp. Cell Res.*, 98, 369, 1976.
22. **Bhorjee, J. S. and Pederson, T.,** Chromatin: its isolation from cultured mammalian cells with particular reference to contamination by nuclear ribonucleoprotein particles, *Biochemistry*, 12, 2766, 1973.
23. **Bresnick, E.,** Alterations in liver nucleoprotein after administration of 3-methylcholanthrene (3MC), *Fed. Proc. Fed. Am. Soc. Exp. Biol.*, 29, 2778, 1970.
24. **Busch, H., Ballal, N. R., Busch, R. K., Samal, B., Choi, Y. C., Davis, F. M., Kunkle, H. M., Nall, K., Parker, D., Rao, M. S., Rothblum, L. I., and Takami, H.,** in *Differentiation and Development*, Academic Press, New York, 1978, 187.
25. **Cameron, I. L. and Jeter, J. R., Eds.,** *Acidic Proteins of the Nucleus*, Academic Press, New York, 1974.
26. **Cesarone, C. F., Bolognesi, C., and Santi, L.,** Improved microfluorometric DNA determination in biological material using 33258 Hoechst, *Anal. Biochem.*, 100, 188, 1979.
27. **Cholon, J. J. and Studzinski, G. P.,** Effect of aminonucleoside on serum stimulation of nonhistone nuclear protein and DNA synthesis in normal and SV-40 transformed human fibroblasts, *Cancer Res.*, 34, 588, 1974.
28. **Coggin, J. H. and Anderson, N. G.,** Cancer, differentiation and embryonic antigens: some central problems, *Adv. Cancer Res.*, 19, 105, 1974.
29. **Craddock, V. M. and Henderson, A. R.,** *De novo* and repair replication of DNA in liver of carcinogen-treated animals, *Cancer Res.*, 38, 2135, 1978.
30. **Deringer, M. K.,** Occurrence of tumors, particularly mammary tumors in agent-free strain C3HeB mice, *J. Natl. Cancer Inst.*, 22, 995, 1959.
31. **Elgin, S. C. R. and Weintraub, H.,** Chromosomal proteins and chromatin structure, *Ann. Rev. Biochem.*, 44, 725, 1975.
32. **Felsenfeld, G.,** Chromatin, *Nature (London)*, 271, 115, 1978.
33. **Feigelson, P., Murthy, L. R., Sippel, A. E., and Morris, H. P.,** Absence of specific messenger-RNA species in hepatoma-cells, *Proc. Am. Assoc. Cancer Res.*, 16, 26, 1975.
34. **Forger, J. M., Choie, D. D., and Freidberg, E. C.,** Nonhistone chromosomal proteins of chemically transformed neoplastic cells in tissue culture, *Cancer Res.*, 36, 258, 1976.
35. **Gilmour, R. S. and Paul, J.,** Role of non-histone components in determining organ specificity of rabbit chromatins, *FEBS Letts.*, 9, 242, 1970.
36. **Giometti, C. S., Anderson, N. G., Tollaksen, S. L., Edwards, J. J., and Anderson, N. L.,** Analytical techniques for cell fractions. XXVII. Use of heart proteins as reference standards in two dimensional electrophoresis, *Anal. Biochem.*, 102, 47, 1980.
37. **Gonzalez-Mujica, F. and Mathias, A. P.,** Proteins from different classes of liver nuclei in normal and thioacetamide-treated rats, *Biochem. J.*, 133, 441, 1973.
38. **Gray, H. B., Jr., Ostrander, D. A., Hodnett, J. L., Legerski, R. J., and Robberson, D. L.,** Extracellular nuclease of pseudomonas BaL 31. Characterisation of single strand-specific deoxyriboendonuclease and double strand deoxyriboexonuclease activities, *Nucleic Acids Res.*, 9(2), 1459, 1981.
39. **Greuander, V.,** *Regular Structures: Lectures in Pattern Theory*, Vol. 3, Springer-Verlag, Basel, 1981.
40. **Gronow, M.,** Nuclear proteins and chemical carcinogenesis, *Chem. Biol. Interact.*, 29, 1, 1980.
41. **Gronow, M. and Thackrah, T.,** Nuclear protein changes during the nitrosamine-induced carcinogenesis of rat liver, *Chem. Biol. Interact.*, 9, 225, 1974.
42. **Gronow, M. and Thackrah, T.,** Changes in the composition of rat liver chromatin fractions during nitrosamine carcinogenesis, *Eur. J. Cancer*, 10, 21, 1974.
43. **Guevara, J., Jr., Johnston, D. A., Ramagli, L. S., Martin, B. A., Capetillo, S., and Rodriguez, L. V.,** Quantitative aspects of silver deposition in proteins resolved in complex polyacrylamide gels, *Electrophoresis*, 3, 197, 1982.
44. **Hart, R. W., Setlow, R. B., and Woodhead, A. D.,** Evidence that pyrimidine dimers in DNA can give rise to tumors, *Proc. Natl. Acad. Sci. U.S.A.*, 74, 5574, 1977.
45. **Heston, W. E.,** Brief communication: testing for possible effects of cedar wood shavings and diet occurrence of mammary gland tumors and hepatomas in C3H-$A^{vy}$ and C3H-$A^{vy}$fB mice, *J. Natl. Cancer Inst.*, 54, 1011, 1975.

46. **Heston, W. E. and Vlahakis, G.,** High incidence of spontaneous hepatomas and the increase of this incidence with urethan in C3H, C3Hf and C3He male mice, *J. Natl. Cancer Inst.*, 24, 425, 1960.
47. **Heston, W. E. and Vlahakis, G.,** C3H-$A^{vy}$ — a high hepatoma and high mammary tumor strain of mice, *J. Natl. Cancer Inst.*, 40, 1161, 1968.
48. **Hirsch, F. W., Hall, K. N., Busch, F. N., Morris, H. P., and Busch, H.,** Comparison of abundant cytosol proteins of rat liver, Novikoff and Morris hepatomas by two-dimensional gel electrophoresis, *Cancer Res.*, 38, 1514, 1978.
49. **Huang, R. C., Bonner, J., and Murray, K.,** Physical and biological properties of soluble nucleohistones, *J. Mol. Biol.*, 34, 305, 1968.
50. **Huberman, E., Mager, R., and Sachs, L.,** Mutagenesis and transformation of normal cells by chemical carcinogens, *Nature (London)*, 264, 360, 1976.
51. **Hurley, P. M., Catsimpoolas, N., and Wogan, G. N.,** Bidimensional electrophoresis of nuclear chromosmal proteins from Aflatoxin induced liver cancer, in *Electrophoresis '78*, Catsimpoolas, N., Ed., Elsevier/North Holland, Amsterdam, 1978, 283.
52. **Johns, E. W. and Forrester, S.,** Studies on nuclear proteins. The binding of extra acidic proteins to deoxyribonucleoprotein during the preparation of nuclear proteins, *Eur. J. Biochem.*, 8, 547, 1969.
53. **Johnston, D. A., Capetillo, S., Ramagli, L. S., Guevara, J., Jr., Gersten, D. M., and Rodriguez, L. V.,** Standardization of protein position in silver-stained two-dimensional polyacrylamide gel electrophoresis, *Electrophoresis*, 5, 110, 1984.
54. **Jungman, R. A. and Schweppe, J. S.,** Binding of chemical carcinogens to nuclear proteins of rat liver, *Cancer Res.*, 32, 952, 1972.
55. **Klose, J.,** Protein mapping by combined isoelectric focusing and electrophoresis of mouse tissue. A novel approach to testing for induced point mutations in mammals, *Humangenetik*, 26, 211, 1975.
56. **Kornberg, R.,** Structure of chromatin, *Ann. Rev. Biochem.*, 36, 931, 1977.
57. **Kostraba, N. C. and Wang, T. Y.,** Transcriptional transformation of Walker tumor chromatin by nonhistone proteins, *Cancer Res.*, 32, 2348, 1972.
58. **Levy, S., Simpson, R. T., and Sober, H. H.,** Fractionation of chromatin components, *Biochemistry*, 11, 1547, 1972.
59. **Lowry, O. H., Rosenbrough, J. H., Farr, A. L., and Randall, R. J.,** Protein measurement with the Folin phenol reagent, *J. Biol. Chem.*, 193, 265, 1951.
60. **Markert, C. L.,** Neoplasia: a disease of cell differentiation, *Cancer Res.*, 28, 1908, 1968.
61. **McConkey, E. H.,** Double-label autoradiography for comparison of complex protein mixtures after gel electrophoresis, *Anal. Biochem.*, 96, 39, 1979.
62. **Mintz, B. and Illemsee, K.,** Normal genetically mosaic mice produced from malignant teratocarcinoma cells, *Proc. Natl. Acad. Sci. U.S.A.*, 72, 3585, 1975.
63. **Montagna, R. A., Maizel, A. L., Becker, F. F., and Rodriguez, L. V.,** Chromatin conformation modulates repair of single strand interruptions by polynucleotide ligase [$^3$H] AMP, *Chem. Biol. Interact.*, 33, 149, 1981.
64. **Montagna, R. A., Rodriguez, L. V., and Becker, F. F.,** A comparative study of the nonhistone proteins of rat liver euchromatin and heterochromatin, *Arch. Biochem. Biophys.*, 179, 617, 1977.
65. **Morrissey, J. H.,** Silver stain for proteins in polyacrylamide gels: a modified procedure with enhanced uniform sensitivity, *Anal. Biochem.*, 117, 307, 1981.
66. **Moule, Y. and Chauveau, J.,** Mise en evidence dans le foie de rat de particules nuclehires contenant un arn rapidement, *Cent. Res. Acad. Sci. (Paris)*, 263, 75, 1966.
67. **O'Farrell, P. H.,** High resolution two-dimensional electrophoresis of proteins, *J. Biol. Chem.*, 250, 4007, 1975.
68. **O'Farrell, P. H. and O'Farrell, P. Z.,** Two dimensional polyacrylamide gel electrophoretic fractionation, in *Methods in Cell Biology*, Prescott, D. M., Ed., Vol. 16, Academic Press, New York, 1977, 407.
69. **Patel, G.,** Isolation of the nuclear acidic proteins, their fractionation, and some general characteristics, in *Acidic Proteins of the Nucleus*, Cameron, I. V. and Jeter, J. R., Jr., Eds., Academic Press, New York, 1974, 29.
70. **Paul, J. and Gilmour, R. A.,** Organ-specific restriction of transcription in mammalian chromatin, *J. Mol. Biol.*, 34, 305, 1968.
71. **Peters, K. E. and Comings, D.,** Two-dimensional gel electrophoresis of rat liver nuclear washes, nuclear matrix, and hnRNA proteins, *J. Cell Biol.*, 86, 135, 1980.
72. **Peterson, J. L. and McConkey, E. H.,** Non-histone chromosomal proteins from HeLa cells. A survey by high resolution, two-dimensional electrophoresis, *J. Biol. Chem.*, 251, 548, 1976.
73. **Rabinowitz, Z. and Sachs, L.,** Control of the reversion of properties in transformed cells, *Nature (London)*, 225, 136, 1970.
74. **Ramagli, L. S., Capetillo, S., Becker, F. F., and Rodriguez, L. V.,** Tumor associated alterations of nonhistone chromosomal proteins during rat hepatocarcinogenesis induced by diverse acting carcinogens, submitted.

75. **Reeder, R. H.,** Transcription of chromatin by bacterial RNA polymerase, *J. Mol. Biol.*, 80, 229, 1973.
76. **Rodriguez, L. V.,** Unpublished observations.
77. **Rodriguez, L. V. and Becker, F. F.,** Rat liver chromatin: fractionation into eu- and heterochromatin with localization of ribosomal genes, *Arch. Biochem. Biophys.*, 173, 428, 1976.
78. **Rodriguez, L. V. and Becker, F. F.,** Rat liver chromatin: distribution of histone and nonhistone proteins in eu- and heterochromatin, *Arch. Biochem. Biophys.*, 173, 438, 1976.
79. **Rodriguez, L. V., Klein, K. K., Armoruso, M., and Becker, F. F.,** Analysis of chromosomal proteins of fractionated chromatin from rat liver and transplantable hepatocellular carcinomas, *Int. J. Cancer*, 24, 490, 1979.
80. **Rodriguez, L. V. and Becker, F. F.,** Alterations in chromosomal proteins in C3H/HeN mice with spontaneous primary hepatocellular carcinomas, *Int. J. Cancer*, 24, 806, 1979.
81. **Rodriguez, L. V., Lapeyre, J.-N., Robberson, D. L., Maizel, A. L., and Becker, F. F.,** Hydrodynamic shearing by VirTis blending conserves nucleosome structure in rat liver chromatin, *Arch. Biochem. Biophys.*, 200, 116, 1980.
82. **Rosenfeld, A. and Kak, A. C.,** *Digital Picture Processing*, Academic Press, New York, 1976, 280.
83. **Rosenfeld, A. and Kak, A. C.,** *Digital Picture Processing*, Academic Press, New York, 1976, 296.
84. **Rouslahti, E., Oh, E., and Jalanko, H.,** Differences in the nuclear proteins of normal and malignant liver tissues, *Oncodev. Biol. Med.*, 1, 17, 1980.
85. **Sabine, J. R., Horton, B. J., and Wicks, M. B.,** Spontaneous tumors in C3H-$A^{vy}$ and C3H-$A^{vy}$fB mice: high incidence in the United States and low incidence in Australia, *J. Natl. Cancer Inst.*, 50, 1237, 1973.
86. **Sales-Martinez, V., Gabaldon, M., and Baguena, J.,** Nonhistone protein changes during the diethylnitrosamine-induced carcinogenesis of rat liver, *Cancer Res.*, 41, 1187, 1981.
87. **Sanders, M. M., Groppi, V. E., Jr., and Browning, E. G.,** Resolution of basic cellular proteins including histone variants by two dimensional gel electrophoresis: evaluation of lysine to arginine ratios and phosphorylation, *Anal. Biochem.*, 103, 157, 1980.
88. **Scoggin, C. H., Gabrielson, E., Davidson, J. N., Jones, C., Pattern, D., and Puck, T. T.,** Two-dimensional electrophoresis of human-CHO cell hybrids containing human chromosome 11, *Somatic Cell Genet.*, 7, 389, 1981.
89. **Shaw, L. M. J. and Huang, R. C.,** A description of two procedures which avoid the use of extreme pH conditions for the resolution of components isolated from chromatins prepared from pig cerebellar and pituitary nuclei, *Biochemistry*, 9, 4530, 1976.
90. **Sinclair, J. and Rickwood, D.,** Two-dimensional gel electrophoresis, in *Gel Electrophoresis of Proteins: A Practical Approach*, Hames, B. D. and Rickwood, D., Eds., IRL Press, London, 1981, 189.
91. **Sippel, A. E., Kurtz, D. T., Morris, H. P., and Fiegelson, P.,** Comparison of *in vivo* translation rates and messenger RNA levels of $\alpha_{2\mu}$-globulin in rat liver and Morris hepatoma 5123D, *Cancer Res.*, 36, 3588, 1976.
92. **Smith, G. J.,** Chromatin-associated proteins as markers of neoplastic transformation in mammalian cells, *Cell Biol. Int. Rep.*, 5, 635, 1981.
93. **Stein, G. S. and Kleinsmith, L. J., Eds.,** *Chromosomal Proteins and Their Role in the Regulation of Gene Expression*, Academic Press, New York, 1975.
94. **Stein, G. S., Spelsburg, T. C., and Kleinsmith, L. J.,** Nonhistone chromosomal proteins and gene regulation, *Science*, 183, 817, 1974.
95. **Stout, D. L. and Becker, F. F.,** Occurrence of progressive DNA damage coincident with the appearance of foci of altered hepatocytes, *Carcinogenesis*, 3, 599, 1982.
96. **Sudhakar, S., Johnston, D. A., Becker, F. F., and Rodriguez, L. V.,** Phosphoprotein kinases and nonhistone chromosomal phosphoproteins during hepatocarcinogenesis induced by diverse acting carcinogens, *Cell. Molec. Biol.*, 30, 225, 1984.
97. **Teebor, J. W. and Becker, F. F.,** Regression and persistence of hyperplastic nodules induced by N-2-fluorenylacetamide and their relationships to hepatocarcinogenesis, *Cancer Res.*, 31, 1, 1971.
98. **Teng, C. S., Teng, C. T., and Allfrey, V. G.,** Studies of nuclear acid proteins. Evidence for their phosphorylation, tissue specificity, selective binding to deoxyribonucleic acid, and stimulatory effects on transcription, *J. Biol. Chem.*, 246, 3597, 1971.
99. **Teplitz, R. L., Barr, K. J., and Lawce, H. J.,** Karyological and biochemical evidence for chromosomal dedifferentiation in neoplasia, *In Vitro*, 7, 195, 1972.
100. **Ts'o, P. O. P., Ed.,** *The Molecular Biology of the Genetic Apparatus*, Vol. 50, North Holland, Amsterdam, 1977, 241.
101. **Tracy, R. P., Currie, R. M., Kyle, R. A., and Young, D. S.,** Two-dimensional gel electrophoresis of serum specimens from patients with monoclonal gammopathies, *Clin. Chem.*, 28, 900, 1982.
102. **Tsanev, R. and Hadjiolov, D.,** Chromosomal proteins in hepatocarcinogenesis, *Z. Krebforsch.*, 91, 237, 1978.

103. **Van Keuren, M. L., Goldman, D., and Merril, C. R.,** Detection of radioactively labeled proteins is quenched by silver staining methods: quenching is minimal for $^{14}C$ and partially reversible for $^{3}H$ with a photochemical stain, *Anal. Biochem.*, 116, 248, 1981.
104. **Walker, M. S., Becker, F. F., and Rodriguez, L. V.,** *In vivo* binding of N-2-acetylaminofluorene and its N-hydroxy derivative to the DNA of fractionated rat liver chromatin, *Chem. Biol. Interact.*, 27, 177, 1979.
105. **Ward, J. M. and Viahakis, G.,** Evaluation of hepatocellular neoplasms in mice, *J. Natl. Cancer Inst.*, 61, 807, 1978.
106. **Wilhelm, J. A., Groves, C. M., and Hnilica, L. S.,** Lack of major cytoplastic protein contamination of rat liver nuclear chromatin, *Experientia*, 28, 514, 1972.
107. **Wray, W., Conn, P. M., and Wray, V. R.,** Isolation of nuclei using hexylene glycol, in *Methods in Cell Virology*, Vol. 16, Prescott, D., Ed., Academic Press, New York, 1977, 69.
108. **Wu, B. C., Spohn, W. H., and Busch, H.,** Two-dimensional gel electrophoresis of nuclear phosphoproteins of Novikoff hepatoma and regenerating liver, *Physiol. Chem. Phys.*, 12, 11, 1980.
109. **Zhao, M. S. and Liew, C. C.,** Production of antibody to phosphoprotein associated with nucleosome structure, *Can. J. Biochem.*, 60, 356, 1982.

Chapter 11

# EPSTEIN-BARR VIRUS NUCLEAR ANTIGEN

**Thomas B. Sculley and Thomas C. Spelsberg**

## TABLE OF CONTENTS

# I. INTRODUCTION

Epstein-Barr virus (EBV) was isolated in 1964 from a cell line established from an African Burkitt's lymphoma biopsy.[1] This virus has been recognized as the etiologic agent of infectious mononucleosis, and has been implicated as the causative agent of nasopharyngeal carcinoma and African Burkitt's lymphoma.[2] EBV is a member of the human herpesviruses, and contains a linear double-stranded DNA molecule with a molecular weight of $100 \times 10^6$ daltons.[5,6] Bone marrow-derived lymphocytes (B-lymphocytes) are the natural host of the virus, for, apart from the fact that these cells possess receptors for the virus, they have been shown to carry the viral genome in vivo.[7]

During the course of a productive infection of B-lymphocytes by EBV, a variety of different viral-induced antigens can be detected: EBV nuclear antigen (EBNA);[8] membrane antigen,[9] which has been subtyped into early and late components;[10,11] early antigens of which there are two types, restricted and diffuse;[12] and viral capsid antigen.[13] All of these antigens were originally defined using naturally occurring human antibodies. In addition, another antigen, the lymphocyte detected membrane antigen (LYDMA) was discovered in the course of cytotoxicity tests using killer thymus-dependent lymphocytes.[14,15]

However, in the case of nonproductive infection of cells, resulting in malignant transformation, only two viral-induced antigens have been detected, EBNA and LYDMA. EBNA can first be detected about 12 hr postinfection.[16] LYDMA presumably either preceeds or accompanies EBNA. Following the expression of EBNA, blastogenesis occurs by about 24 hr and cellular DNA synthesis at about 36 hr.[16] After expression of EBNA, the viral DNA is replicated and some viral DNA molecules became integrated into the host cell DNA.[17]

While the presence of EBNA in cells may not necessarily result in transformation,[18] its early appearance after EBV infection, prior to cellular DNA synthesis, and its association with the chromatin suggest that it may play an essential role in the transformation event.

# II. BIOCHEMICAL CHARACTERIZATION OF EBNA

EBV genome-positive cells possess a soluble complement-fixing antigen (S-antigen) present in high-speed centrifugation supernatants of extracts from these cells.[19-21] The presence of the S-antigen in EBV genome-positive but not EBV genome-negative lymphoblastoid cell lines and its identification with sera containing antibodies to other EBV-induced antigens but not with antibody-negative sera suggested that the S-antigen was induced by EBV.[20-22]

Recent evidence indicates that S-antigen and EBNA are one and the same molecule. A correlation exists between anti-EBNA antibody titers and anti-S-antigen titers in human sera from EBV-infected individuals.[23] Utilizing isoelectric focusing, Pickler et al.[24] obtained a single peak for both S-antigen and EBNA at pH 4.6. Also, partially purified preparations of S-antigen have been shown to specifically absorb antibodies to EBNA,[25] while Lenoir et al.[26] demonstrated that samples of S-antigen could inhibit the anticomplement immunofluorescence reaction used to detect EBNA. Therefore, both of these will be assumed to be equivalent and will be referred to as EBNA in this review.

EBNA has been partially purified from the Raji cell line (a nonproducer EBV genome-positive lymphoblastoid line) by Baron et al.[27] Employing ammonium sulfate precipitation, chromatography on DEAE-cellulose and poly-C-cellulose, they obtained an approximate 100-fold purification with the antigen exhibiting a molecular weight of 130,000. Lenoir et al.[26] employed sucrose gradient centrifugation, gel filtration, and ion exchange chromatography to isolate EBNA from extracts of Raji cells. The sedimentation coefficient of EBNA was estimated to be 8.5S corresponding to a molecular weight of 180,000. This molecular weight was confirmed by gel filtration on Sephadex® G-200 and Sepharose® 6B. Also using the Raji cell line, Matsuo et al.[28] further characterized EBNA. They were able to show that EBNA was heat stable, retaining its antigenic activity even after exposure to 80°C. Analysis of the antigen by isoelectric focusing and crossed immunoelectrophoresis revealed an isoelectric point of 4.8. By chromatography on Sephadex® G-200 and Sepharose® 6B, the molecular weight of the antigen was estimated to be between 220,000 and 240,000.

By applying an extract of Raji cells to chromatography on DNA-cellulose, Ohno et al.[29] partially purified EBNA 100- to 200-fold. The molecular weight of the antigen was estimated to be 174,000 ± 15,000 as determined by both sucrose gradient centrifugation and gel filtration, though chromatography on Sephadex® G-200 in buffer containing 0.5 or 1.0 *M* NaCl reduced the molecular weight of EBNA to 98,000 ± 8,000. This data was interpreted to indicate that the antigen existed as a dimer in low salt buffers and as a monomer in high salt buffers.

Further characterization of EBNA was performed by Baron and Strominger.[30] The antigen was purified 85-fold from a nuclear pellet derived from either RPMI 4265 or Raji cells. The purification employed heat treatment, ammonium sulfate precipitation, preparative ultracentrifugation, and chromatography on DNA-cellulose. Analyses of the antigen by sodium dodecyl sulfate polyacrylamide gel electrophoresis (SDS-PAGE) revealed 5 to 7 bands ranging in molecular weight between 12,000 and 70,000, with the most prominent band having a molecular weight of 49,000. They also showed that treatment of EBNA with DNase but not RNase caused an approximate 60% loss of antigenic activity indicating that EBNA may exist within the cell complexed to DNA.

Two groups have reportedly purified EBNA to homogeneity. Matsuo et al.[31] utilized ammonium sulfate precipitation, chromatography on DNA-cellulose, and immunoaffinity columns. These authors achieved a 3000-fold purification with a yield of 8% from Raji cell extracts. Analysis of the purified antigen by SDS-PAGE revealed the presence of three protein bands, having approximate molecular weights of 100,000, 70,000, and 50,000. Additional experiments indicated that the 70,000-dalton component was heat-labile since this protein was lost if the cell extract was heated prior to purification of EBNA. Based upon previously reported molecular weight estimations of EBNA in low salt buffer, these authors suggested that in its native state, EBNA was composed of a complex of each of the three proteins. Luka et al.[32] employed chromatography on DNA-cellulose, blue dextran-agarose, hydroxylapatite, and Sephadex® G-200 to purify EBNA from extracts of Raji and Namalwa cells. This purification scheme resulted in a 1200-fold purification with a yield of 45%. The molecular weight of the antigen was calculated to be between 170,000 and 200,000 by chromatography on Sephadex® G-200 and sucrose gradient centrifugation and 48,000 by SDS-PAGE. These results were interpreted to indicate that EBNA had a monomer molecular weight of 48,000 and existed in its native form as a tetramer.

Additional evidence that EBNA possessed a subunit molecular weight of about 50,000 came from Hentzen et al.[33] These workers subjected extracts containing EBNA to chromatography on CL-Sepharose®-6B under both nondenaturing and denaturing conditions. EBNA exhibited a molecular weight of 180,000 under nondenaturing conditions, but behaved with a molecular weight of approximately 50,000 in the presence of SDS or 6 *M* guanidine hydrochloride.

In summary, therefore, past studies on EBNA have indicated that the native molecular weight of the antigen is between 130,000 and 240,000, that it exists as a tetramer in low salt buffers and as a dimer in high salt buffers, and that the antigen has a monomer molecular weight of approximately 50,000. Recently, however, reports have been published suggesting that the monomer molecular weight of EBNA may be higher than the 50,000 previously reported. Utilizing fluoroimmunoelectrophoresis and radioimmunoelectrophoresis, Strnad et al.[34] identified a 65,000 mol wt antigen in Raji cells. The identification of this antigen as EBNA was based upon its detection with anti-EBNA positive sera but not anti-EBNA negative sera and its presence in Raji cells but not three EBV genome-negative human B cell lines. The antigen eluted with EBNA during partial purification by chromatography on DNA-Sepharose® and Blue Dextran-Sepharose®, and was able to absorb anti-EBNA antibodies.

In addition to the 65,000-dalton antigen, an 81,000-dalton antigen was also detected in both Raji cells and B95-8 cells. Antigens with molecular weights of 74,000, 70,000, and 73,000 were identified in B95-8, P3HR-1, and Namalwa cells, respectively.

Spelsberg et al.[35] identified a new species of EBNA termed Class II EBNA. This antigen was shown to be more tightly bound to chromatin than the classical EBNA termed Class I EBNA, and to have an apparent pI of between 8.5 and 9.2 as opposed to 4.6 determined for Class I EBNA. The new antigen was also shown to differ immunologically from Class I EBNA. The molecular weight of Class II EBNA was estimated to be between 60,000 and 70,000 by gel filtration. By subjecting samples of $NC_{37}$ cells to two-dimensional gel electrophoresis and then employing radioimmunoelectrophoresis, it was revealed that both Class I and Class II EBNA exhibited monomer molecular weights of approximately 70,000.

The identification of 70,000 mol w antigens in $NC_{37}$ cells by radioimmunoelectrophoresis does not necessarily indicate that these were the only EBV-induced antigens present. Other antigens such as the 48,000-dalton protein purified by Luka et al.[32] may have also been present but could have been irreversibly denatured upon electrophoresis on SDS polyacrylamide gels and, therefore, not detected by radioimmunoelectrophoresis. To investigate this possibility, an extract was prepared from $NC_{37}$ cells by suspending the cells in 0.4 *M* NaCl, 20 m*M* Tris-HCl (pH7.4), 1 m*M* phenylmethylsulfonyl fluoride (PMSF), 0.5% Triton® X-100. Insoluble material was removed by centrifugation and the supernatant dialyzed against the same buffer without Triton® X-100. This extract contains class I EBNA (as defined by Spelsberg et al.[35]) and should also contain the 48,000-dalton EBNA component (Luka et al.[32]) based upon procedures reported to extract this antigen. The solution was chromatographed on a column of phosphocellulose equilibrated in 0.4 *M* NaCl, 20 m*M* Tris-HCl (pH 7.4), 1 m*M* PMSF. Protein was eluted from the column with a linear gradient of sodium chloride 0.4 to 1.0 *M* containing 20 m*M* Tris pH 7.4, 1 m*M* PMSF, and EBNA was located using the $^{51}Cr$ complement fixation assay of Sever.[36] The majority of the protein passed through the column without binding while EBNA was eluted between 0.6 and 0.8 *M* NaCl (Figure 1A). Samples from every tenth fraction were subjected to electrophoresis on 10% polyacrylamide SDS gels and the proteins were either stained with Coomassie blue (Figure 1B) or transferred to nitrocellulose and incubated with an anti-EBNA positive serum, then $^{125}I$-labeled protein A[37] (Figure 1C). The anti-EBNA positive serum reacted with a protein having an approximate molecular weight of 70,000 which was present in the same fractions that reacted in the $^{51}Cr$ complement fixation assay.

The results indicated that only one EBNA antigen was present and that this antigen had a monomer molecular weight of 70,000. However, it was still possible that the 48,000-dalton EBNA happened to co-elute from the phosphocellulose column together with the 70,000-dalton EBNA but may not have been identified by radioimmunoelectrophoresis. To determine whether this was the case, the peak of complement fixing activity eluted from the phosphocellulose was pooled, lyophilized, redissolved in 6 *M* guanidine-HCl, 0.1 *M* sodium acetate (pH 6.0), 1% 2-mercaptoethanol and chromatographed on a column of CL-Sepharose®-6B equilibrated in the same buffer. Under these conditions, the antigen would be completely denatured and should elute from the column at its monomer molecular weight. All detectable complement fixing activity eluted from the Sepharose® column as a single peak portraying a molecular weight of between 65,000 and 75,000 (Figure 2). No antigenic activity was present in the 45,000- to 53,000-molecular weight range.

If the monomer molecular weight of EBNA is truly between 65,000 and 75,000, then it is likely that the 48,000-dalton antigen studied by Hentzen et al.[33] and purified by Luka et al.[32,38] was a proteolytic degradation product of EBNA. If this were the case, it might explain the data of Matsuo et al.[31] These authors were able to purify EBNA more than 3000-fold with the final preparation yielding three proteins having molecular weights of 100,000, 70,000, and 50,000. If the cell extract was heated prior to purification, only the 100,000- and 50,000-dalton proteins were recovered. Assuming that the 70,000-dalton protein was

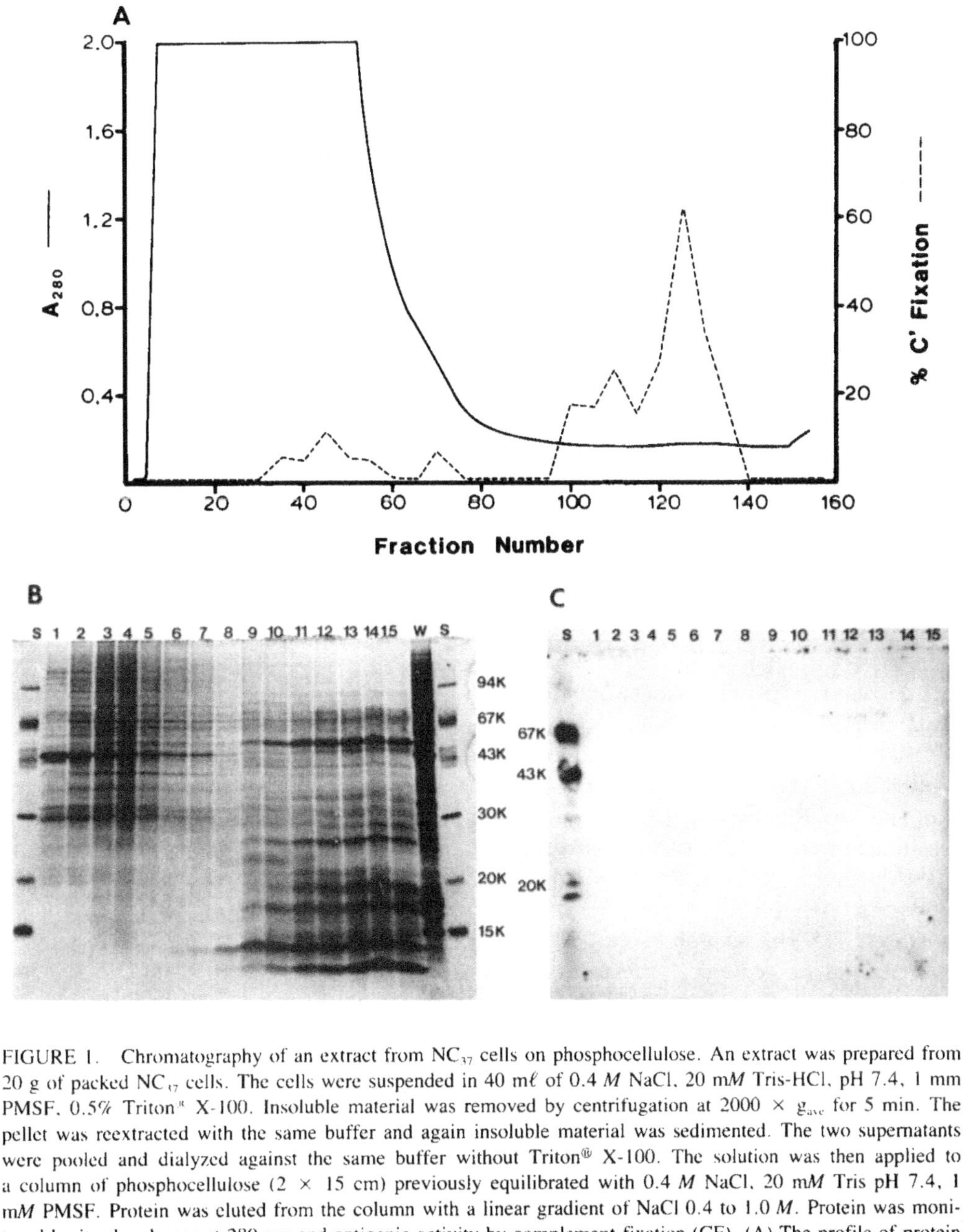


FIGURE 1. Chromatography of an extract from $NC_{37}$ cells on phosphocellulose. An extract was prepared from 20 g of packed $NC_{37}$ cells. The cells were suspended in 40 mℓ of 0.4 *M* NaCl, 20 m*M* Tris-HCl, pH 7.4, 1 mm PMSF, 0.5% Triton® X-100. Insoluble material was removed by centrifugation at 2000 × $g_{ave}$ for 5 min. The pellet was reextracted with the same buffer and again insoluble material was sedimented. The two supernatants were pooled and dialyzed against the same buffer without Triton® X-100. The solution was then applied to a column of phosphocellulose (2 × 15 cm) previously equilibrated with 0.4 *M* NaCl, 20 m*M* Tris pH 7.4, 1 m*M* PMSF. Protein was eluted from the column with a linear gradient of NaCl 0.4 to 1.0 *M*. Protein was monitored by its absorbance at 280 nm and antigenic activity by complement fixation (CF). (A) The profile of protein (———) and CF activity (— — —); (B) samples from every tenth fraction were subjected to electrophoresis on a 10% polyacrylamide SDS slab gel and protein was visualized by staining with Coomassie Blue, (C) samples from every tenth fraction were subjected to radioimmunoelectrophoresis and EBNA located with an anti-EBNA positive serum. The molecular weights of standard proteins and $^{125}$I-labeled standard proteins are indicated.

EBNA, then heating the cell extract would encourage proteolysis and would reduce EBNA to the 50,000-dalton component.

## III. BIOLOGICAL ACTIVITY OF EBNA

All of the studies thus far reported on the biological activity of EBNA have employed partially purified samples of the antigen. Therefore, one must be aware of the possibility

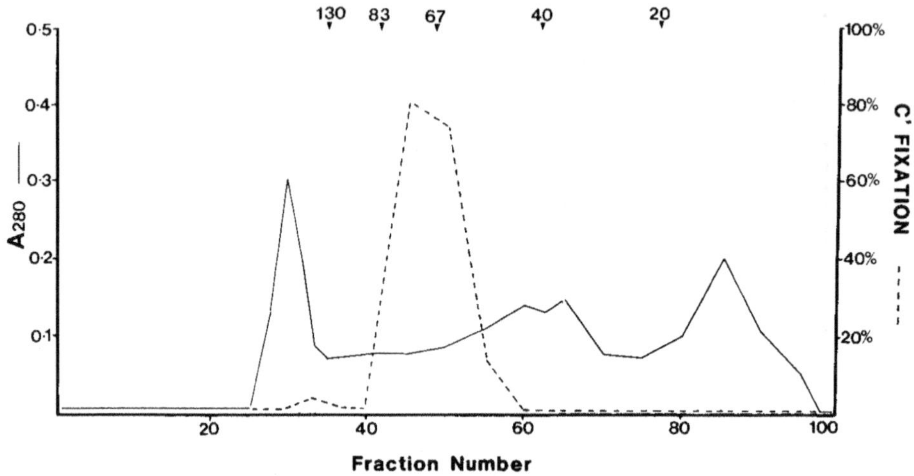


FIGURE 2. Analysis of EBNA by chromatography on CL-Sepharose®-6B. The peak of antigenic activity eluted from the phosphocellulose column was pooled, dialyzed against 0.1 m*M* PMSF, and lyophilized. The protein was dissolved in 5 mℓ of GM guanidine-HCl, 100 m*M* sodium acetate pH 6.0, 1% 2-mercaptoethanol and applied to a column of -CL-Sepharose®-6B (2.5 × 95 cm) equilibrated with the same buffer. The elution of proteins with known molecular weights are indicated. The elution of protein was monitored by its absorbance at 280 n*M* (———) and antigens by CF (— — —).

that the observed activity may be due to an EBV-induced cellular protein or a normal cellular protein associated with EBNA.

One of the first activities observed for EBNA was its ability to function as a DNA-binding protein. Luka et al.[25] solubilized EBNA from the nuclei of two EBV-transformed cell lines, Raji and AW-Ramos, using a high concentration of NaCl. The DNA binding properties of the antigen were studied using chromatography on DNA-cellulose columns. They found that EBNA bound to both double-and single-stranded DNA based upon the concentration of salt required to elute EBNA from the DNA-cellulose column. This affinity of EBNA for double-stranded DNA was also observed by Lenoir et al.[39] and Baron et al.[27]

If some of the biological effects of EBNA arise from its ability to bind to DNA, then it would be logical to assume that EBNA might interact with specific DNA sequences. Hirsch et al.[40] studied the affinity of partially purified samples of EBNA for various DNAs utilizing the acid-fixed nuclear binding assay. It was found that EBV DNA was twice as efficient as herpes simplex virus DNA and about three times more efficient than various cellular DNAs in eliminating the ability of EBNA to convert acid-fixed nuclei to the EBNA-positive form. Even so, the authors did not consider the differences to be convincing enough to conclude that EBNA binds to specific EBV DNA sequences.

A possible function of EBNA, which may be related to its ability to bind to DNA, was reported by Kamata et al.[42] It was found that the addition of a partially purified preparation of EBNA to chromatin of human diploid cells caused stimulation of the chromatin template activity for transcription. The stimulation of transcription was determined by measuring the ethidium bromide-binding capacity of chromatin and the incorporation of [$^3$H]UTP into RNA. That this stimulation was a consequence of EBNA was supported by the fact that the stimulation was inhibited with anti-EBNA positive sera but not with anti-EBNA negative sera. Also, extracts from EBV genome-negative BJAB cells failed to cause any stimulation. As cited earlier, EBNA is expressed in cells shortly after EBV infection, but prior to cellular DNA synthesis. If EBNA does have the capacity to stimulate transcription, it might serve a function in initiating cellular DNA synthesis.

Spelsberg et al.[35] identified two different species of EBNA in $NC_{37}$ cells. The class I EBNA was loosely associated with the chromatin and was extracted with relatively low ionic strength buffers. All studies on the biological activity of EBNA thus far reported would have employed the class I EBNA. The class II EBNA, however, was found to be tightly associated with chromatin requiring high salt-urea or guanidine-HCl for extraction. While little or no specificity in the DNA-binding properties of the class I EBNA have been observed, the tight association of the class II EBNA with the chromatin suggests that its primary function may be to interact with DNA. Purification of the class II EBNA and analysis of its interaction with DNA may be more rewarding in demonstrating an association of EBNA with specific DNA sequences.

Many of the characteristics of EBNA show similarities to the T antigens of papova- and adenoviruses. Work by Kamata et al.[43] indicated that EBNA could possess protein kinase activity, a property also possessed by T antigens of SV-40.[44,45] These workers partially purified EBNA by chromatography on Sephadex® G-200, DNA-cellulose, DEAE-Sephadex® A-50, and heparin-Sepharose®. The specific activity of the protein kinase rose concurrently with the rise in specific activity of EBNA during the purification. The kinase activity was immunoprecipitated with anti-EBNA positive sera, but not anti-EBNA negative sera and was detected by immunoprecipitation in Raji cells but not EBV genome-negative Ramos or BJAB cells. Analysis of in vitro phosphorylated products in the immunoprecipitates by gel electrophoresis showed phosphorylation of a 50,000-dalton protein which the authors assumed to be EBNA. This protein was found to be phosphorylated at both serine and threonine residues.

If, as assumed in the previous section, EBNA has a subunit molecular weight of between 65,000 and 75,000, then the phosphorylated 50,000-dalton protein identified by immunoprecipitation with anti-EBNA positive sera could be complexed with EBNA, Luka et al.[38] presented data suggesting that EBNA was associated with a 53,000-dalton cellular protein. Immunoprecipitation of their 48,000-dalton EBNA component resulted in the simultaneous precipitation of a 53,000-dalton protein. Analysis of the amino acid compositions and peptide maps of these proteins revealed that they were not related to each other. And the observation that the 53,000-dalton protein was also present in two EBV genome-negative cell lines indicated that it was not an EBV encoded protein. Whether this 53,000-dalton protein is also associated with the 65,000- to 75,000-molecular-weight species of EBNA has yet to be determined.

## IV. CONCLUSION

Utilizing fluoro-and radioimmunoelectrophoresis, EBNA has been identified in a variety of different EBV-genome positive cell lines. The antigen portrayed a subunit molecular weight of between 65,000 and 75,000 depending upon the cell line examined.[34] Also, employing radioimmunoelectrophoresis, Spelsberg et al.[35] identified 70,000-mol w EBNA components in $NC_{37}$ cells. As yet the relationship between the 65,000- to 75,000-antigens and the 48,000-dalton EBNA component purified by Luka et al.[32] is unknown. It is possible that the 48,000-dalton protein is an additional EBNA antigen, undetectable by radioimmunoelectrophoresis. Though partial purification of EBNA from an extract of $NC_{37}$ cells identified the antigen as having a molecular weight of 70,000 both by radioimmunoelectrophoresis and gel filtration on CL-Sepharose®-6B, no other EBNA components were detected either by radioimmunoelectrophoresis or by complement fixation. In addition, EBNA has been partially purified from extracts of Raji cells. This antigen also portrayed a subunit molecular weight of between 65,000 and 70,000.[45] These results suggest that the 48,000-dalton antigen might be a degraded form of EBNA. Ultimately, purification of EBNA to homogeneity and comparison of its amino acid composition and peptide map to those of the 48,000-dalton antigen should resolve this question.

While the studies responsible for attributing specific biological functions to EBNA are obviously hampered by the fact that the antigen has only been partially purified, there are other problems also involved. It would be helpful to know whether EBNA has maintained its molecular weight throughout the purification scheme employed, or if the final preparation contains a degraded form of the antigen. Also, there is increasing evidence that EBV genome-positive cells contain multiple EBNA components. Strnad et al.[34] identified multiple EBNA antigens in a number of cell lines, while Spelsberg et al.[35] demonstrated the presence of two distinct EBNAs in $NC_{37}$ cells. Therefore, it would be of value to determine if multiple EBNA antigens were present in the cell, and if so, which component(s) are responsible for the observed biological activity.

## REFERENCES


1. **Epstein, M. A., Achong, B. G., and Barr, Y. M.,** Virus particles in cultured lymphoblasts from Burkitt's lymphoma, *Lancet*, 1, 702, 1964.
2. **Klein, G.,** The Epstein-Barr virus, in *The Herpes Viruses*, Kaplan, A. S., Ed., Academic Press, New York, 1973, 521.
3. **Epstein, M. A. and Achong, B. C.,** Recent progress in Epstein-Barr virus research, *Ann. Rev. Microbiol.*, 31, 421, 1977.
4. **de Thé, G.,** Role of Epstein-Barr virus in human diseases: infectious mononucleosis, Burkitt's lymphoma and nasopharyngeal carcinoma, in *Viral Oncology*, Klein, G., Ed., Raven Press, New York, 1980, 769.
5. **Zur Hausen, H. and Schulte-Holthausen, H.,** Presence of EB virus nucleic acid in a virus-free line of Burkitt tumor cells, *Nature (London)*, 227, 254, 1970.
6. **Nonoyama, M. and Pagano, J.,** Complementary RNA specific to the DNA of Epstein-Barr virus: detection of EB viral genomes in nonproductive cells, *Nature (London) New Biol.*, 233, 103, 1971.
7. **Crawford, D. H., Rickison, A. B., Finerty, S., and Epstein, M. A.,** Epstein-Barr (EB) virus genome-containing EB nuclear antigen-negative B-lymphocyte populations in blood in acute infectious mononucleosis, *J. Gen. Virol.*, 38, 449, 1978.
8. **Reedman, B. M. and Klein, G.,** Cellular localization of an Epstein-Barr Virus (EBV)-associated complement-fixing antigen in producer and non-producer lymphoblastoid cell lines, *Int. J. Cancer*, 17, 499, 1973.
9. **Klein, G., Clifford, P., Klein, E., and Stjernsward, J.,** Search for tumor-specific immune reactions in Burkitt's lymphoma patients by the membrane immunofluorescence reaction, *Proc. Natl. Acad. Sci. U.S.A.*, 55, 1628, 1966.
10. **Ernberg, I., Klein, G., Kourilsky, F. M., and Silvestre, D.,** Differentiation between early and late membrane antigen on human lymphoblastoid cell lines infected with Epstein-Barr virus. I. Immunofluorescence, *J. Natl. Cancer Inst.*, 53, 61, 1974.
11. **Silvestre, D., Ernberg, I., Neauport-Sautes, C., Kourilsky, F. M., and Klein, G.,** Differentiation between early and late membrane antigen on human lymphoblastoid cell lines infected with Epstein-Barr virus. II. Immunoelectron microscopy, *J. Natl. Cancer Inst.*, 53, 67, 1974.
12. **Henle, G., Henle, W., and Klein, G.,** Demonstration of two distinct components in the early antigen complex of Epstein-Barr virus-infected cells, *Int. J. Cancer*, 8, 272, 1971.
13. **Henle, G. and Henle, W.,** Immunofluorescence in cells derived from Burkitt's lymphoma, *J. Bacteriol.*, 91, 1248, 1966.
14. **Svedmyr, E. and Jondal, M.,** Cytotoxic effects on cells specific for B cell lines transformed by Epstein-Barr virus are present in patients with infectious mononucleosis, *Proc. Natl. Acad. Sci. U.S.A.*, 72, 1622, 1975.
15. **Bakacs, T., Svedmyr, E., Klein, E., Rombo, L., and Weiland, O.,** EBV-related cytotoxicity of Fe-receptor negative T lymphocytes separated from the blood of infectious mononucleosis patients, *Cancer Lett.*, 4, 185, 1978.
16. **Takada, K. and Osato, T.,** Analysis of the transformation of human lymphocytes by Epstein-Barr virus. I. Sequential occurrence from the virus-determined nuclear antigen synthesis, to blastogenesis, to DNA synthesis, *Intervirology*, 11, 39, 1978.
17. **Glaser, R., Nonoyama, M., Shows, T. B., Henle, G., and Henle, W.,** Epstein-Barr virus: studies on the association of virus genome with human chromosomes in hybrid cells, in *Oncogenesis and Herpes Viruses II*, Part 1, de Thé, G., Epstein, M. A., and Zur Hausen, H., Eds., International Agency for Research Against Cancer, 1976, 457.
18. **Pope, J. H.,** Transformation by the virus in vitro, in *The Epstein-Barr Virus*, Epstein, M. A. and Achong, B. G., Eds., Springer-Verlag, New York, 1979, 205.

19. **Armstrong, D., Henle, G., and Henle, W.,** Complement-fixation tests with cell lines derived from Burkitt's lymphoma and acute leukemias, *J. Bacteriol.*, 91, 1257, 1966.
20. **Pope, J. H., Horne, M. K., and Wetters, E. J.,** Significance of a complement-fixing antigen associated with herpes-like virus and detected in the Raji cell line, *Nature (London)*, 222, 166, 1969.
21. **Vonka, V., Benyesh-Melnick, M., Lewis, R. T., and Wimberly, I.,** Some properties of the soluble (S) antigen of cultured lymphoblastoid cell lines, *Arch. Ges. Virusforsch.*, 31, 118, 1970.
22. **Vonka, V., Porter, D. D., McCombs, R. M., and Benyesh-Melnick, M.,** Studies with viral and soluble antigens from cultured Burkitt lymphoma cells, *Bacteriol., Proc.*, 112, 154, 1969.
23. **de Thé, G., Ho, J. H. C., Ablashi, D. V., Day, N. E., Macario, A. J. L., Martin-Benthelon, M. C., Pearson, G., and Sohier, R.,** Nasopharyngeal carcinoma. IX. Antibodies to EBNA and correlation with response to other EBV antigens in Chinese patients, *Int. J. Cancer*, 16, 713, 1974.
24. **Pickler, G. M., Spelsberg, T. C., and Pearson, G. R.,** Isolation of the Epstein-Barr virus nuclear antigen from chromatin preparations, in *Proc. 3rd Int. Symp. Oncogenesis and Herpes-Viruses*, de Thé, G., Henle, W., and Rapp, F., Eds., International Agency for Research Against Cancer, Lyon, 1978, 243.
25. **Luka, J., Siegert, W., and Klein, G.,** Solubilization of the Epstein-Barr virus-determined nuclear antigen and its characterization as a DNA binding protein, *J. Virol.*, 22, 1, 1977.
26. **Lenoir, G., Berthelon, M. C., Favre, M. C., and de Thé, G.,** Characterization of Epstein-Barr virus antigens. I. Biochemical analysis of the complement-fixing soluble antigen and relationship with Epstein-Barr virus-associated nuclear antigen, *J. Virol.*, 17, 672, 1976.
27. **Baron, D., Benz, W. C., Carmichael, G., Yocum, R. R., and Strominger, J. L.,** Assay and partial purification of Epstein-Barr virus nuclear antigen, in *Proc. Int. Workshop on the Production, Concentration and Purification of Epstein-Barr Virus*, International Agency for Research Against Cancer, Lyon, 1975, 2547.
28. **Matsuo, T., Nishi, G., Hirai, H., and Osato, T.,** Studies of Epstein-Barr virus-related antigens. II. Biochemical properties of soluble antigen in Raji Burkitt lymphoma cells, *Int. J. Cancer*, 19, 364, 1977.
29. **Ohno, S., Luka, J., Lindahl, T., and Klein, G.,** Identification of a purified complement-fixing antigen as the Epstein-Barr virus-determined nuclear antigen (EBNA) by its binding to metaphase chromosomes, *Proc. Natl. Acad. Sci. U.S.A.*, 74, 1605, 1977.
30. **Baron, D. and Strominger, J. L.,** Partial purification and properties of the Epstein-Barr virus associated nuclear antigen, *J. Biol. Chem.*, 253, 2875, 1978.
31. **Matsuo, T., Hibi, N., Nishi, S., Hirai, H., and Osato, T.,** Studies on Epstein-Barr virus-related antigens. III. Purification of the virus-determined nuclear antigen (EBNA) from nonproducer Raji cells, *Int. J. Cancer*, 22, 747, 1978.
32. **Luka, J., Lindahl, T., and Klein, G.,** Purification of the Epstein-Barr virus determined nuclear antigen from Epstein-Barr virus-transformed human lymphoid cell lines, *J. Virol.*, 27, 604, 1978.
33. **Hentzen, D., Lenoir, G. M., Berthleon, M. C., and Daillie, J.,** Epstein-Barr virus (EBV) antigenic determinants on subunits of the EBV determined nuclear antigen (EBNA), *Biochem. Biophys. Res. Commun.*, 96, 425, 1980.
34. **Strnad, B. C., Schuster, T. C., Hopkins, R. F., Neubauer, R. H., and Rabin, H.,** Identification of an Epstein-Barr virus nuclear antigen by fluoroimmunoelectrophoresis and radioimmunoelectrophoresis, *J. Virol.*, 38, 996, 1981.
35. **Spelsberg, T. C., Sculley, T. B., Pikler, G. M., Gilbert, J. A., and Pearson, G. R.,** Evidence for two classes of chromatin-associated Epstein-Barr virus-determined nuclear antigen, *J. Virol.*, 43, 555, 1982.
36. **Sever, J. L.,** Application of a microtechnique to viral serological investigations, *J. Immunol.*, 88, 320, 1962.
37. **Burnette, W.N.,** "Western blotting": electrophoretic transfer of protein from SDS-polyacrylamide gels to unmodified nitrocellulose and radiographic detection with antibody and radioiodinated protein A, *Anal. Biochem.*, 112, 195, 1981.
38. **Luka, J., Jornvall, H., and Klein, G.,** Purification and biochemical characterization of the Epstein-Barr virus (EBV)-determined antigen (EBNA) and an associated protein with a 53K subunit, *J. Virol.*, 35, 592, 1980.
39. **Lenoir, G., Berthelon, M. C., Favre, M. D., and de Thé, G.,** Characterization of Epstein-Barr Virus (EBV) antigens. III. Detection of early antigen using anti-complementary immunofluorescence (ACIF) and complement fixation texts, *Biomedicine*, 23, 461, 1975.
40. **Hirsch, I., Kuchlerova, L., Brichacek, B., Schankova, A., and Vonka, V.,** Blocking of acid-fixed nuclear binding of Epstein-Barr virus nuclear antigen (EBNA) by different DNA species, *J. Gen. Virol.*, 44, 840, 1979.
41. **Vonka, V. and Hirsch, I.,** Epstein-Barr-Virus nuclear antigen, *Prog. Med. Virol.*, 28, 145, 1982.
42. **Kamata, T., Tanaka, S., Aikawa, S., Hinuma, Y., and Watanabe, Y.,** A possible function of Epstein-Barr virus-determined nuclear antigen (EBNA): stimulation of chromatin template activity *in vitro*, *Virology*, 95, 222, 1979.

43. **Kamata, T., Takaki, K., Hinuma, Y., and Watanabe, Y.,** Protein kinase activity associated with Epstein-Barr virus-determined nuclear antigen, *Virology*, 113, 512, 1981.
44. **Griffen, J. D., Spangler, G., and Livingston, D. M.,** Protein kinase activity associated with simian virus 40 T antigen, *Proc. Natl. Acad. Sci. U.S.A.*, 76, 2610, 1979.
45. **Sculley, T. B., Kreofsky, T., Pearson, G. R., and Spelsberg, T. C.,** Partial purification of the Epstein-Barr virus nuclear antigen(s), *J. Biol. Chem.*, 258, 3974, 1983.
46. **Tjian, R. and Robbins, A.,** Enzymatic activities associated with a purified simian virus 40 T-antigen-related protein, *Proc. Natl. Acad. Sci. U.S.A.*, 76, 610, 1979.

Chapter 12

# NONRANDOM BINDING OF CHEMICAL CARCINOGENS IN MAMMALIAN NUCLEAR SUBFRACTIONS

**Carol J. Mirell, Rebecca Trafzer, Joon Rho, and Isaac Bekhor***

## TABLE OF CONTENTS



* We would like to dedicate this chapter to the memory of Dr. Charles Heidelberger, a pioneer in chemical carcinogenic studies, a brilliant educator, and above all, a fine human being. We have all benefited much from him.

# I. INTRODUCTION

The search for the key to cancer has enticed many investigators, yet the causes and mechanisms of this group of disorders eludes us. Since the time that coal tar was shown to be carcinogenic,[57] the interest in chemicals as carcinogens has grown dramatically. Many classes of carcinogens have been investigated including polycyclic hydrocarbons, azo-dyes, acetyl amino fluorenes, nitrosamines, nitrosamides, thioacetamides, ethionine, and others (for a review, refer to Gronow[26]).

It has generally been postulated that carcinogenesis is actually a mutagenic event. In fact, it has become commonplace to test the oncogenicity of various compounds by monitoring point mutations and reversions in bacterial populations.[1] Direct attack on DNA has been thought to be the mechanism that could lead to neoplastic growth.[13,31,40,41] The binding of ultimate carcinogens to DNA has been suggested to be the most critical event in the initiation of carcinogenesis.[15,27,29,30] There have been extensive reports evaluating DNA-carcinogen adducts from numerous laboratories. In a study with mice, Eastman and Bresnick[20a] have shown that the degree of persistence of DNA-methylcholanthrene adducts correlates directly with susceptibility to neoplastic transformation. Arrand and Murray[3] have demonstrated that the DNA of human lung epithelium (a target tissue) binds benzopyrene metabolites to a much greater extent than does nontarget fibroblast cells. In another study, Shoyab[60] showed that reiterated DNA sequences preferentially bind dimethylbenzanthracene when the carcinogen was administered at low dosages. Studies by Walker et al.[67] indicate random distribution of *N*-2-acetylaminofluorene and its *N*-hydroxy derivative to DNA when the chromatin is fractionated on the basis of hydrodynamic shearing or DNAase sensitivity. However, it has been reported that preferential binding does occur in micrococcal nuclease sensitive sites.[32,45,67]

In some instances, there is a direct relationship between mutagenesis and carcinogenesis. For example, mutations that block induction of aryl hydrocarbon hydroxylase make mice less sensitive to long-term carcinogenicity of certain polycyclic aromatic hydrocarbons (PAH).[38,50] In the case of skin cancers, there is strong evidence that the initiating event is the production of pyrimidine dimers by UV irradiation.[28] Cairns[16] has pointed out that mutagens can cause increased sister chromatid exchanges and has suggested that transposition of DNA may be crucial in leading to some human cancers. It has been shown, for example, that the integration of tumor viruses can be associated with transpositions in adjoining regions of host DNA.[8]

With the demonstration of the generalized existence of cellular oncogenes,[6] the possibility that chemical carcinogens induce enhanced expression of viral-like particles has spurred many investigators to speculate on the implications in neoplastic processes. Whitmire et al.[70] showed that methylcholanthrene induced the production of C-type particles in C57/Blk mice, although these were nontumorigenic upon reinjection into other animals. Freeman et al.[23] also obtained C-type particles from methylcholanthrene-induced hamster tumors. In a study by Getz et al.[25] it was shown that both a nontransformed cell line C3H/10T$^1/_2$ (clone 8) and its methylcholanthrene-transformed derivative contained murine leukemic viral RNA sequences in quantitatively similar amounts; it was therefore concluded that the transcription of DNA into RNA in this case is not related to the maintenance of the transformed state. In studies by Tennant et al.[63] chemical carcinogens were shown to mediate viral induction, but at frequencies vastly exceeding mutational rates. It seems likely that these studies indicate that chemical carcinogens may cause viral induction, but that induction in itself is not necessarily responsible for transformation.

With the rapid technological advances made in DNA sequencing and genetic engineering, it has been easy to overlook the significance of other macromolecules in the cell. There are numerous instances of altered gene expression that cannot be explained by sequence alterations.[2,19,21,51,69] In cancer research, the emphasis has likewise been placed on DNA modification. Aberration in gene expression may not be caused exclusively by DNA alteration.

The patterns of many cancers, leukemias in particular, suggest that the disease involves stem cell changes which are a reflection of altered gene expression.[7,22,44] Changes in development may be the consequence of chemical carcinogens interacting with other macromolecules. It has been shown that carcinogens bind to RNA and proteins as well as to DNA.[39,71] Morin et al.[48] propose that the macromolecular complexes produced by carcinogens may alter the structure of chromosomes and result in nondisjunction during mitosis, possibly resulting in malignant transformation.

In view of the fact that nonhistone chromosomal (NHC) proteins are regarded as possible regulators of gene activity, it seems reasonable to analyze the potentially significant interaction of carcinogens with these proteins. In most cases, it has been found that all carcinogens react with proteins as well as with DNA, and in many cases the degree of binding to protein is higher than the binding to DNA. There have been some reports, however, that indicate that the majority of the carcinogen is bound to DNA.[37] Jungmann and Schweppe[33] measured the binding of radiolabeled *N*-hydroxy-*N*-2-fluoroenylacetamide, *p*-dimethylaminobenzene, and 7,12-dimethylbenzanthracene to chromosomal proteins from rat liver. These investigators demonstrated stable interactions of these carcinogens with both histone and NHC proteins. It was also shown that the degree of binding to a particular protein was dependent on the carcinogen studied. In addition, there seems to be some degree of saturability since pretreatment with a nonradioactive carcinogen reduces the measureable radioactivity observed. Analyses of carcinogen binding to histones has shown that high levels of the metabolites are associated with $H_1$, low levels with $H_4$, and moderate levels with $H_2$ and $H_3$.[12,56] Vaught and Bresnick[66] have shown that when rat liver nuclei were incubated with benzo(*a*)pyrene (BAP) the histones bound twofold and the nonhistones bound fourfold the amount of carcinogen as did DNA. Studies by Spelsberg's group[49] have shown that transcriptionally active regions of the AKR mouse embryo nucleus binds methylcholanthrene preferentially. Further studies[71] have analyzed the active fraction for BAP binding. It was found that in this fraction, 94% of the BAP was bound to proteins, 5% to RNA, and 1% to DNA.

Transcriptionally active regions of chromatin contain hyperacetylated core histones.[17] Koostra,[35] however, has recently reported that hyperacetylation of core histones had no effect on the extent of binding of BAP to chromatin. It has also been shown that chemical carcinogens alter transcriptional activity.[10] If modification of core histones does not influence the binding of chemical carcinogens, it seems that modification of nonhistones may be of some relevance in the observed changes in transcription.

Recent studies point to the significance of those proteins remaining after high salt treatment of the nucleus.[42,55,59] The proteins which aggregate with DNA (that is, those that cannot be readily separated from DNA under high salt conditions) contain tissue-specific antigens,[18,58] show preference for actively replicating[55] and transcribing sites,[46,68] and bind steroid-receptor complexes.[61,68] It has been reported that carcinogens, like hormones, bind to receptor proteins in the cytosol and are translocated to the nucleus by a temperature-dependent process.[65] In our laboratory we also have examined the effect of 2 *M* NaCl extraction on chromatin preparations. We extract chromatin with 2 *M* NaCl and then separate the DNA on the basis of protein-DNA interactions.[4,47,52] This results in two DNA subfractions: (1) DNA-P, the minor fraction which is tightly complexed with salt-resistant proteins, and (2) DNA-S, the major fraction which is protein depleted. We are then able to analyze the protein and DNA components of both fractions. Our work has suggested that the DNA-P is enriched in tissue-specific genes, for example, globin in reticulocytes,[4] lactogen in human placenta,[52] and ovalbumin in oviduct.[47]

Since the literature and our own data point to the significance of the salt-resistant fraction, we felt that this fraction should be investigated for its possible role in carcinogenesis. We therefore began an analysis of fractionated nuclei for carcinogen binding sites.

In the experiments described below, we have studied the binding affinities of methylcholanthrene and BAP in fractionated mammalian nuclei. We have analyzed rat liver nuclei for methylcholanthrene binding; however, the majority of the data was obtained from a transformable mouse fibroblast line, C3H/10T$^1/_2$, generously supplied by Dr. Charles Heidelberger. We have shown by radiolabeling that the DNA-P fractions show a much higher binding capacity for both BAP and methylcholanthrene. In addition, fluorometry data with BAP is presented which clearly indicates that the covalently bound adduct is BAP-diol-epoxide Proteinase K digests and guanidinium chloride dissociation data indicate that the majority of the carcinogen is bound to protein rather than to DNA. Although such dramatic nonrandom binding does not prove that protein-carcinogen interactions are responsible for transformation, the overwhelming enrichment of both BAP and methylcholanthrene in the DNA-P fraction argues in favor of proteins playing a significant role in neoplastic processes.

## II. EXPERIMENTAL PROCEDURES

### A. Animals

Male Sprague Dawley rats obtained from Simonsen Laboratories were used. To induce mixed function oxidase activity, the rats were injected i.p. with 3-methylcholanthrene in corn oil at a dose of 40 mg/kg. The rats were sacrificed by decapitation 2 days following induction, the livers excised and sliced for in vitro incubation. Livers (20 g of tissue in 100 mℓ media) were incubated at 37°C in Modified F-10 (GIBCO) in the presence of $^3$H-3-methylcholanthrene at 4 mCi/mℓ (specific activity = 26 Ci/mmol). Incubations were conducted at various times to establish maximal incorporation. Following incubation, nuclei were prepared as previously described.[4] The purified nuclei were then analyzed for DNA content by diphenylamine assay[14] and for radioactivity using a Beckman 8100 Scintillation Counter.

Nuclei were then fractionated into DNA-P and DNA-S fractions as described below. The fractions were then analyzed for radioactivity.

### B. Tissue Culture

C3H/10T$^1/_2$ (clone 8) cells were kindly provided by Dr. Charles Heidelberger, LAC-USC Cancer Center, Los Angeles. The cells were grown in Eagle's basal medium (GIBCO Laboratories) supplemented with 10% heat-inactivated fetal calf serum, moxalactam (30 μg/mℓ), and gentamycin (30 μg/mℓ). The cells were maintained at 37°C in a 5% $CO_2$-air humidified incubator, either in tissue culture flasks or roller bottles. Cells were routinely passed using 0.1% trypsin and harvested manually with a rubber policeman. No spontaneous transformation was observed in control cultures.

### C. Radiolabeling

For experiments involved with carcinogen labeling, the cells were induced with benzanthracene (3 μg/mℓ) 48 hr before adding isotope. In the case of BAP 5 μCi/mℓ of tritiated isotope (specific activity = 18 Ci/mmol) was added along with 2.5 μg/mℓ of unlabeled BAP 48 hr prior to harvest. In experiments involving radioactive methylcholanthrene, 1 μCi/mℓ (specific activity = 26 Ci/mmol) of tritiated isotope was added along with 5 μg/mℓ of unlabeled methylcholanthrene 48 hr prior to harvest. For tracing proteins, either 4 μCi/mℓ$^3$ H-leucine (5 mCi/mmol), 0.2 μCi/mℓ $^{14}$C-leucine (356 Ci/mmol), or 0.5 μCi/mℓ $^{35}$S-methionine (560 Ci/mmol) were used. DNA was labeled with $^3$H-thymidine (20 Ci/mmol) at 1.6 μCi/mℓ.

### D. Preparation of Nuclei, Chromatin, and Chromatin Subfractions

At confluency, cells were harvested in 0.14 *M* NaCl, 1 m*M* ethylenediaminetetraacetic acid (EDTA), TPD (10 m*M* Tris-HCl, pH 8.3, 0.1 m*M* PMSF, 0.2 m*M* dithiothreitol). After a low-speed centrifugation, the cells were lysed in TPD containing 5 m*M* $MgCl_2$, 0.1 *M* NaCl, 0.2% Triton® X-100. The crude nuclei were then washed repeatedly with TDP, 5

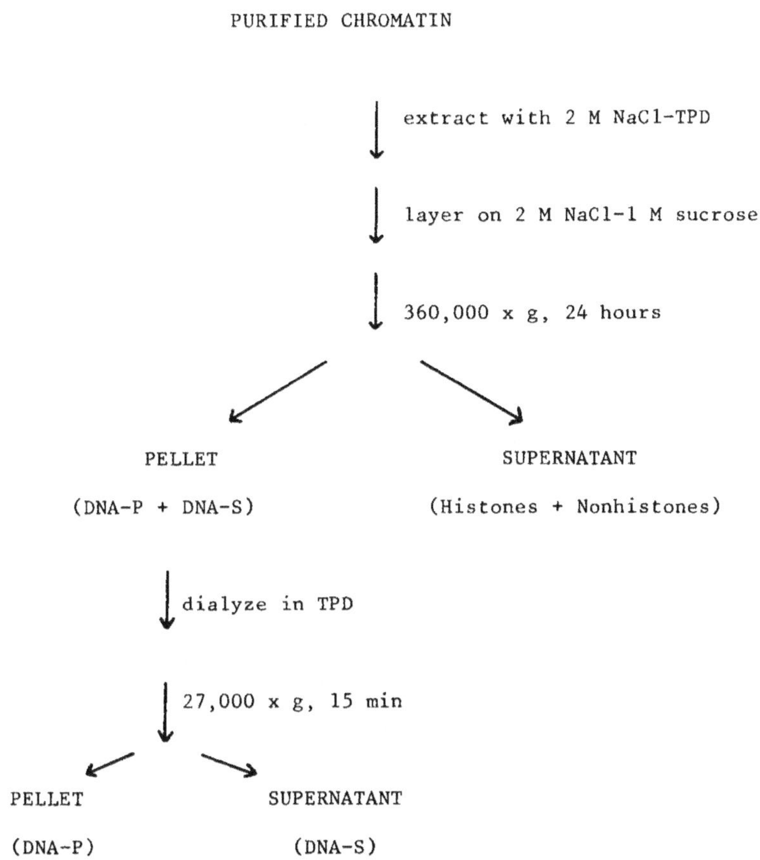


FIGURE 1. Fractionation scheme for the separation of soluble DNA (DNA-S) from protein bound DNA (DNA-P).

m*M* $MgCl_2$, 0.1 *M* NaCl to remove the Triton® X-100. Nuclei were observed microscopically to ensure removal of cytoplasmic tags. Crude nuclei were then suspended in 2.1 *M* sucrose, layered over 2.2 *M* sucrose, and centrifuged for 90 min at 26 Krpm in a Beckman SW27 rotor. The nuclear pellets were homogenized in TPD, and the chromatin collected by centrifuging the homogenate through 1.7 *M* sucrose. The chromatin was then extracted with 2 *M* NaCl as previously described.[4,47,52] This was followed by dialysis and low-speed centrifugation to separate the DNA-P aggregate from the solubilized DNA-S (see Figure 1).

### E. Removal of Noncovalently Linked Benzo(*a*)pyrene or Methylcholanthrene

The noncovalently linked carcinogen in our various chromatin fractions is routinely removed by solubilization in ethyl acetate. One volume of sample is vortexed for 1 min at 25°C with two volumes of ethyl acetate. The aqueous phase is recovered following low-speed centrifugation. It was found that six extractions were required to remove all of the soluble carcinogen. The remaining bound carcinogen is presumed to be covalently linked either to protein, DNA, or RNA. Trace amounts of ethyl acetate are subsequently removed by gassing with nitrogen.

### F. Fluorometry

We have developed a highly sensitive fluorescence instrument which allows us to ascertain the structural identity of PAH. This instrument can detect concentrations as low as 1.0 pmol/mℓ. When operated in the three-dimensional mode, the instrument provides spectral infor-

**Table 1**
**INCORPORATION OF 3 MCA BY NUCLEI FOLLOWING IN VITRO INCUBATION OF RAT LIVER SLICES**

| Time (min) | cpm MCA/mg DNA |
|---|---|
| 30 | 830 |
| 60 | 3200 |
| 120 | 560 |

mation that appears as a series of isointensity contour lines when monitored by an X-Y recorder. The DNA-P fraction was analyzed for BAP adducts using 0.1-mℓ of sample per assay. The excitation wavelength ranged from 250 to 500 nm and the emission wavelength ranged from 300 to 600 nm.

### G. Density Gradient Separation of Proteinase K Digested Samples

DNA-P and -S samples containing approximately 15 μg/mℓ of DNA were digested with Proteinase K at a concentration of 50 μg/mℓ in 1% sarcosyl. Digestion proceeded for 2 hr at 37°C. The digests were then made 57 *M* in CsCl. After being placed into centrifuge tubes, the samples were overlaid with mineral oil and centrifuged at 30 Krpm for 40 hr in a SW60 Beckman rotor at 20°C. The gradients were either hand fractionated or fractionated with an ISCO Density Gradient Fractionator equipped with a U.V. monitor. Fractions were then suspended in scintillation cocktail and counted in a Beckman 8100 Scintillation Spectrometer.

### H. Density Gradient Separation Following Guanidinium Chloride Dissociation

Samples of DNA-P containing approximately 10 μg of DNA were incubated at 56°C for 45 min in 8.0 *M* guanidinium chloride, 2% mercaptoethanol, and 10 m*M* EDTA. Following incubation, the samples were layered over 2 volumes of 4 *M* CsCl and centrifuged for 48 hr at 25°C at 35 Krpm in a Beckman SW60 rotor. The gradients were then hand fractionated. Each fraction was individually dialyzed and assayed for protein, DNA, and carcinogen content.

### I. Other Assays

DNA was assayed by diphenylamine[14] or diaminobenzoic acid.[64] Protein was assayed by the method of Lowry et al.[43] or Bradford.[9]

## III. RESULTS

The fractionation scheme used (see Figure 1) to analyze carcinogen binding sites was previously used to study chromatins from reticulocytes,[4] liver,[4] placenta,[52] and oviduct[47] tissues. We have found that salt extraction removes approximately 90% of the nonhistones and all of the histone chromosomal proteins. The proteins remaining seem to aggregate with a minor fraction of DNA shown to contain nonrandomly distributed sequences. Other investigators have shown that carcinogens bind to NHC proteins,[33,66] and it is our interest to determine whether this binding could be correlated with our findings of nonrandom DNA distribution with respect to protein aggregation.

The initial work was conducted with rat liver, a tissue that is not readily transformed by the carcinogens studied. It was determined that maximum incorporation of methylcholanthrene (MCA) by rat liver slices occurred after 60 min of incubation (see Table 1). Further

**Table 2**
**DISTRIBUTION OF 3-MCA IN NUCLEAR SUBFRACTIONS ISOLATED FROM RAT LIVER**

| Fraction | DNA yield mg | DNA yield % | MCA (cpm) | cpm MCA/mg DNA |
|---|---|---|---|---|
| Chromatin | 2.2 | 100.0 | N.D. | N.D. |
| DNA-P | 0.051 | 2.3 | 605 | 11,700 |
| DNA-S | 1.780 | 82.7 | 320 | 180 |
| NaCl extract | 0.320 | 14.9 | 1,510 | 4,720 |

**Table 3**
**DNA AND PROTEIN YIELDS FROM FRACTIONATED $10T^1/_2$ NUCLEI**

| Fraction | DNA yield mg | DNA yield % | Protein yield mg | Protein yield % | Protein/DNA |
|---|---|---|---|---|---|
| DNA-P | 0.19 | 7.5 | 1.2 | 12.6 | 6.3 |
| DNA-S | 2.03 | 79.9 | 0.97 | 10.2 | 0.48 |
| 2 *M* NaCl extract | 0.32 | 12.6 | 7.3 | 77.1 | 22.8 |

analysis was conducted with the 60-min samples. Nuclei were extracted with 2 *M* NaCl, and the fractions were analyzed for DNA and MCA. Table 2 shows that approximately 15% of the DNA is solubilized in the 2 *M* NaCl, 82.7% is found in the DNA-S, and 2.3% in the DNA-P fraction. We have expressed our results with respect to DNA; however, it is not possible from this experiment to determine what proportion of the MCA is protein and/or DNA bound. The extensive labeling of DNA-P and the NaCl extract suggests that MCA is preferentially found in protein enriched fractions, 25% in DNA-P and 62% in the NaCl extract. The enormous enrichment of carcinogen in the DNA-P fraction (65-fold with respect to DNA-S) is of particular interest because of previous studies indicating the importance of this fraction in both replicative[55] and transcriptional[46,68] processes. In order to evaluate further the binding sites of carcinogens, we switched to a transformable cell line, a tissue which more readily binds MCA and BAP metabolites.

Fractionation of fibroblast cells was performed on chromatin isolated from purified nuclei. The DNA and protein composition of each fraction is shown in Table 3. The DNA distribution in these cells is similar to other tissues studied (2 to 15% of the DNA being found in the DNA-P fraction). The primary distinction is the elevated levels of protein found in these fibroblast preparations. In most chromatin the protein to DNA ratio approximates two,[5] whereas these cells show a protein to DNA ratio of 3.5—4.0. All three fractions show elevated levels of protein.

To determine whether the primary carcinogen binding target is protein or DNA, a guanidinium dissociation of DNA-P from cells exposed to BAP was performed. As can be seen in Figure 2, CsCl density gradient centrifugation resolved the guanidinium treated DNA-P into two major peaks. Peak A contains all of the protein. The bulk of the DNA is found in peak B, although low levels were also detected in peak A. The BAP metabolite was found primarily in peak A, with low levels also appearing in peak B. Although protein dissociation is incomplete as evidenced by some DNA remaining in peak A, it can be calculated that low levels of BAP metabolites are linked to the DNA, but that the majority of the carcinogen is protein bound.

Further efforts to determine carcinogen binding sites was performed using Proteinase K

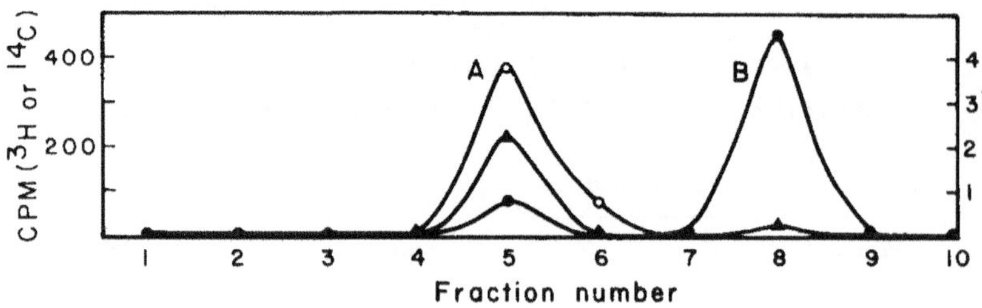


FIGURE 2. Dissociation of DNA-P labeled with BAP by guanidinium chloride. Fractionation of the CsCl density gradient resulted in two major peaks. Peak A contains all of the protein, minor amounts of DNA, and most of the BPDE; peak B contains most of the DNA and minor amounts of BPDE. (○) $^{14}$C-leucine, (●) $^3$H-thymidine, (▲), BPDE, quantitated by fluorometry.

digestion followed by density gradient centrifugation. A reference sample of chromatin, isolated from cells incubated with $^3$H-thymidine and $^{35}$S-methionine, was used to determine the effect of Proteinase K digestion on DNA and protein banding patterns (see Figure 3). In the undigested sample, approximately 65% of the DNA was found in fraction 1, density = 1.43, and 35% was found in fraction 14, density = 1.68. Most of the protein bands on top of the gradient, although about 2% migrates to a density of 1.68. Proteinase K digestion results in a shift of virtually all of the DNA into fraction 15, density = 1.7, indicating relatively complete separation of DNA from protein. There is virtually no protein detectable throughout the gradient; all measureable $^{35}$S counts were found either on top of the gradient or in the dialysate. Fibroblasts labeled with MCA were fractionated into DNA-P and -S; each fraction was Proteinase K digested and and fractionated following density gradient centrifugation. In a nondigested sample of DNA-P (Figure 4a), approximately 55% of the MCA was found on top of the gradient, 40% at a density of 1.56, and a minor amount at the bottom of the gradient, presumably complexed with RNA.

Following digestion (Figure 4b), approximately 45% of the MCA was found on top of the gradient, 15% at a density of 1.7, 35% liberated into the dialysate, and the remaining at the bottom of the gradient. Since we have shown that Proteinase K effectively removes DNA from the top of the gradient, it can be inferred that virtually all of the MCA found in the top fraction following digestion is associated with protein. This combined with the MCA liberated into the dialysate shows that approximately 80% of the MCA is protein bound.

When DNA-S was treated with Proteinase K, there was no detectable shift in the MCA distribution. Since DNA-S has such a low content of proteins, this lack of redistribution, in response to Proteinase K, is not unexpected (Figure 5). Thus, the observation that MCA distribution in DNA-S is unaffected by Proteinase K digestion is consistent with our conclusion that chemical carcinogens are preferentially bound to proteins.

Metabolic activation is required for many carcinogens, and identification of the ultimate carcinogen has been the subject of numerous reports. Previous studies utilizing fluorescence spectra and HPLC have shown that metabolism of both methylcholanthrene and BAP results in many hydroxylated metabolites.[20,24] In a sample of DNA-P isolated from cells exposed to BAP, multiple metabolic forms were found (see Figure 6a), some of which were hydrophobically linked and some of which were apparently covalently linked to the protein-DNA complex. Extensive extractions with ethyl acetate removes hydrophobically linked residues, allowing an analysis of the putatively covalently linked metabolites. Comparison with known standards (see Figure 7a and b) shows that both nonmetabolized and multiple forms of metabolized BAP can be found in the nonextracted sample. Washing results in a profile that is characteristic of benzo(*a*)pyrene-7,8-diol-9,10-epoxide (BPDE) with a strong fluorescence peak at 415 nm with excitation maximum at 340 nm and a weak secondary fluorescence peak at 475 nm with excitation maximum of 395.

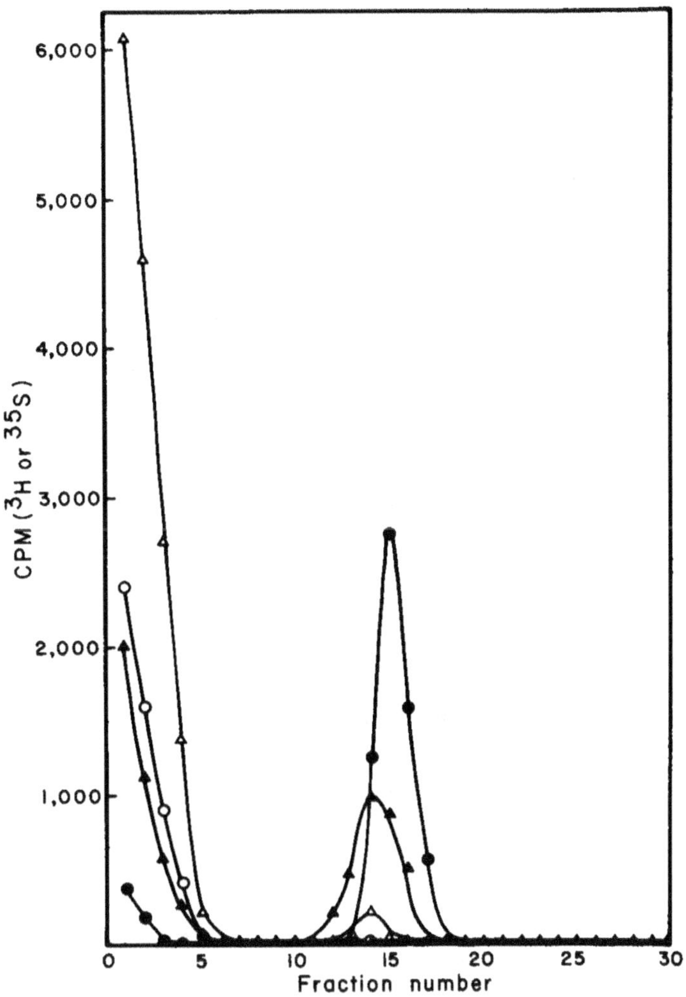


FIGURE 3. The effect of Proteinase K digestion of radiolabeled chromatin on the distribution of DNA and protein in a CsCl density gradient. (▲) $^3$H-thymidine, control; (●) $^3$H-thymidine, digested; (△) $^{35}$S-methionine, control; (○) $^{35}$S-methionine, digested.

An analysis of the DNA-P and DNA-S fractions from cells treated either with BAP or MCA shows substantial enrichment of the carcinogen in the DNA-P fractions. In the case of BAP, DNA-P shows a 40-fold enrichment with respect to DNA-S, and in the case of MCA, DNA-P shows a 6.5-fold enrichment with respect to DNA-S (see Table 4). As we have shown, approximately 80% of the carcinogen in the DNA-P fraction is protein bound, the remaining being bound to DNA (approximately 15%) and RNA (approximately 5%). The distribution of carcinogens among these macromolecules is similar with other reports.[71] However, the new discovery that the carcinogen is highly enriched in the DNA-P fraction, both in rat liver and in fibroblasts, reinforces our hypothesis that the proteins in this fraction are highly significant in development, the range of activity being expanded to include carcinogenesis as well as the normal transcriptional and replicative functions.

## IV. DISCUSSION

The molecular targets of chemical carcinogens include DNA, RNA, and proteins. Various

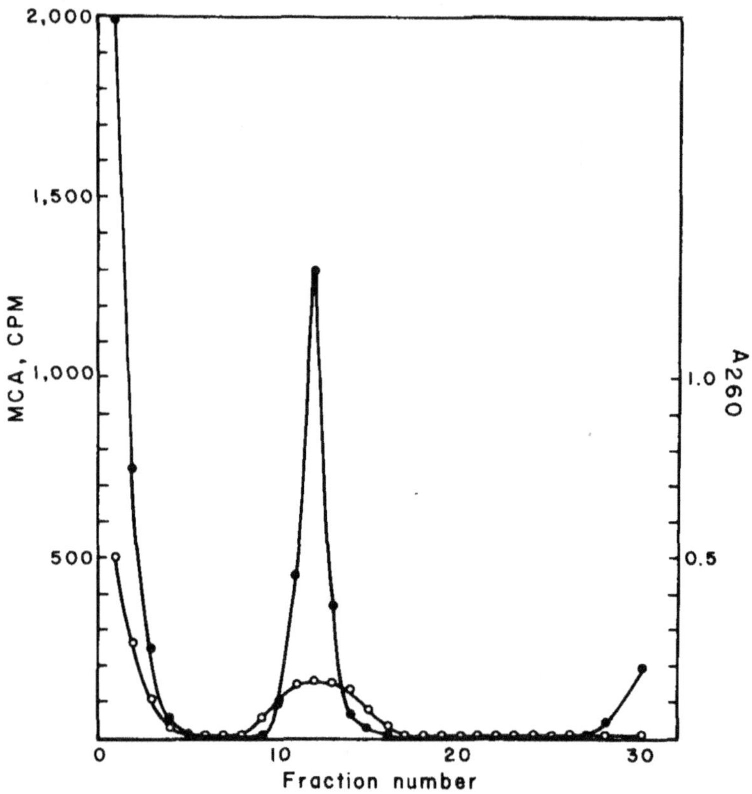


A

FIGURE 4. (A) Control sample of DNA-P labeled with MCA and centrifuged through CsCl; (●) $^3$H-MCA; (○) $A_{260}$. (B) Proteinase K digested DNA-P, labeled with MCA and centrifuged through CsCl; (●) $^3$H-MCA. (○) $A_{260}$.

sources have reported that chemical carcinogens (1) are randomly distributed throughout DNA,[67] (2) bind selectively to internucleosomal linker regions,[32,45,67] (3) bind preferentially to nucleosomes containing high mobility group (HMG) proteins,[36] (4) bind to active regions of the genome,[3,49] (5) bind to histones,[12,56,66] and (6) bind to nonhistone proteins.[66,71] The theory that carcinogenesis is a mutagenic process has been expanded to include large changes in chromosomal structure such as genetic transposition[8,16] and nondisjunction during cell division.[48] In addition, it has been argued that changes in the DNA may interfere with repair enzymes and thus result in alterations in transcriptional read-out.[11] In studies aimed at determining the significance of proteins in oncogenesis, it was found that both RNA polymerase and kinase activity can be greatly enhanced over control levels when MCA was injected. It was concluded from this study that these proteins may be the target molecules of MCA.[62] In a study evaluating the persistence of carcinogen-nucleoside adducts in mouse tissues, it was found that most of the carcinogen did not cochromatograph as nucleoside-bound adducts,[20a] adding further evidence that other nuclear components bind significant quantities of PAH.

In our own studies, we have found that both MCA and BAP are nonrandomly distributed in nuclear fractions. The data from both rat liver and fibroblasts indicate that relative to DNA content, small amounts of carcinogen are found in the DNA-S fraction, the majority either released with 2 *M* NaCl or found in the residual DNA-P fraction. An analysis of the density gradient data of MCA-treated cells shows that the DNA band (density = 1.7) from

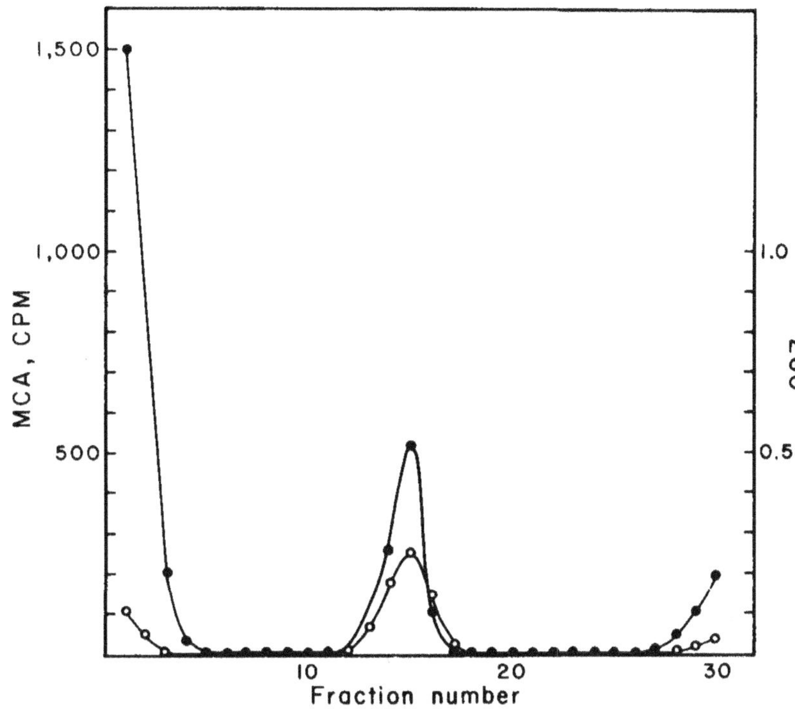


FIGURE 4B

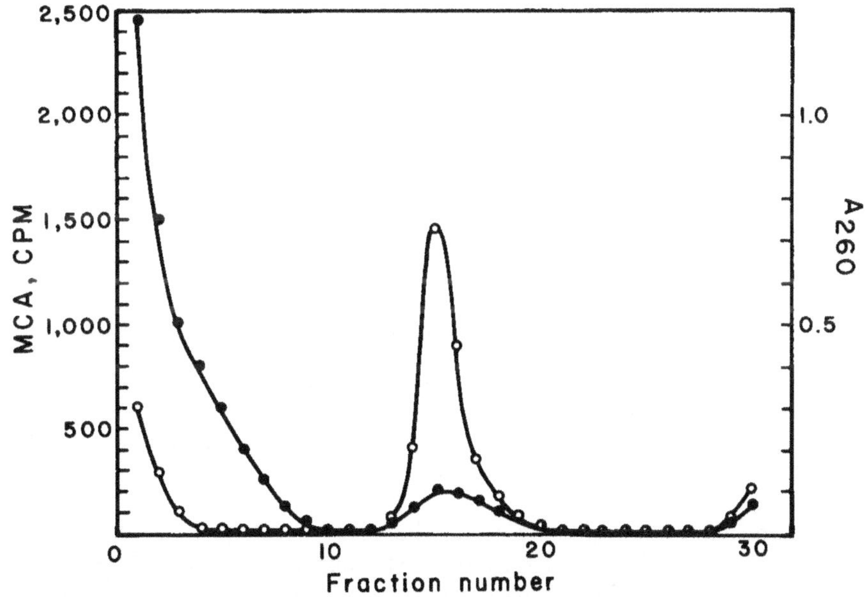


FIGURE 5. DNA-S, labeled with $^3$H-MCA. The curves of Proteinase K digested and nondigested samples are coincident. (●) $^3$H-MCA; (○) $A_{260}$.

Proteinase K digested DNA-P (Figure 4b) contains approximately twice as much carcinogen as the DNA band (density = 1.68) from DNA-S (Figure 5). However, an analysis of undigested samples shows a 6.5-fold enrichment of MCA in the DNA-P fraction (see Table

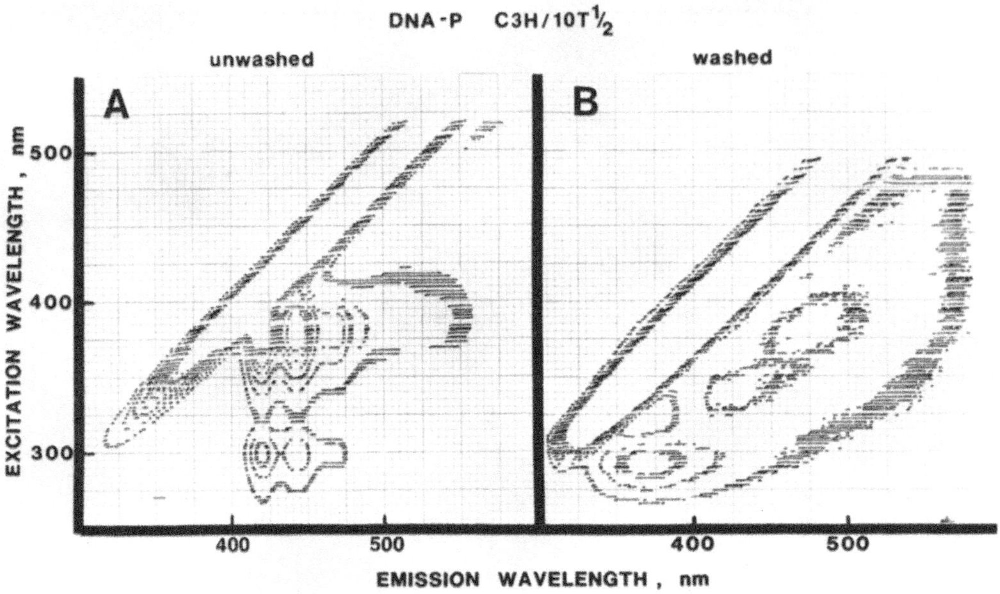


FIGURE 6. DNA-P analyzed by fluorometry. Unwashed sample (A) sample repeatedly extracted with ethyl acetate (B).

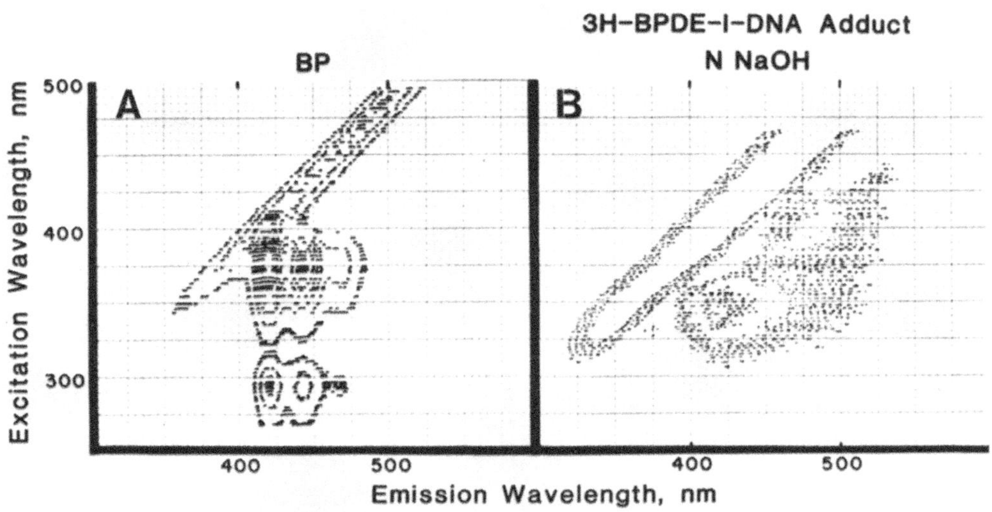


FIGURE 7. Fluorometric tracings of standards. (A) A mixture of benzopyrene and its metabolites; (B) benzo(*a*) pyrene-7,8,-diol-9,10-epoxide, the covalently linked adduct.

4). Obviously the unequal partitioning of the MCA between DNA-P and -S is due primarily to protein-carcinogen binding. We have found a striking proportion of the carcinogen bound to proteins. Both the guanidinium dissociation studies and the Proteinase K digestion experiments indicate that greater than 80% of the carcinogen is protein bound.

Many functions have been ascribed to nonhistone proteins (for a review, refer to Chapter 4). Of particular interest is the finding that many cytosolic proteins act as carriers to transport various ligands to the nucleus; it is interesting to note that various genes have been induced presumably as a result of the interaction of a receptor-ligand complex with the nucleus, such as metalliothionein by receptor-glucocorticoids[34] and ovalbumin by receptor-estrogen.[54] The

**Table 4**
**DISTRIBUTION OF CARCINOGENS IN FRACTIONATED 10T$^1/_2$ NUCLEI**

| Fraction | pg BPDE/μg DNA | pg MCA/μg DNA |
|---|---|---|
| DNA-P | 200 | 71 |
| DNA-S | 5 | 11 |

specificity of the induction suggests that some factors in the nucleus determine the susceptibility of various genes. It has been reported that steroid-receptor complexes seem to fractionate with a subset of NHC proteins,[61,68] and it is thus possible to speculate that the factors mediating receptor-ligand activity may reside within this subset. In addition to reports of hormone translocation, there have been studies indicating that carcinogens are translocated to the nucleus by receptor proteins.[53,65] It is possible then that some carcinogens may mediate gene activity by receptor-ligand interaction with a subset of nonhistone proteins.

We have previously reported that the DNA-P fraction contains active genes. In addition, studies on the nuclear matrix, a protein-DNA complex isolated after salt extraction and therefore similar to DNA-P, have shown that the insoluble proteins show preference for actively replicating DNA sites.[55] The preferential binding of the carcinogen in this fraction suggests that both transcriptional and replicative activity may be affected. It is conceivable that the modification by carcinogens is specific as in the case of some steroid modified expression. It is also conceivable that macromolecular complexes containing polycyclic hydrocarbons, protein, and DNA could act in a more generalized way, severely altering the conformational state of chromatin, thereby causing drastic changes in both replication and transcription.

## ACKNOWLEDGMENTS

This work was in part supported by NIH grants CA-19974 (to Joon Rho), DE-04031 (to Isaac Bekhor), and Training Grant DE-07006. We also wish to acknowledge the encouragement given by Robert Siev to Carol J. Mirell throughout this work.

## REFERENCES

1. **Ames, B. N., Durston, W. E., Yamasaki, E., and Lee, F. D.,** Carcinogens are mutagens: a simple test system combining liver homogenates for activation and bacteria for detection, *Proc. Natl. Acad. Sci. U.S.A.*, 70, 2281, 1973.
2. **Ananiev, E. V. and Ilyin, Y. V.,** A comparative study of the location of mobile disperse genes in salivary gland and midgut polytene chromosomes of Drosophila melanogaster, *Chromosoma*, 82, 429, 1981.
3. **Arrand, J. E. and Murray, A. M.,** Benzopyrene groups bind preferentially to the DNA of active chromatin in human lung cells, *Nucleic Acids Res.*, 10, 1547, 1982.
4. **Bekhor, I. and Mirell, C. J.,** Simple isolation of DNA hydrophobically complexed with presumed gene regulatory proteins ($M_3$), *Biochemistry*, 18, 609, 1979.
5. **Birnie, G. D.,** *Subnuclear Components*, Butterworths, London, 1976.
6. **Bishop, J. M.,** Oncogenes, *Sci. Am.*, 246, 80, 1982.
7. **Boettinger, D. and Durban, E. M.,** Progenitor-cell populations can be infected by RNA tumor viruses, but transformation is dependent on the expression of specific differentiated functions, *Cold Spring Harbor Symp. Quant. Biol.*, 44, 1249, 1979.
8. **Botchan, M., Stringer, J., Mitchison, T., and Sambrook, J.,** Integration and excision of SV40 DNA from the chromosome of a transformed cell, *Cell*, 20, 143, 1980.
9. **Bradford, N. M.,** A rapid and sensitive method for the quantitation of microgram quantities of protein utilizing the principle of protein-dye binding, *Anal. Biochem.*, 72, 248, 1976.

10. **Bresnick, E., Brand, R., and Knight, J. A.,** Ribonucleic acid biosynthesis in methylcholanthrene treated rats, *Biochim. Biophys. Acta*, 114, 227, 1966.
11. **Bresnick, E. and Eastman, A.,** Alkylation of mammalian cell DNA, persistence of adducts, and relationship to carcinogenesis, *Drug Metab. Rev.*, 13, 189, 1982.
12. **Bresnick, E., Vaught, J. B., Chuang, A. H. L., Stroming, T. A., Bockman, E., and Mukhtar, H.,** Nuclear aryl hydrocarbon hydroxylase and interaction of polycyclic hydrocarbons with nuclear components, *Arch. Biochem. Biophys.*, 181, 257, 1977.
13. **Brookes, P.,** Covalent interaction of carcinogens with DNA, *Life Sci.*, 16, 331, 1975.
14. **Burton, K.,** A study of the conditions and mechanisms of the diphenylamine reaction for the colorimetric estimation of deoxyribonucleic acid, *Biochem. J.*, 62, 315, 1956.
15. **Buty, S. G., Thompson, S., and Slaga, T. J.,** The role of epidermal aryl hydrocarbon hydroxylase in the covalent binding of polycyclic hydrocarbons to DNA and its relationship to tumor initiation, *Biochem. Biophys. Res. Commun.*, 70, 1102, 1976.
16. **Cairns, J.,** The origin of human cancers, *Nature (London)*, 289, 353, 1981.
17. **Chahal, S. S., Matthews, H. R., and Bradbury, E. M.,** Acetylation of histone $H_4$ and its role in chromatin structure and function, *Nature (London)*, 287, 76, 1980.
18. **Chytil, F. and Spelsberg, T. C.,** Tissue differences in antigenic properties of non-histone protein-DNA complexes, *Nature (London), New Biol.*, 233, 215, 1971.
19. **Cochet, M., Gannon, F., Hen, R., Maroteaux, L., Perrin, F., and Chambon, P.,** Organisation and sequence studies of the 17-piece chicken conalbumin gene, *Nature (London)*, 282, 567, 1979.
20. **Eastman, A., and Bresnick, E.,** Metabolism and DNA binding of 3-methylcholanthrene, *Cancer Res.*, 39, 4316, 1979.

20a. **Eastman, A. and Bresnick, E.,** Persistent binding of 3-methylcholanthrene to mouse lung DNA and its correlation with susceptibility to pulmonary neoplasia, *Cancer Res.*, 39, 2400, 1979.

21. **Engel, J. D. and Dodgson, J. B.,** Analysis of the adult and embryonic chicken globin genes in chromosomal DNA, *J. Biol. Chem.*, 253, 8239, 1978.
22. **Fialkow, P. J., Gartler, S. M., and Yoshida, A.,** Clonal origin of chronic myelocytic leukemia in man, *Proc. Natl. Acad. Sci. U.S.A.*, 58, 1468, 1967.
23. **Freeman, A. E., Kelloff, G. J., Vernon, M. L., Lane, W. T., Capps, W. I., Bumgarner, S. D., Turner, H. C., and Huebner, R. J.,** Prevalence of endogenous type-C virus in normal hamster tissues and hamster tumors induced by chemical carcinogens, Simian virus 40, and polyoma virus, *J. Natl. Cancer Inst.*, 52, 1469, 1974.
24. **Gehly, E. H., Fahl, W. E., Jefcoate, C. R., and Heidelberger, C.,** The metabolism of benzo(a)pyrene by cytochrome P-450 in transformable and nontransformable C3H mouse fibroblasts, *J. Biol. Chem.*, 254, 5041, 1979.
25. **Getz, M. J., Elder, P. K., and Moses, H. L.,** Equivalent expression of endogenous murine leukemia virus-related genes in C3H/10T$^1/_2$ cells and chemically transformed derivative cells, *Cancer Res.*, 38, 566, 1978.
26. **Gronow, M.,** Nuclear proteins and chemical carcinogenesis, *Chem. Biol. Interact.*, 29, 1, 1980.
27. **Haddow, A.,** Molecular repair, wound healing, and carcinogenesis: tumor production a possible over-healing? *Adv. Cancer Res.*, 16, 181, 1972.
28. **Hart, R. W., Setlow, R. B., and Woodhead, A. D.,** Evidence that pyrimidine dimers in DNA can give rise to tumors, *Proc. Natl. Acad. Sci. U.S.A.*, 74, 5574, 1977.
29. **Heidelberger, C.,** Chemical oncogenesis in culture, *Adv. Cancer Res.*, 18, 317, 1973.
30. **Huberman, E. and Sachs, L.,** DNA binding and its relationship to carcinogenesis by different polycyclic hydrocarbons, *Int. J. Cancer*, 19, 122, 1976.
31. **Irving, C. C.,** Interaction of chemical carcinogens with DNA, in *Methods in Cancer Research*, Vol. 7, Busch, H., Ed., Academic Press, New York, 1973, chap. 5.
32. **Jahn, C. L. and Litman, G. W.,** Distribution of covalently bound benzo(a)pyrene in chromatin, *Biochem. Biophys. Res. Commun.*, 76, 534, 1977.
33. **Jungmann, R. A. and Schweppe, J. S.,** Binding of chemical carcinogens to nuclear proteins of rat liver, *Cancer Res.*, 33, 952, 1972.
34. **Karin, M. and Herschman, H. R.,** Glucocorticoid hormone receptor mediated induction of metallothionein synthesis in HeLa cells, *J. Cell Physiol.*, 103, 35, 1980.
35. **Koostra, A.** Effect of histone acetylation on the formation and removal of B(a)P chromatin adducts, *Nucleic Acids Res.*, 10, 2775, 1982.
36. **Koostra, A., Shah, Y. B., and Slaga, T. J.,** Binding of B(a) P diol-epoxide to nucleosomes containing high mobility group proteins, *FEBS Lett.*, 116, 62, 1980.
37. **Koostra, A., Slaga, T. J., and Olins, D. E.,** Interaction of benzo(a)pyrene diolepoxide with nuclei and isolated chromatin, *Chem. Biol. Interac.*, 28, 225, 1979.

38. **Kouri, R. E., Ratrie, H., and Whitmire, C. E.,** Evidence of a genetic relationship between susceptibility to 3-methyl-cholanthrene-induced subcutaneous tumors and inducibility of aryl hydrocarbon hydroxylase, *J. Natl. Cancer Inst.*, 51, 197, 1973.
39. **Kuroki, T. and Heidelberger, C.,** The binding of polycyclic aromatic hydrocarbons to the DNA, RNA, and proteins of transformable cells in culture, *Cancer Res.*, 31, 2168, 1971.
40. **Lawley, P. D.,** Carcinogenesis by alkylating agents, in *Chemical Carcinogenesis*, Searle, C. E., Ed., American Chemical Society Monogr. 173, American Chemical Society, Washington, D.C., 1976, 83.
41. **Loeb, L. A., Springgote, C. F., and Battula, N.,** Errors in DNA replication as a basis of malignant changes, *Cancer Res.*, 34, 2311, 1974.
42. **Long, B. H., Huang, C. Y., and Pogo, A. O.,** Isolation and characterization of the nuclear matrix in Friend erythroleukemia cells: chromatin and hnRNA interactions in the nulear matrix, *Cell*, 18, 1079, 1979.
43. **Lowry, O. H., Rosebrough, N. I., Farr, A. L., and Randall, R. J.,** Protein measurement with the folin phenol reagent, *J. Biol. Chem.*, 193, 265, 1951.
44. **Markert, C. L.,** Neoplasia a disease of cell differentiation, *Cancer Res.*, 28, 1908, 1968.
45. **Metzger, G., Wilhelm, F. X., and Wilhelm, M. L.,** Non-random binding of a chemical carcinogen to the DNA in chromatin, *Biochem. Biophys. Res. Commun.*, 75, 703, 1977.
46. **Miller, T. E., Huang, C. Y., and Pogo, A. O.,** Rat liver nuclear skeleton and ribonucleoprotein complexes containing hnRNA, *J. Cell Biol.*, 76, 675, 1978.
47. **Mirell, C. J. and Bekhor, I.,** The effect of estrogen on the concentration of ovalbumin gene sequence in the 2 *M* NaCl residual fraction of oviduct chromatin, *FEBS Lett.*, 150, 117, 1982.
48. **Morin, N. R., Zeldin, P. E., Kubinski, Z. O., Bhattacharya, P. K., and Kubinski, H.,** Macromolecular complexes produced by chemical carcinogens and ultraviolet radiation, *Cancer Res.*, 37, 3802, 1977.
49. **Moses, H. L., Webster, R. A., Martin, G. D., and Spelsberg, T. C.,** Binding of polycyclic aromatic hydrocarbons to transcriptionally active nuclear subfractions of AKR mouse embryo cells, *Cancer Res.*, 36, 2905, 1976.
50. **Nebert, D. W.,** Pharmacogenetics: an approach to understanding chemical and biological aspects of cancer, *J. Natl. Cancer Inst.*, 65, 1279, 1980.
51. **Nguyen-Huu, M. C., Stratmann, M., Grover, B., Wurtz, T., Land, H., Giesecke, K., Sippel, A. E., and Schultz, G.,** Chicken lysozyme gene contains several intervening sequences, *Proc. Natl. Acad. Sci. U.S.A.*, 76, 76, 1979.
52. **Norman, G. L. and Bekhor, I.,** Enrichment of selected active human gene sequences in the placental deoxyribonucleic acid fraction associated with tightly bound nonhistone chromosomal proteins, *Biochemistry*, 20, 3568, 1981.
53. **Okey, A. B., Bondy, G. P., Mason, M. E., Nebert, D. W., Forster-Gibson, C. J., Muncan, J., and Defresne, M. J.,** Temperature-dependent cytosol-to-nuclear translocation of the Ah receptor for 2,3,7,8-tetrachlorodibenzo-p-dioxin in continuous cell culture lines, *J. Biol. Chem.*, 255, 11415, 1980.
54. **O'Malley, B., Spelsberg, T. C., Schrader, W. T., Chytil, F., and Steggles, A. W.,** Mechanisms of interaction of a hormone receptor complex with the genome of a eukaryotic target cell, *Nature (London)*, 235, 141, 1972.
55. **Pardoll, D. M., Vogelstein, B., and Coffey, D. S.,** A fixed site of DNA replication in eucaryotic cells, *Cell*, 19, 527, 1980.
56. **Pezzuto, J. M., Lea, M. A., and Yang, C. S.,** Binding of metabolically activated benzo(a)pyrene to DNA and histones of rat liver, lung and regenerating liver, *Life Sci.*, 22, 105, 1978.
57. **Pott, P.,** Chirurgical observations relative to the cancer of the scrotum, 1775, *Natl. Cancer Inst. Monogr.*, 10, 7, 1963.
58. **Pumo, D. E., Wierzbicki, R., and Chiu, J.-E.,** Chicken reticulocyte nuclear antigen: its identification and relation to transcriptive activity in erythropoietic cells, *Biochemistry*, 19, 2362, 1980.
59. **Rackowicz-Sculczynska, E. M. and Horst, A.,** Non-histone proteins in fractionated chromatin before and after DNAase II treatment, *Biochim. Biophys. Acta*, 653, 69, 1981.
60. **Shoyab, M.,** Dose-dependent preferential binding of polycyclic aromatic hydrocarbons to reiterated DNA of murine skin cells in culture, *Proc. Natl. Acad. Sci. U.S.A.*, 73, 5841, 1978.
61. **Spelsberg, T. C., Webster, R., Pikler, G., Thrall, C., and Weiss, D.,** Role of nuclear proteins as high affinity sites (acceptors) for progesterone in the avian oviduct, *J. Steroid Biochem.*, 7, 1091, 1976.
62. **Szeberenyi, J., Kleeberg, U., and Tigyi, A.,** Effect of 3-methylcholanthrene on RNA polymerase and protein kinase activities of transcriptionally active chromatin subfractions in rat liver, *Experentia*, 38, 687, 1982.
63. **Tennant, R. W., Otten, J. A., Myer, F. E., and Rascati, R. J.,** Induction of retrovirus gene expression in mouse cells by some chemical mutagens, *Cancer Res.*, 42, 3050, 1982.
64. **Thomas, P. S. and Farquhar, M. N.,** Specific measurement of DNA in nuclei and nucleic acids using diaminobenzoic acid, 89, 35, 1978.
65. **Tierny, B., Weaver, D., Heintz, N. H., Schaeffer, W. I., and Bresnick, E.,** The identity and nuclear uptake of a cytosolic binding protein for 3-methylcholanthrene, *Arch. Biochem. Biophys.*, 200, 513, 1980.

66. **Vaught, J. and Bresnick, E.,** Binding of polycyclic hydrocarbons to nuclear components in vitro, *Biochem. Biophys. Res. Commun.*, 69, 587, 1976.
67. **Walker, M. S., Becker, F. F., and Rodriguez, L. V.,** In vivo binding of N-2-acetylaminofluorene and its N-hydroxy derivative to the DNA of fractionated rat liver chromatin, *Chem. Biol. Interact.*, 27, 177, 1979.
68. **Wang, T. Y.,** The role of nonhistone chromosomal proteins in the interaction of prostate chromatin with androgen-receptor complex, *Biochim. Biophys. Acta*, 518, 81, 1978.
69. **Weinstock, R., Sweet, R., Weiss, M., Cedar, H., and Axel, R.,** Intragenic spacers interrupt the ovalbumin gene, *Proc. Natl. Acad. Sci. U.S.A.*, 75, 1299, 1978.
70. **Whitmire, C. E., Salerno, R. A., Rabstein, L. S., Hueaner, R. J., and Turner, H. C.,** RNA tumor virus antigen expression in chemically induced tumors. Virus-genome-specified common antigens detected by complement fixation in mouse tumors induced by 3-methylcholanthrene, *J. Natl. Cancer Inst.*, 47, 1255, 1971.
71. **Zytkovicz, T. H., Moses, H. L., and Spelsberg, T. C.,** Covalent binding of benzo(a)pyrene metabolites to DNA, RNA and chromatin proteins in the AKR mouse embryo cell line, *Chem. Biol. Interact.*, 35, 39, 1981.

# INDEX

## A



## B



## C

## D

## E

## F



## G



## H

## I



## L



## M

## N

## O



## P

## R



## S

## T



## U



## V



## W



## X



## Y



## Z

For Product Safety Concerns and Information please contact our EU representative GPSR@taylorandfrancis.com Taylor & Francis Verlag GmbH, Kaufingerstraße 24, 80331 München, Germany

**Batch number: 10731074**

Printed by Printforce, the Netherlands